T0203047

Python For ArcGIS

Laura Tateosian

Python For ArcGIS

Laura Tateosian
North Carolina State University
Raleigh, NC, USA

ISBN 978-3-319-79250-7   ISBN 978-3-319-18398-5   (eBook)
DOI 10.1007/978-3-319-18398-5

Springer Cham Heidelberg New York Dordrecht London
© Springer International Publishing Switzerland 2015
Softcover reprint of the hardcover 1st edition 2015

Printed on acid-free paper

Springer International Publishing AG Switzerland is part of Springer Science+Business Media (www.springer.com)

Preface

You know... this would be a lot faster with a script!

Imagine... You've just begun a new job as a GIS specialist for the National Park Service. Your supervisor has asked you to analyze some wildlife data. She gives you a specific example to start with: One data table (Bird Species) contains a list of over 600 bird species known to be native to North Carolina. Another data table (Bird Inventory) contains over 5,000 records, each corresponding to a sighting of a particular bird. Your task is to clean the data, reformat the file for GIS compatibility, and summarize the data by determining what percent of the native Bird Species appear in the inventory and map the results. Once you complete this, she would like you to repeat this process for historical datasets for the last 10 years of monthly records. After that, the next assignment will be to answer the same question based on monthly species inventory datasets for fish and invertebrates.

Performing this process manually for one dataset could be time consuming and error prone. Performing this manually for numerous datasets is completely impractical. Common GIS tasks such as this provide a strong motivation for learning how to automate workflows.

Python programming to the rescue! This analysis can be performed in single Python script. You could even use Python to automatically add the sighting locations to ArcMap® and create a report with the results. You could also add a graphical user interface (GUI) so that colleagues can run the analysis themselves on other datasets. This book will equip you to do all of these things and more.

This book is aimed primarily at teaching GIS Python scripting to GIS users who have little or no Python programming experience. Topics are organized to build upon each other, so for this audience, reading the chapters sequentially is recommended. Computer programmers who are not familiar with ArcGIS can also use this book to learn to automate GIS analyses with the ArcGIS Python API, accessible through the arcpy package. These readers can quickly peruse the general Python concepts and focus mainly on the sections describing arcpy package functionality.

The book comes with hundreds of example Python scripts. Download these along with the sample datasets so that you can run the examples and modify them to experiment as you follow the text. Where and how to download the supplementary material is explained in the first chapter. Rolling your sleeves up and trying things yourself will really boost your understanding as you use this book. Remember, the goal is to become empowered to save time by using Python and be more efficient in your work. Enjoy!

Raleigh, NC   Laura Tateosian

Acknowledgements

Thank you to the numerous people who have helped with this project. Over the years, many students have studied from drafts of this material and made suggestions for improvements. They have also expressed enthusiasm about how useful this knowledge has been for their work, which encouraged me to enable a wider distribution by writing this book.

Special thanks to those who have reviewed the book, Pankaj Chopra of Emory University, Brian McLean of the North Carolina Department of Agriculture and Consumer Services, Holly Brackett of Wake Technical Community College, Rahul Bhosle of GIS Data Resources, Inc., and Makiko Shukunobe of North Carolina State University. Thank you to Sarah Tateosian for the artwork. Thank you to Tom Danninger, Michael Kanters, Joe Roise Justin Shedd, and Bill Slocumb of North Carolina State University and Cheryl Sams at the National Park Service for assistance with datasets. I also would like to thank Mark Hammond and Kiriakos Vlahos, the authors of PythonWin and PyScripter, respectively. Finally, I would like to express my gratitude for mentorship from Hugh Devine, Christopher Healey, Helena Mitasova, Sarah Stein, and Alan Tharp.

Contents

Chapter 1
Introduction

Abstract Geospatial data analysis is a key component of decision-making and planning for numerous applications. Geographic Information Systems (GIS), such as ArcGIS®, provide rich analysis and mapping platforms. Modern technology enables us to collect and store massive amounts of geospatial data. The data formats vary widely and analysis requires numerous iterations. These characteristics make computer programming essential for exploring this data. Python is an approachable programming language for automating geospatial data analysis. This chapter discusses the capabilities of scripting for geospatial data analysis, some characteristics of the Python programming language, and the online code and data resources for this book. After downloading and setting up these resources locally, readers can walk through the step-by-step code example that follows. Last, this chapter presents the organization of the remainder of the book.

Chapter Objectives

After reading this chapter, you'll be able to do the following:

- Articulate in general terms, what scripting can do for GIS workflows.
- Explain why Python is selected for GIS programming.
- Install and locate the sample materials provided with the book.
- Contrast the view of compound GIS datasets in Windows Explorer and ArcCatalog™.
- Run code in the ArcGIS® Python Window.

© Springer International Publishing Switzerland 2015
L. Tateosian, *Python For ArcGIS*, DOI 10.1007/978-3-319-18398-5_1

Geographic data analysis involves processing multiple data samples. The analysis may need to be repeated on multiple fields, files, and directories, repeated monthly or even daily, and it may need to be performed by multiple users. Computer programming can be used to automate these repetitive tasks. Scripting can increase productivity and facilitate sharing. Some scriptable tasks involve common data management activities, such as, reformatting data, copying files for backups, and searching database content. Scripts can also harness the tool sets provided by Geographic Information Systems (GIS) for processing geospatial data, i.e., *geoprocessing*. This book focuses on the Python scripting language and geoprocessing with ArcGIS software.

Scripting offers two core capabilities that are needed in nearly any GIS work:

- Efficient batch processing.
- Automated file reading and writing.

Scripts can access or modify GIS datasets and their fields and records and perform analysis at any of these levels. These automated workflows can also be embellished with GUIs and shared for reuse for additional economy of effort.

1.1 Python and GIS

The programming language named Python, created by Guido van Rossum, Dutch computer programmer and fan of the comedy group Monty Python, is an ideal programming language for GIS users for several reasons:

- **Python is easy to pick up**. Python is a nice 'starter' programming language: easy to interpret with a clean visual layout. Python uses English keywords or indentation frequently where other languages use punctuation. Some languages require a lot of set-up code before even creating a program that says 'Hello.' With Python, the code you need to print Hello is `print 'Hello'`.
- **Python is object-oriented**. The idea of object-oriented programming (OOP) was a paradigm shift from functional programming approach used before the 1990s. In functional programming, coding is like writing a collection of mathematical functions. By contrast, object-oriented coding is organized around objects which have properties and functions that do things to that object. OOP languages share common conventions, making it easier for those who have some OOP experience in other languages. There are also advantages to programmers at any level such as context menus which provide cues for completing a line of code.

- **Python help abounds**. Another reason to use Python is the abundance of resources available. Python is an open-source programming language. In the spirit of open-source software, the Python programming community posts plenty of free information online. 'PythonResources.pdf', found with the book's Chapter 1 sample scripts (see Section 1.2), lists some key tutorials, references, and forums.
- **GIS embraces Python**. Due to many of the reasons listed above, the GIS community has adopted the Python programming language. ArcGIS software, in particular has embraced Python and expands the Python functionality with each new release. Python scripts can be used to run ArcGIS geoprocessing tools and more. The term *geoprocessing* refers to manipulating geographic data with a GIS. Examples of geoprocessing include calculating buffer zones around geographic features or intersecting layers of geographic data. The Esri® software, ArcGIS Desktop, even provides a built-in Python command prompt for running Python code statements. The 'ArcGIS Resources' site provides extensive online help for Python, including examples and code templates. Several open-source GIS programs also provide Python programming interfaces. For example, GRASS GIS includes an embedded Python command prompt for running GRASS geoprocessing tools via Python. QGIS and PostGreSQL/PostGIS commands can be also run from Python. Once you know Python for ArcGIS Desktop, you'll have a good foundation to learn Python for other GIS tools.
- **Python comes with ArcGIS**. Python is installed automatically when you install ArcGIS. To work with this book, you need to install ArcGIS Desktop version 10.1 or higher. The example in Section 1.4 shows how to use Python inside of ArcGIS. From Chapter 2 onward, you'll use PythonWin or PyScripter software, instead of ArcGIS, to run Python. Chapter 2 explains the installation procedure for these programs, which only takes a few steps.

1.2   Sample Data and Scripts

The examples and exercises in this book use sample data and scripts available for download from http://www.springer.com/us/book/9783319183978. Click on the 'Supplementary Files' link to download 'gispy.zip'. Place it directly under the 'C:\' drive on your computer. Uncompress the file by right-clicking and selecting 'extract here'. Once this is complete, the resources you need for this book should be inside the 'C:\gispy' directory. Examples and exercises are designed to use data under the 'gispy' directory structured as shown in Figure 1.1.

The download contains sample scripts, a scratch workspace, and sample data:

- **Sample scripts** correspond to the examples that appear in the text. The 'C:\gispy\sample_scripts' directory contains one folder for each chapter. Each time a sample script is referenced by script name, such as 'simpleBuffer.py', it appears in the corresponding directory for that chapter.
- **Scratch workspace** provides a sandbox. 'C:\gispy\scratch' is a directory where output data can be sent. The directory is initially empty. You can run scripts that generate output, check the results in this directory, and then clear this space before starting the next example. This makes it easy to check if the desired output was created and to keep input data directories uncluttered by output files.
- **Sample data** for testing the examples and exercises is located in 'C:\gispy\data'. There is a folder for each chapter. You will learn how to write and run scripts in any directory, but for consistency in the examples and exercises, directories in 'C:\gispy' are specified throughout the book.

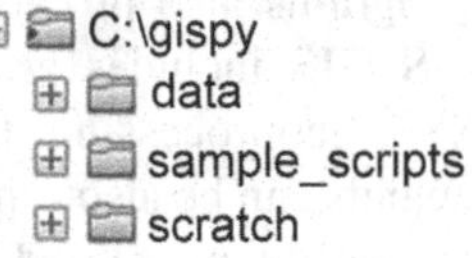

                                      □ ▣ C:\gispy
                                          ⊞ ▤ data
                                          ⊞ ▤ sample_scripts
                                          ⊞ ▤ scratch

Figure 1.1 Examples in this book use these directories.

1.3 GIS Data Formats

Several GIS data formats are used in this book, including compound data formats such as GRID rasters, geodatabases, and shapefiles. In Windows Explorer, you can see the file components that make up these compound data formats. In ArcCatalog™, which is designed for browsing GIS data formats, you see the software interpretation of these files with both geographic and tabular previews of the data. This section looks at three examples (GRID rasters, Shapefiles, and Geodatabase) comparing the Windows Explorer data representations with the ArcCatalog ones. We will also discuss two additional data types (dBASE and layer files) used in this book that consist of only one file each, but require some explanation.

1.3.1 GRID Raster

A *GRID raster* defines a geographic space with an array of equally sized cells arranged in columns and rows. Unlike other raster formats, such as JPEG or PNG, the file name does have a file extension. The file format consists of two directories, each of which contains multiple files. One directory has the name of the raster and contains '.adf' files which store information about extent, cell resolution, and so forth; the other directory, named 'info', contains '.dat' and '.nit' files which store file organization information. Figure 1.2 shows a GRID raster named 'getty_rast' in Windows Explorer (left) and in ArcCatalog (right). Windows Explorer, displays the two directories, 'getty_rast' and 'info' that together define the raster named 'getty_ rast'. The ArcCatalog directory tree displays the same GRID raster as a single item with a grid-like icon.

1.3.2 Shapefile

A *shapefile* (also called a stand-alone feature class), stores geographic features and their non-geographic attributes. This is a popular format for storing GIS vector data. *Vector data* stores features as sets of points and lines as opposed to rasters which store data in grid cells. The vector features, consisting of geometric primitives (points, lines, or polygons), with associated data attributes stored in a set of supporting files. Though it is referred to as a shapefile, it consists of three or more files, each with different file extensions. The '.shp' (shapefile), '.shx' (header), and '.dbf' (associated database) files are mandatory. You may also have additional files such as '.prj' (projection) and '.lyr' (layer) files. Figure 1.5 shows the shapefile named 'park' in Windows Explorer (which lists multiple files) and ArcCatalog (which displays only a single file). Shapefiles are often referred to with their '.shp' extension in Python scripts.

1.3.3 dBASE Files

One of the shapefile mandatory file types ('.dbf') can also occur as a stand-alone database file. The '.dbf' file format originated with a database management system named 'dBASE'. This format for storing tabular data is referred to as a dBASE file.

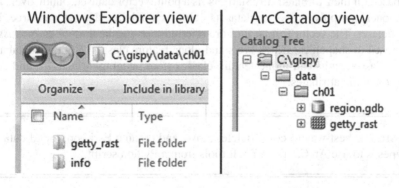

Figure 1.2 Windows Explorer and ArcCatalog views of an Esri GRID raster dataset, 'getty_rast'.

If a dBASE file appears in a directory with a shapefile by the same (base) name, it is associated with the shapefile. If no shapefile in the directory has the same name, it is a stand-alone file.

1.3.4 Layer Files

A '.lyr' file can be used along with a shapefile to store visualization metadata. Usually when a shapefile is viewed in ArcGIS, the *symbology* (the visual representation elements of features, such as color, outline, and so forth) is randomly assigned. Each time it is previewed in ArcCatalog a polygon shapefile might be displayed with a different color, for example. The symbology can be edited in ArcMap® and then a layer file can store these settings. A layer file contains the specifications for the representation of a geographic dataset (a shapefile or raster dataset) and must be stored in same directory as the geographic data. A common source of confusion is another use of the term 'layer'. Data that is added to a map is referred to as a layer (of data). This is not referring to a file, but rather an attribute of the map, data which it displays.

1.3.5 Geodatabase

Esri has three geodatabase formats: file, personal, and ArcSDE™. A *geodatabase* stores a collection of GIS datasets. Multiple formats of data (raster, vector, tabular, and so forth) can be stored together in a geodatabase. Figure 1.3 shows a *file geodatabase*, 'regions.gdb' in Windows Explorer and in ArcCatalog. The left image shows that 'region.gdb' is the name of a directory and inside the directory is a set of files associated with each of the datasets (with extensions .freelist, .gdbindexes, .gdbtable, .gdbtablx, and so forth), only a few of which are shown in Figure 1.3. The ArcCatalog view in Figure 1.3 shows the five datasets (four vector and one raster) in this geodatabase. The geodatabase has a silo-shaped icon. Clicking the geodatabase name expands the tree to show the datasets stored in the geodatabase. The dataset icons vary based on their formats: 'fireStations' is a point vector dataset, 'landCover' and 'workzones' are polygon vector datasets, 'trail' is a polyline dataset, and 'tree' is a raster dataset. The vector format files are referred to as geodatabase *feature classes*, as opposed to shapefiles, which are stand-alone feature classes. Both geodatabases feature classes and stand-alone feature classes store geographic features and their non-geographic attributes.

Note The best way to copy, delete, move, and rename Esri compound data types is to use ArcCatalog or call tools from a Python script.

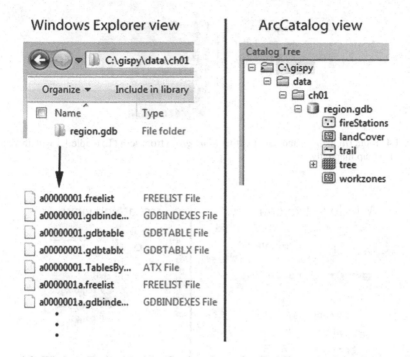

Figure 1.3   Windows Explorer and ArcCatalog views of an Esri file geodatabase, 'region.gdb'.

1.4   An Introductory Example

Example 1.1: This Python script calls the Buffer (Analysis) tool.

```
# simpleBuffer.py
import arcpy
# Buffer park.shp by 0.25 miles. The output buffer erases the
# input features so that the buffer is only outside it.
# The ends of the buffers are rounded and all buffers are
# dissolved together as a single feature.
arcpy.Buffer_analysis('C:/gispy/data/ch01/park.shp',
                      'C:/gispy/scratch/parkBuffer.shp',
                      '0.25 miles', 'OUTSIDE_ONLY', 'ROUND', 'ALL')
```

The ArcGIS Buffer (Analysis) tool, creates polygon buffers around input geographic features (e.g., Figure 1.4). The buffer distance, the side of the input feature to buffer, the shape of the buffer, and so forth can be specified. Buffer analysis has many applications, including highway noise pollution, cell phone tower coverage, and proximity of public parks, to name a few. To get a feel for working with

Figure 1.4 Input (*light gray*) and buffer output (*dark gray*) from script Example 1.1 with the input added to the map manually.

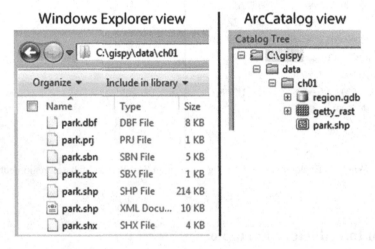

Figure 1.5 Windows Explorer and ArcCatalog views of an Esri shapefile, 'park.shp'.

the sample data and scripts, use a line of Python code to call the Buffer tool and generate buffers around the input features with the following steps:

1. Preview 'park.shp' in 'C:/gispy/data/ch01' using both Windows Explorer and ArcCatalog, as shown in Figure 1.5.
2. When you preview the file in ArcCatalog, a lock file appears in the Windows Explorer directory. Locking interferes with geoprocessing tools. To unlock the file, select another directory in ArcCatalog and then refresh the table of contents (F5). If the lock persists, close ArcCatalog.
3. Open ArcMap. Open the ArcGIS Python Window by clicking the Python button on the standard toolbar, as shown in Figure 1.6 (this window can also be opened from the Geoprocessing Menu under 'Python').
4. Open Notepad (or Wordpad) on your computer.

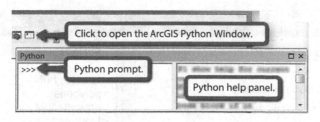

Figure 1.6 The ArcGIS Python Window embedded in ArcMap.

5. In Notepad, browse to the sample scripts for Chapter 1 ('C:\gispy\sample_
   scripts\ch01') and open 'simpleBuffer.py'. It should look like the script shown
   in Example 1.1.
6. Copy the last three lines of code from 'simpleBuffer.py' into the ArcGIS Python
   Window.

Be sure to copy the entirety of all three lines of code, starting with arcpy.
Buffer, all the way through the right parenthesis. It should look like this:

```
arcpy.Buffer_analysis('C:/gispy/data/ch01/park.shp',
                      'C:/gispy/scratch/parkBuffer.shp',
                      '0.25 miles', 'OUTSIDE_ONLY', 'ROUND', 'ALL')
```

7. Press the 'Enter' key and you'll see messages that the buffering is occurring.
8. When the process completes, confirm that an output buffer file has been created
   and added to the map (Figure 1.4 displays the output that was automatically
   added to the map in dark gray and the input which was added to the map by
   hand in light gray). The feature color is randomly assigned, so your buffer color
   may be different.
9. Confirm that you see a message in the Python Window giving the name of the
   result file. If you get an error message instead, then the input data may have
   been moved or corrupted or the Python statement wasn't copied correctly.
10. You have just called a tool from Python. This is just like running it from the
    ArcToolbox™ GUI such as the one shown in Figure 1.7 (You can launch this
    dialog with ArcToolbox > Analysis tools > Proximity > Buffer). The items in the
    parentheses in the Python tool call are the parameter values, the user input.
    Compare these values to the user input in Figure 1.7. Can you spot three differ-
    ences between the parameters used in the Python statement you ran and those
    used in Figure 1.7?

The ArcMap Python Window is good for running simple code and testing code
related to maps (we'll use it again in Chapter 24); However, Chapter 2 introduces
other software which we'll use much more frequently to save and run scripts. Before
moving on to the organization of the book, let's answer the question posed in step
10. The three parameter value differences are the output file names (C:/gispy/
scratch/parkBuffer.shp versus C:\gispy\data\ch01\park_Buff.shp), the buffer distances (0.25 miles versus 5 miles), and the dissolve types (ALL versus NONE).

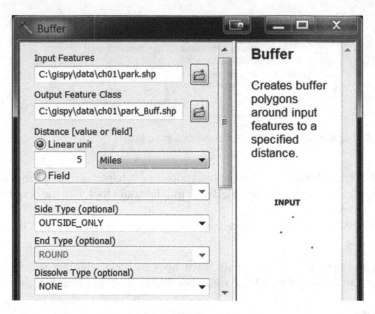

Figure 1.7 Each input slot in the graphical user interface (GUI) for running the buffer tool corresponds to a parameter in the Python code blueprint.

1.5   Organization of This Book

This book focuses on automatically reading, writing, analyzing, and mapping geospatial data. The reader will learn how to create Python scripts for repetitive processes and design flexible, reusable, portable, robust GIS processing tools. The book uses Python to work with the ArcGIS arcpy™ package, as well as HTML, KML, and SQL. Script tool user-interfaces, Python toolboxes, and the arcpy mapping module are also discussed. Expositions are accompanied by interactive code snippets and full script examples. Readers can run the interactive code examples while reading along. The script examples are available as Python files in the chapter 'sample scripts' directory downloadable from the Springer Web site (as explained in Section 1.2). Each chapter begins with a set of learning objectives and concludes with a list of 'key terms' and a set of exercises.

The chapters are designed to be read sequentially. General Python concepts are organized to build the Python skills needed for the ArcGIS scripting capabilities. General Python topic discussions are presented with GIS examples. Chapter 2 introduces the Python programming language and the software used to run Python scripts. Chapters 3 and 4 discuss four core Python data types. Chapters 5 and 6 cover ArcGIS tool help and calling tools with Python using the arcpy package. Chapter 7 deals with getting input from the user. Chapter 8 introduces programming control structures. This provides a backdrop for the decision-making and looping syntax discussed in Chapters 9 and 10. Python can use special arcpy functions to describe data (used for decision-making in Chapter 9) and list GIS data. Batch geoprocessing is performed on lists of GIS data in Chapter 11. Chapter 12 highlights some additional useful list manipulation techniques.

As scripts become more complex, debugging (Chapter 13) and handling errors (Chapter 14) become important skills. Creating reusable code by defining functions (Chapter 15) and modules (Chapter 16) also becomes key. Next, Chapter 17 discusses reading and writing data records using arcpy cursors. In Chapter 18, another Python data structure, called a dictionary, is introduced. Dictionaries can be useful during file reading and writing, which is discussed in Chapter 19. Chapter 20 explains how to access online data, decompress files, and read, write, and parse markup languages. Another code reuse technique, the user-defined class, is presented in Chapter 21. Chapters 22 and 23 show how to create GUIs for file input or other GIS data types with Script Tools and Python toolboxes. Finally, Chapter 24 uses the arcpy mapping module to perform mapping tasks with Python.

1.6   Key Terms

Geoprocessing   Geodatabase
GRID rasters   Feature class
Vector data   Window Explorer vs. ArcCatalog
Symbology   Buffer (Analysis) tool
dBASE file   ArcGIS Python Window
Layer file

Chapter 2
Beginning Python

Abstract Before you can create GIS Python scripts, you need to know where to write and run the code and you need a familiarity with basic programming concepts. If you're unfamiliar with Python, it will be worthwhile to take some time to go over the basics presented in this chapter before commencing the next chapter. This chapter discusses Python development software for Windows® operating systems, interactive mode and scripting, running scripts with arguments, and some fundamental characteristics of Python, including comments, keywords, indentation, variable usage and naming, traceback messages, dynamic typing and built-in modules, functions, constants, and exceptions.

Chapter Objectives
After reading this chapter, you'll be able to do the following:

- Test individual lines of code interactively in a code editor.
- Run Python scripts in a code editor.
- Differentiate between scripting and interactive code editor windows.
- Pass input to a script.
- Explain the advantages of using an integrated development environment, over a general purpose text editor.
- Match code text color with code components.
- Define eight fundamental components of Python code.

2.1   Where to Write Code

Python scripts can be written and saved in any text editor; when a file is saved with a '.py' extension, the computer interprets it as a Python script. However, an *integrated development environment* (IDE), a software application designed for computer programming, is a better choice than a general purpose text editor, because it is tailored for programming. The *syntax* of a programming language is the set of rules that define how to form code statements that the computer will be able to interpret in that language. An IDE can check code syntax, highlight special code

© Springer International Publishing Switzerland 2015   13
L. Tateosian, *Python For ArcGIS*, DOI 10.1007/978-3-319-18398-5_2

statements, suggest ways to complete a code statement, and provide special tools, called debuggers, to investigate errors in the code.

The introductory example in Chapter 1 used the ArcMap Python window. The Python window embedded in ArcGIS desktop applications has some IDE functionality, such as Help pages and automatic code completion. It allows the user to save code (right-click>Save as) or load a saved script (right-click>Load), but it is missing some of the functionality of a stand-alone IDE. For example, it provides no means to pass input into a script and it doesn't provide a debugger. Stand-alone IDE's are also lightweight and allow scripts to be run and tested outside of ArcGIS software. For these reasons, we will mainly use a popular stand-alone IDE called PythonWin.

The PythonWin IDE provides two windows for two modes of executing code: an interactive environment and a window for scripts (Figure 2.1). The interactive environment works like this:

1. The user types a line of code in the interactive window (for example, **print** `'Hello'`).
2. The user presses 'Enter' to indicate that the line of code is complete.
3. The single line of code is run.

The interactive mode is a convenient way to try out small pieces of code and see the results immediately. The code written in the interactive window is not saved

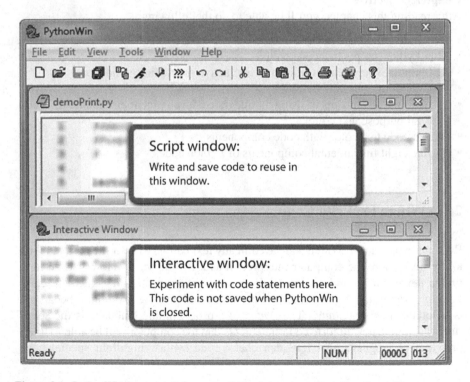

Figure 2.1 PythonWin has two windows: one for scripts and one for interactive coding.

Figure 2.2 The Interactive window opens when PythonWin is launched.

when the IDE is closed. Often we want to save lines of related code for reuse, in other words, we want to save scripts. A *Python script* is a program (a set of lines of Python code) saved in a file with the '.py' extension. A script can later be reopened in an IDE and run in its entirety from there.

Python is installed automatically when ArcGIS is installed, but PythonWin is not. To follow along in the rest of the chapter, install PythonWin and PyScripter based on the instructions in Exercise 1. Then launch PythonWin and locate the Python prompt, as shown in Figure 2.2.

PyScripter is also a good choice as a Python IDE. PyScripter's equivalent of PythonWin's Interactive Window is the Python Interpreter window. PyScripter is more complex than PythonWin, but has additional functionality, such as interactive syntax checking, window tabs, variable watch tools, and the ability to create projects.

2.2 How to Run Code in PythonWin and PyScripter

Once you have installed PythonWin, you'll need to understand how to use the Interactive Window and script windows. When PythonWin is first opened, only the Interactive Window is displayed, because you're not editing a script yet. Python statements in the window labeled 'Interactive Window' are executed as soon as you finish typing the line of code and press the 'Enter' key. The >>> symbol in the Interactive Window is the *Python prompt*. Just like the prompt in the ArcGIS Python Window, it indicates that Python is waiting for input.

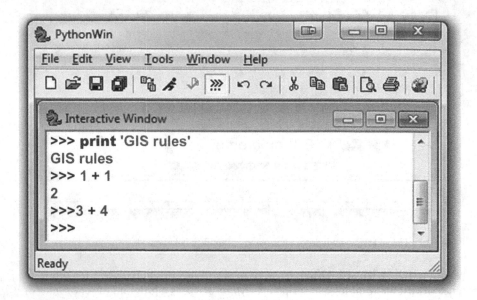

Type something at the Python prompt and press the 'Enter' key. On the following line, Python displays the result. Enter print "GIS rules" and it displays 'GIS rules'. Enter 1 + 1 and Python displays 2. Backspace and then enter 3 + 4. Python doesn't display 7 because there is no space between the prompt and the Python code. There must be exactly one space between the prompt and the input. This problem only occurs if you 'Backspace' too far. When you press the 'Enter' key, the cursor is automatically placed in the correct position, ready for the next command. PyScripter avoids this issue by not allowing you to remove that space after the prompt.

Instead of showing screen shots of the Interactive Window, this book usually uses text to display these interactions. For example, the screen shot above would be replaced with this text:

```
>>> print 'GIS rules.'
GIS rules.
>>> 1 + 1
2
>>> 3 + 4
>>>
```

Table 2.1 PythonWin buttons, keyboard shortcuts, and menu instructions.

Action	Button	Keyboard shortcut	Menu
Create a new script		Ctrl + n, Enter	File > New > OK
Open a script		Ctrl + o	File > Open...
Save a script		Ctrl + s	File > Save
Close script			File > Close
Run a script		Ctrl + r, Enter	File > Run > OK
Tile the windows			Window > Tile
Show line numbers			View > Options > Editor > Set Line numbers > – 30
Shift focus between windows		Ctrl + Tab	
Open/close Interactive Window	≫		View > Interactive Window
Clear the window		Ctrl + a, Delete	Edit > Select All, Edit > Delete
Toggle whitespace		Ctrl + w	View > Options > Tabs and whitespace > check 'View whitespace'

> **Note** Interactive Window examples are given as text with Python prompts. To try the examples yourself, note that the prompt indicates that you should type what follows on that line in the Interactive Window, but don't type the >>> marks.

The interactive mode is helpful for testing small pieces of code and you'll continue to use this as you develop code, but ultimately you will also be writing Python scripts. Editing scripts in an IDE is similar to working with a text editor. You can use buttons (or menu selections or keyboard shortcuts) for creating, saving, closing, and opening scripts. Table 2.1 shows these options. If you're unfamiliar with PythonWin, it would be useful to walk through the following example, to practice these actions.

Create a new blank script as follows:

1. Choose File > new or press **Ctrl + n** or click the 'New' button ⬜.
2. Select 'Python script'.
3. Click 'OK'.

A new blank script window with no >>> prompt appears. Next, organize the display by tiling the windows and turning on line numbers. To tile the windows as shown in Figure 2.1, click in the script window, then select: Window menu>Tile. When you have two or more windows open in PythonWin, you need to be aware of the focus. The *focus* is the active window where you've clicked the mouse most recently. If you click in the Interactive Window and tile the windows again, the Interactive Window will be stacked on top, because the 'Tile' option places the focus window on top. To display the line numbers next to the code in the script window, select the View menu>Options>Editor and set 'Line numbers' to 30. This sets the width of the line numbers margin to 30 pixels. Next, add some content to the script, by typing the following lines of code in the new window:

```
print 'I love GIS...'
print 'and soon I will love Python!'
```

Text font and color in the script window is used to differentiate code elements as we will discuss in an upcoming section. Next, save the Python script in 'C:/gispy/scratch'. To save the script:

1. Click File>Save or press **Ctrl+s** or click the 'Save' button 💾.
2. A 'Save as' window appears. Browse to the 'C:\gispy\scratch' directory.
3. Set the file name to 'gisRules.py'.
4. Click 'Save'.

Beware of your focus; if you select **Ctrl+s** while your focus is in the Interactive Window, it will save the contents of the Interactive Window instead of your script. Confirm that you can view the 'gisRules.py' file name in the 'C:\gispy\scratch' directory in ArcCatalog and the file extension in Windows Explorer. If not, see the box titled "Listing Python scripts in ArcCatalog and Windows Explorer."

Listing Python Scripts in ArcCatalog and Windows Explorer

I. By default, ArcCatalog does not display scripts in the TOC. To change this setting, use the following steps:

1. Customize menu>ArcCatalog Options>File Types tab>New Type
2. Enter *Python* for Description and *py* for File extension
3. Import File Type From Registry...
4. Click 'OK' to complete the process.

It may be necessary to restart ArcCatalog in order to view the python files.

II. By default, Windows® operating system users can see Python scripts in Windows Explorer, but the file extension may be hidden. If you don't see the '.py' extension on Python files, change the settings under the Windows Explorer tools menu. The procedure varies depending on the Windows® operating system version. For example, in Windows 7, follow these instructions:

1. Tools>Folder Options…>View.
2. Uncheck 'Hide extensions for known file types'.
3. Click 'Apply to All Folders'.

and in Windows 8, click View, then check 'File name extensions'.

Back in PythonWin, run the program:

1. Select File>Run or press **Ctrl+r** or click the 'Run' button .
2. A 'Run Script' window appears. Click 'OK'.

PythonWin will run the script and you should see these results in the Interactive Window:

```
>>> I love GIS...
and soon I will love Python!
```

Code from the script prints output in the Interactive Window. PythonWin prints the first line in black and the other in teal; in this case, the text coloring is inconsequential and can be ignored.

With the focus in the 'gisRules.py' script window, select File>Close or click the X in the upper right corner of the window to close the script. Next, we'll reopen 'gisRules.py' to demonstrate running a saved script. To open and rerun it:

1. Select File>Open or press **Ctrl+o** or click the 'Open' button .
2. Browse to the file, 'gisRules.py' in C:\gispy\scratch.
3. Select File>Run or press **Ctrl+r** or click the 'Run' button .
4. A 'Run Script' window appears. Click 'OK'.

You will see the same statements printed again in the Interactive Window. Clearing the Interactive Window between running scripts to make it easier to identify new output. To do so, click inside the Interactive Window to give it focus. Then select all the contents (**Ctrl+a**) and delete them. PythonWin allows you to open multiple script windows, so you can view more than one script at a time, but it only opens one Interactive Window. To open and close the Interactive Window, click on the button that looks like a Python prompt . Table 2.1 summarizes the actions described in this example.

The buttons and keyboard shortcuts are similar in PyScripter. PyScripter uses Ctrl+n, Ctrl+o, Ctrl+s, and Ctrl+Tab in the same way. One notable difference is that running a script is triggered by the green arrow button ▷ and keyboard short-cut, Ctrl+F9 (instead of Ctrl+r). Closing a window is Ctrl+F4. Additional options can be configured by going to Tools>Options>IDE Shortcuts.

2.3   How to Pass Input to a Script

Now you know how to run lines of code directly in the Interactive Window and how to create and run Python scripts in PythonWin. You also need to know how to give input to a script in PythonWin. Getting input from the user enables code reuse without editing the code. User input, referred to as *arguments*, is given to the script in PythonWin through the 'Run Script' window. To use arguments, type them in the 'Run Script' window text box labeled 'Arguments'. To use multiple arguments, separate them by spaces.

This example runs a script 'add_version1.py' with no arguments and then runs 'add_version2.py', which takes two arguments:

1. In PythonWin open (**Ctrl+o**) 'add_version1.py', which looks like this:

```
# add_version1.py: Add two numbers.
a = 5
b = 6
c = a + b
# format(c) substitutes the value of c for {0}.
print 'The sum is {0}.'.format(c)
```

2. Run the script (**Ctrl+r>OK**). The output in the Interactive Window should look like this:

```
>>> The sum is 11.
```

3. 'add_version1.py' is so simple that it adds the same two numbers every time it is run. 'add_version2.py' instead adds two input numbers provided by the user. Open 'add_version2.py', which looks like this:

```
# add_version2.py: Add two numbers given as arguments.
# Use the built-in Python sys module.
import sys
# sys.argv is the system argument vector.
# int changes the input to an integer number.
a = int(sys.argv[1])
b = int(sys.argv[2])
c = a + b
print 'The sum is {0}.'.format(c)
```

4. This time, we will add the numbers 2 and 3. To run the script with these two arguments, select **Ctrl + r** and in the 'Run Script' window 'Arguments' text box, place a 2 and 3, separated by a space.

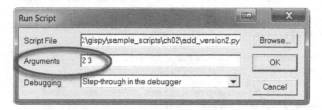

5. Click '**OK**'. The output in the Interactive Window should look like this:

```
>>> The sum is 5.
```

The beginning of 'add_version2.py' differs from 'add_version1.py' so that it can use arguments. The new lines of code in 'add_version2.py' use the system argument vector, `sys.argv`, to get the values of the arguments that are passed into the script. `sys.argv[1]` holds the first argument and `sys.argv[2]` holds the second one. We'll revisit script arguments in more depth in an upcoming chapter. For now, you know enough to run some simple examples.

2.4 Python Components

The color, font, and indentation of Python code, as it appears in an IDE, highlights code components such as comments, keywords, block structures, numbers, and strings. This special formatting, called *context highlighting*, provides extra cues to help programmers. For example, the keyword `import` appears as bold blue in a script in PythonWin, but if the word is misspelled as 'improt', it will have no special formatting. The script, 'describe_fc.py', in Figure 2.3, shows several highlighted code components. 'describe_fc.py' prints basic information about each feature class in a workspace. To try this example, follow steps 1–5:

1. In ArcCatalog, preview the two feature classes ('park.shp' and 'fires.shp') in 'C:/gispy/data/ch02'. Before moving on to step 2, click on the ch02 directory in the ArcCatalog table of contents and select F5 to refresh the view and release the locks on the feature classes.
2. Open the 'describe_fc.py' in PythonWin.
3. Launch the 'Run Script' window (**Ctrl + r**).
4. Type "C:/gispy/data/ch02" in the Arguments text box. The script uses this argument as the data workspace.
5. Click 'OK' and confirm that the output looks like the output shown in Figure 2.4.

```
1    # describe_fc.py
2    # Purpose: Print information about each feature class in a workspace.
3    # Usage: Workspace
4    # Example input: C:/gispy/data/ch02
5    # Output: A list of basic information about each feature class.
6    # Author: Lou Lou Who 7/20/2055
7
8    import arcpy, sys
9
10   # GET the input workspace from the user.
11   arcpy.env.workspace = sys.argv[1]
12
13   # GET a list of the feature classes in the workspace.
14   fcs = arcpy.ListFeatureClasses()
15
16   # PRINT basic information about each feature class in the folder.
17   print 'Feature classes in folder {0}:'.format(arcpy.env.workspace)
18  -for fc in fcs:
19       desc = arcpy.Describe(fc)
20       print 'Name:          {0}'.format(fc)
21       print 'Data type:     {0}'.format(desc.dataType)
22       print 'Data class:    {0}'.format(desc.dataSetType)
23       print 'Type:          {0}'.format(desc.featureType)
24       print 'Shape type:    {0}'.format(desc.shapeType)
25       print 'Has M:         {0}'.format(desc.hasM)
26       print 'Has Z:         {0}'.format(desc.hasZ)
27       print
28   print 'Feature class list complete.'
29
```

Figure 2.3 Python script 'describe_fc.py' as it appears in an IDE.

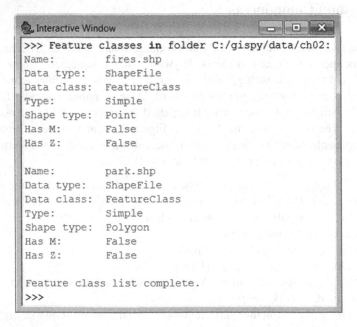

```
Interactive Window                            _  □  ✕
>>> Feature classes in folder C:/gispy/data/ch02:
Name:          fires.shp
Data type:     ShapeFile
Data class:    FeatureClass
Type:          Simple
Shape type:    Point
Has M:         False
Has Z:         False


Name:          park.shp
Data type:     ShapeFile
Data class:    FeatureClass
Type:          Simple
Shape type:    Polygon
Has M:         False
Has Z:         False


Feature class list complete.
>>>
```

Figure 2.4 Output printed in the PythonWin Interactive Window when 'describe_fc.py' (Figure 2.2) runs.

Figure 2.3 shows the script as it is displayed with the default settings in PythonWin. The italic text, bold text, and indentation correspond to comments, keywords, and block structures, respectively. These components along with variables and assignment statements are discussed next.

2.4.1 Comments

Text lines shown in italics by the IDE are *comments*, information only included for humans readers, not for the computer to interpret. Anything that follows one or more hash sign (#) on a line of Python code is a comment and is ignored by the computer. By default, PythonWin displays comments as italicized green text when the comment starts with one hash sign and italicized gray text when the comment starts with two or more consecutive hash signs. Comments have several uses:

- Provide metadata—script name, purpose, author, data, usage (input), sample input syntax, expected output. These comments are placed at the beginning of the script (lines 1–6 in 'describe_fc.py').
- Outline—for the programmer to fill in the code details and for the reader to glean the overall workflow. For example, an outline for 'describe_fc.py' looks like this:

```
# GET the input workspace from the user.
# GET a list of the feature classes in the workspace.
# PRINT basic information about each feature class in the folder.
```

- Clarify specific pieces of code—good Python code is highly readable, but sometimes comments are still helpful. Skilled Python programmers use expository commenting selectively.
- Debug—help the programmer isolate mistakes in the code. Since comments are ignored by the computer, creating 'commented out' sections, can help to focus attention on another section of code. Comment out or uncomment multiple lines of code in PythonWin by selecting the lines and clicking **Alt + 3**/**Alt + 4** (or right-click and choose **Source code > Comment out region/Uncomment region**).

2.4.2 Keywords

Python keywords, words reserved for special usage, are shown by default in bold blue in PythonWin. 'describe_fc.py' in Figure 2.3 uses keywords `import`, `print`, `for`, and `in`. Table 2.2 gives a list of Python keywords. Python is case sensitive; keywords must be written in all lower case.

Table 2.2 Python 2.7 keywords.

and	del	from	not	while
as	elif	global	or	with
assert	else	if	pass	yield
break	except	import	print	
class	exec	in	raise	
continue	finally	is	return	
def	for	lambda	try	

2.4.3 Indentation

Indentation is meaningful in Python. Notice that lines 19–27 are indented to the same position in 'describe_fc.py.' The code in line 18, which starts with the for keyword, tells Python to repeat what follows for each feature class listed in the input directory. Lines 19–27 are indented because they are the block of code that gets repeated. The Name, Data type, and so forth are printed for each feature class. The for keyword structure is an example of a Python *block structure*, a set of related lines of code (a block of code). Block structures will be discussed in more detail later, but for now, it's useful to have some understanding of the significance of indentation in Python. Items within a block structure are sequential code statements indented the same amount to indicate that they are related. The first line of code dedented (moved back a notch) after a block structure does not belong to the block structure. 'describe_fc.py' prints 'Feature class list complete' only once, because line 28 is dedented (the opposite of indented). Python does not have mandatory statement termination keywords or characters such as 'end for' or curly brackets to end the block structure; indentation is used instead. For example, if lines 20–27 were dedented, only one feature class description would be printed, the last one in the list. Indentation and loops will be discussed in more detail in an upcoming chapter.

2.4.4 Built-in Functions

'describe_fc.py' uses the print keyword to print output to the Interactive Window. This works because print is both a keyword and a built-in function. A *function* is a named set of code that performs a task. A *built-in function* is a function supplied by Python. Use it by typing the name of the function and the input separated by commas, usually inside parentheses. A code statement that uses a built-in function has this general format:

```
functionName(argument1, argument2, argument3,...)
```

The built-in `print` function, discussed in more detail in Chapter 3, is an exception to this rule; it does not require parentheses. In Python 2.7, which is the version of Python currently used by ArcGIS, the parentheses are optional in print statements. In Python 3.0 and higher, they are required.

Programming documentation uses special terminology related to dealing with functions, such as 'calling functions', 'passing arguments', and 'returning values'. These terms are defined here using built-in function examples:

- A code statement that invokes a function is referred to as a *function call*. We say we are 'calling the function', because we call upon it by name to do its work. Think of a function as if it's a task assigned to a butler. If you want him to make tea, you need to call on the butler to do so. You don't need to know any more details about how the tea is made, because he does it for you. There is no Python `make_tea` function, so we'll look at the built-in `round` function instead. The following line of code is a function call that calls the `round` function to round a number:

```
>>> round(1.7)
2.0
```

- Providing input for the function is referred to as *passing arguments* into the function.
- When you call the butler to make tea, you need to tell him if you want herbal or green tea or some other kind. The type of tea would be provided as an *argument*. The term 'parameter' is closely related to arguments. *Parameters*, are the pieces of information that can be specified to a function. The `round` function has one required parameter, a number to be rounded. The specific values passed in when calling the function are the arguments. The number `1.7` is used as the argument in the example above. An argument is an expression used when calling the function. The difference between the terms 'argument' and 'parameter' is subtle and often these terms are used interchangeably. The following line of code calls the built-in `min` function to find the minimum number. Here, we pass in three comma separated arguments and the function finds the smallest of the three values:

```
>>> min(1, 2, 3)
1
```

- Some functions come up with results from the actions they perform. Others do some action that may make changes or print information, but don't send a result to the caller. Those that do are said to *return a value*. When we ask our butler to do something, he goes away and does his work. Sometimes he returns with a result—like the cup of tea we asked for. Other times, he just performs his duty and there's no value returned—he dimmed the lights, if that's what you requested, but he doesn't bring your slippers if you only asked him to adjust the lighting. The built-in `round` function returns the rounded number and the `min` function returns the minimum value. These return values are printed in the Interactive

Window, but would not be printed in a script. By contrast, the `help` function is designed to print help about an argument, but it returns nothing. The following line of code calls the built-in `help` function to print help documentation for the `round` function. The name of the function is passed as an argument:

```
>>> help(round)
Help on built-in function round in module __builtin__:

round(...)
round(number[, ndigits]) -> floating point number

Round a number to a given precision in decimal digits (default
0 digits).
This always returns a floating point number. Precision may be
negative.
```

The help prints a *signature*, a template for how to use the function which lists the required and optional parameters. The first parameter of the `round` function is the number to be rounded. The second parameter, `ndigits`, specifies the number of digits for rounding precision. The square brackets surrounding `ndigits` mean that it is an optional argument. The arrow pointing to 'floating point number' means that this is the type of value that is returned by the function.

Other sections of this book employ additional built-in functions (e.g., `enumerate`, `int`, `float`, `len`, `open`, `range`, `raw_input`, and `type`). Search online with the terms, 'Python Standard Library built-in functions' for a complete list of built-in functions and their uses.

2.4.5 Variables, Assignment Statements, and Dynamic Typing

Viewing 'describe_fc.py' in PythonWin also shows script elements in cyan-blue and olive-green. By default in PythonWin and PyScripter, numeric data types are colored cyan-blue and string data types are colored olive-green. All objects in Python have a data type. To understand data types in Python you need to be familiar with variables, assignment statements, and dynamic typing. Variables are a basic building block of computer programming. A programming *variable*, is a name that gets assigned a value (it's like an algebra variable, except that it can be given non-numeric values as well as numeric ones). The following statement assigns the value 145 to the variable FID:

```
>>> FID = 145
```

This line of code is called an assignment statement. An *assignment statement* is a line of code (or statement) used to set the value of a variable. An assignment statement consists of the variable name (on the left), a value (on the right) and a single equals sign in the middle.

Assignment Statement

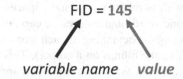

FID = 145

variable name   value

To print the value of a variable inside a script, you need to use the print function. This works in the Interactive Window too, but it is not necessary to use the print function in the Interactive Window. When you type a variable name and press the 'Enter' key in the Interactive Window, Python prints its value.

```
>>> inputData = 'trees.shp'
>>> inputData
'trees.shp'
```

A programming variable is similar to an algebra variable, except that it can be given non- numeric values, so the data type of a variable is an important piece of information. In other programming languages, declaration statements are used to specify the type. Python determines the data type of a variable based on the value assigned to it. This is called *dynamic typing*. You can check the data type of a variable using the built-in type function. The following tells us that FID is an 'int' type variable and inputData is an 'str' type variable:

```
>>> type(FID)
<type 'int'>
>>> type(inputData)
<type 'str'>
```

'int' stands for integer and 'str' stands for string, (variable types which are discussed in detail in the Chapter 3). Due to dynamic typing, the type of a variable can change within a script. These statements show the type of a variable named avg changing from integer to string:

```
>>> avg = 92
>>> type(avg)
<type 'int'>
>>> avg = 'A'
>>> type(avg)
<type 'str'>
```

First, Python dynamically sets the data type of avg as an integer, since 92 is an integer. Python considers characters within quotes to be strings. So when avg is set to 'A', it is dynamically typed to string. Dynamic typing is agile, but beware that if you inadvertently use the same name for two variables, the first value will be overwritten by the second.

2.4.6   Variables Names and Tracebacks

Variable names can't start with numbers nor contain spaces. For names that are a combination of more than one word, underscores or capitalization can be used to break up the words. Capitalizing the beginning of each word is known as camel case (the capital letters stick up like the humps on a camel). This book uses a variation called lower camel case—all lower case for the first word and capitalization for the first letter of the rest. For example, `inputRasterData` is lower camel case.

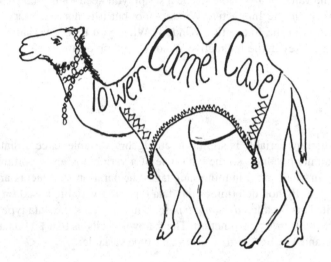

Variable names are case sensitive. Though `FID` has a value of `145` because we assigned it in a previous example, in the following code, Python reports fid as undefined:

```
>>> fid
Traceback (most recent call last):
File "<interactive input>", line 1, in <module>
NameError: name 'fid' is not defined
>>> FID
145
```

When we attempt to check the value of fid, PythonWin prints an error message called a *traceback*. The traceback traces the steps back to where the error occurred. This example was typed in the Interactive Window, so it says 'interactive input', line 1. The last line of the traceback message explains the error. Python doesn't recognize 'fid'. It was never assigned a value, so it is considered to be an undefined object. When Python encounters an undefined object, it calls this kind of error a `NameError`.

> **Tip** Look for an explanation of the error in the last line of a traceback
> message.

Keywords cannot be used as variable names. The following code attempts to use
the keyword **print** as a variable name:

```
>>> print = 'inputData'
Traceback ( File "<interactive input>", line 1
print = 'inputData'
        ^

SyntaxError: invalid syntax
```

Again, a traceback error message is printed, but this time, the message indicates
invalid syntax because the keyword is being used in a way that it is not designed to
be used. This statement does not conform with the rules about how print statements
should be formed, so it reports a SyntaxError.

Python ensures that keywords are not used as variable names by reporting an
error; however, Python does not report an error if you use the name of a built-in
function as variable name. Making this mistake can cause unexpected behavior. For
example, in the code below, the built-in min function is working correctly at the
outset. Then min is used as a variable name in the assignment statement min = 5
Python accepts this with no error. But this makes min into an integer variable
instead of a built-in function, so we can no longer use it to find the minimum value:

```
>>> type(min)
<type 'builtin_function_or_method'>
>>> min(1, 2, 3)
1
>>> min = 5
>>> min(1, 2, 3)
Traceback (most recent call last):
File "<interactive input>", line 1, in <module>
TypeError: 'int' object is not callable
>>> type(min)
<type 'int'>
```

A TypeError is printed the second time we try to use min to find the minimum. min is now an int type so it can no longer be called as a function. To restore
min as a built-in function, you must restart PythonWin.

2.4.7   Built-in Constants and Exceptions

In addition to built-in functions, Python has built-in constants and exceptions. No built-in names should be used as variable names in scripts to avoid losing their special functionality. Built-in constants such as, None, True, and False, are assigned their values as soon as Python is launched, so that value is available to be used anywhere in Python. The built-in constant None is a null value placeholder. The data type of True and False is 'bool' for boolean. Boolean objects can either be True or False :

```
>>> type(True)
<type 'bool'>
>>> True
True
```

The built-in constants True and False will be used in upcoming chapters to set variables that control aspects of the geoprocessing environment. For example, the following lines of code change the environment to allow geoprocessing output to overwrite existing files:

```
>>> import arcpy
>>> arcpy.env.overwriteOutput = True
```

The built-in exceptions, such as NameError or SyntaxError, are created for reporting specific errors in the code. Built-in exceptions are common errors that Python can identify automatically. An exception is *thrown* when one of these errors is detected. This means a traceback message is generated and the message is printed in the Interactive Window; if the erroneous code is in a script, the script will stop running at that point.

New programmers often encounter NameError, SyntaxError, and TypeError exceptions. The NameError usually occurs because of a spelling or capitalization mistake in the code. The SyntaxError can occur for many reasons, but the underlying problem is that one of the rules of properly formed Python has been violated. A TypeError occurs when the code attempts to use an operator differently from how it's designed. For example, code that adds an integer to a string generates a TypeError.

There are many other types of built-in exceptions. The names of these usually have a suffix of 'Error'. A traceback message is printed whenever one of these exceptions is thrown. The final line of traceback error messages states the name of the exception. We'll revisit exceptions and tracebacks in upcoming chapters.

The built-in dir function can be used to print a list of all the built-ins in Python. Type dir(__builtins__) to print the built-ins in PythonWin (there are two underscores on each side).

```
>>> dir(__builtins__)
['ArithmeticError', 'AssertionError', 'AttributeError',
'BaseException', ..., 'str', 'sum', 'super', 'tuple', 'type',
'unichr', 'unicode', 'vars', 'xrange', 'zip']
```

Only a few of the items in the list are printed here, as you'll see when you run the code, there are over 100 built-ins. The built-in constants and built-in exceptions are interspersed in the beginning of this list. The built-in functions are listed next, starting with abs.

2.4.8 Standard (Built-in) Modules

When Python is installed, a library of standard modules is automatically installed. A *module* is a Python file containing related Python code. Python's standard library covers a wide variety of functionality. Examples include math functions, file copying, unzipping files, graphical user interfaces, and even Internet protocols for retrieving online data. To use a module you first use the import keyword. The import statement can be applied to one or more modules using the following format:

```
import moduleName1, moduleName2, ...
```

Once a module is imported, its name can be used to invoke its functionality. For example, the following code imports the standard math module and then uses it to convert 180 degrees to radians:

```
>>> import math
>>> math.radians(180)
3.141592653589793
```

The online 'Python Standard Library' contains documentation of the vast set of standard modules. For now, we'll just highlight two modules (sys and os) that are used in early chapters and introduce others as the need arises.

The sys module gives access to variables that are used by the *Python interpreter*, the program that runs Python. For example, the following code imports the sys module and prints sys.path:

```
>>> import sys
>>> sys.path
['C:\\gispy\\sample_scripts\ch02', u'c:\\program files
(x86)\\ArcGis\\desktop10.3\\arcpy', 'C:\\Windows\\system32\\python27.
zip', 'C:\\Python27\\ArcGIS10.3\\DLLs', 'C:\\Python27\\ArcGIS10.3\\lib',
'C:\\Python27\\ArcGIS10.3\\lib\\plat-win']
```

Notice that `sys.path` is a list of directory paths (only a few are shown here—your list may differ). These are the locations used by the Python interpreter when it encounters an import statement. The Python interpreter searches within these directories for the name of the module being imported. If the module isn't found in one of these directories, the import won't succeed.

The following code prints the file name of the script that was run most recently in PythonWin.

```
>>> sys.argv[0]
'C:\\gispy\\sample_scripts\\ch02\\describe_fc.py'
```

The `sys` module is used to get the workspace from user input on line 11 in Figure 2.3 (`arcpy.env.workspace = sys.argv[1]`). Chapter 7 discusses this useful `sys` module variable in more depth.

The `os` module allows you to access operating system-related methods (`os` stands for operating system). The following code uses the `os` module to print a list of the files in the 'C:/gispy/data/ch02' directory:

```
>>> import os
>>> os.listdir('C:/gispy/data/ch02')
['fires.dbf', 'fires.prj', 'fires.sbn', 'fires.sbx', 'fires.shp',
'fires.shp.xml', 'fires.shx', 'park.dbf', 'park.prj', 'park.sbn',
'park.sbx', 'park.shp', 'park.shp.xml', 'park.shx']
```

2.5 Key Terms

Integrated development environment (IDE)	Dynamic typing
Python script	Block structure, block of code
PythonWin	Dedented
PythonWin Interactive Window	Tracebacks
PythonWin script window	Built-in functions, constants, and exceptions
PyScripter	
Python prompt (>>>)	Function arguments
Python interpreter	Function signatures
Window focus	Function parameters
Script arguments	Exceptions thrown
Context highlighting	`True, False, None`
Code comments	`NameError, SyntaxError,` and `TypeError`
Hash sign	
Python keywords	Module
Variables	Standard modules
Assignment statement	`math, sys, os, shutil`

2.6 Exercises

1. Set up your integrated development environment (IDE) software and check the functionality as described here:

   (a) To install PythonWin,

      1. Browse 'C:\gispy\sample_scripts\ch02\programs\PythonWin'.
      2. Double-click on the executable ('exe') file.
      3. Launch PythonWin. A PythonWin desktop icon should be created during installation. If not, search the machine for 'PythonWin.exe' and create a shortcut to this executable. When you first launch PythonWin, it will display the 'Interactive Window' with a prompt that looks just like the one in the ArcGIS Python window as shown in Figure 2.2.
      4. Test PythonWin. Type `import arcpy` in the Interactive Window and press the 'Enter' key. If no error appears, everything is working. In other words, if you see only a prompt sign on the next line (>>>), then it worked. If, instead, you see red text, this is an error message. Check your ArcGIS software version. This book is designed for ArcGIS 10.1-10.3, which use the 2.7 version of Python. If you are using a different version of ArcGIS, you will need to get a different PythonWin executable. Start by searching online for 'pywin32 download', then navigate to the version you need.

ArcGIS version	Python version
10.1, 10.2, 10.3	2.7
10	2.6
9.3	2.5
9.2	2.4.1
9.1	2.1

   (b) Install PyScripter. To install PyScripter, browse to 'C:\gispy\sample_scripts\ch02\programs\PyScripter'. Double-click on the '.exe' file. Install using the defaults. PyScripter's equivalent of PythonWin's Interactive Window is the Python Interpreter window. Confirm that it is working correctly by typing `import arcpy` in the Python Interpreter window. Press the 'Enter' key. If you see only a prompt sign on the next line (>>>), then it worked. If, instead, you see red text, this is an error message. The ArcGIS Resources Python Forum is a good place for additional trouble-shooting.

2. Type the code statements below in the Interactive Window. Notice that the built-in `type` function gives two different results—first `str` and second `int`. Which

word or phrase in the 'Key terms' list at the end of this chapter explains this
phenomenon?

```
>>> month = 'December'
>>> type(month)
>>> month = 12
>>> type(month)
```

3. Match each key term with the most closely related statement. There are two dis-
tracter statements that should not be used.

Key term	Statement
1. Python keyword	A. This is a built-in boolean constant.
2. IDE	B. If this isn't in the Script Window, you might save the Interactive Window by mistake.
3. Hash sign	
4. Tracebacks	C. Special font formatting (such as color) based on meaning of text in the code.
5. Window focus	D. The controversial debates surrounding Python versus Perl scripting languages.
6. Script arguments	
7. Context highlighting	E. PythonWin shows **print** in blue because it is one of these.
8. Assignment statement	F. If a variable name is misspelled, this can occur.
9. Dynamic typing	G. A specialized editor for developing scripts.
10. False	H. Information passed into a script by the user.
11. NameError	I. Automatically set the data type of a variable when a value is assigned.
12. Dedented code	
13. `math`, `os`, and `sys`	J. `x = 5` is an example of one of these.
	K. Messages printed when exceptions are thrown.
	L. This signals the end of a related block of code.
	M. The shape of a camel's back.
	N. Code following this character is only meant for the human reader.
	O. These are examples of Python standard modules.

4. Type the following lines of code in the Interactive Window. Each line of code should raise a built-in exception. Report the *name* of the built-in exception that is raised (The first row is done for you).

Python code	Exception name
`class = 'combustibles'`	SyntaxError
`'five' + 6`	
`Min`	
`int('five')`	
`5/0`	
`input file = 'park.shp'`	

5. **times.py** Write a script named 'times.py' that finds the product of two input integers and prints the results. Start by opening 'add_version2.py', modifying the last two lines and saving the file with the name 'times.py'. An asterisk is used as the multiplication operator in Python, so you'll use `c = a * b` to multiply. Also change the comment on line 1 to reflect the modified functionality. Run the script using two integer arguments to check that it works. The example below shows the expected behavior. Check your results against this example.

Example input:
2 3

Example output:
```
>>> The product is 6.
```

Chapter 3
Basic Data Types: Numbers and Strings

Abstract All Python objects having a data type. Built-in Python data types, such as integers, floating point values, and strings are the building blocks for GIS scripts. This chapter uses GIS examples to discuss Python numeric data types, mathematical operators, string data types, and string operations and methods.

Chapter Objectives
After reading this chapter, you'll be able to do the following:

- Perform mathematical operations on numeric data types.
- Differentiate between integer and floating point number division.
- Determine the data type of a variable.
- Index into, slice, and concatenate strings.
- Find the length of a string and check if a substring is contained in a string.
- Replace substrings, modify text case in strings, split strings, and join items into a single string.
- Differentiate between string variables and string literals.
- Locate online help for the specialized functions associated with strings.
- Create strings that represent the location of data.
- Format strings and numbers for printing.

3.1 Numbers

Python has four numeric data types: int, long, float, and complex. Examples in this book mainly use int (signed integers) and float (floating point numbers). With dynamic typing, the variable type is decided by the assignment statement which gives it a value. Float values have a decimal point; Integer values don't. In the following example, x is an integer and y is a float:

```
>>> x = 2
>>> y = 2.0
>>> type(x)
```

© Springer International Publishing Switzerland 2015
L. Tateosian, *Python For ArcGIS*, DOI 10.1007/978-3-319-18398-5_3

Table 3.1 Numerical operators.

Operation	Operator	Example	Result
Addition	+	7 + 2	9
Subtraction	−	7 − 2	5
Multiplication	*	7 * 2	14
Division	/	7 / 2	3
Exponentiation	**	7**2	9
Modulus division	%	7 % 2	1

```
<type 'int'>
>>> type(y)
<type 'float'>
```

By default, PythonWin uses cyan-blue to display numeric values (as well as output in the Interactive Window). Table 3.1 shows the operators symbols, +, −, *, **, and / for addition, subtraction, multiplication, exponentiation, and division. Mathematical order of operations is preserved. 2 + 3 * 4 gives 14 not 20, though 2 + (3 * 4) is a clearer way to write it. Be aware that Python integer division and floating point division give different results. If both division operands are integers, the result will be *integer division* with the fractional part truncated; in other words, it always rounds down. If at least one of the operands are float, the result retains higher precision.

```
>>> 8/3
2
>>> 8.0/3
2.6666666666666665
```

In Python versions 3.0 and higher, this behavior is changed so that 8/3 gives the same result as 8.0/3.

3.2 What Is a String?

By default, PythonWin uses olive-green text to display string literals. A *string literal* is a set of characters surrounded by quotation marks. A variable assigned a string literal value is called a *string variable*. The difference between these two terms is important, but both of these items are sometimes referred to simply as 'strings'. The characters inside the quotes in the string literal are interpreted literally. In GIS scripts, we use string literals for things like the name of a workspace, input file, or output file. Meanwhile, the characters that make up a string variable are simply a name for the program to use to represent its value. Variable names should not be surrounded by quotes. The operations and methods described in the next section apply to both string literals and string variables. Both are objects of data type str (for string).

Strings

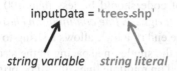

inputData = 'trees.shp'

string variable   *string literal*

The type of quotes used to create a Python string literal is flexible, as long as the opening and closing quotes match. Here are two variations:

```
>>> inputData = 'trees.shp'
>>> inputData
'trees.shp'
>>> inputData = "trees.shp"
>>> inputData
'trees.shp'
```

There is a subtle difference between printing strings with and without the print function in the Interactive Window; the former removes the quotation marks entirely and the latter uses single quotation marks.

```
>>> print inputData
trees.shp
>>> inputData
'trees.shp'
```

String literals in single or double quotes cannot span more than one line. A string literal with no closing quote raises a SyntaxError as shown in the traceback message:

```
>>> output = 'a b c
Traceback ( File "<interactive input>", line 1
output = "a b c
             ^
SyntaxError: EOL while scanning string literal
```

Triple quotes can be used to create a string that spans more than one line:

```
>>> output = """a b c
... d e f"""
>>> print output
a b c
d e f
```

The triple quotes store the carriage return as part of the string. In some cases we don't want a carriage return stored in the string. For example, we might need to store a long file path name. The path won't be interpreted correctly if a carriage return is embedded in it. But very wide lines of Python code are awkward to read. As a rule of thumb, Python lines of code should be less than 100 characters wide, so the reader doesn't need to scroll right. A *line continuation character* ('\'), a backslash embedded in a string at the end of a line, allows a string to be written on more than one line, while preserving the single line spacing in the string literal value. In other words, though it is assigned on multiple lines, it will be printed as a single line variable. Place the character at the end of the line and the string literal can be continued on the next line, but the string value will not contain the carriage return:

```
>>> output = 'a b c \
d e f \
g h i'
>>> print output
a b c d e f g h i
```

Numerical characters surrounded by quotation marks are considered to be strings literals by Python.

```
>>> FID = 145
>>> type(FID)
<type 'int'>
>>> countyNum = '145'
>>> type(countyNum)
<type 'str'>
```

If the variable is a string type, you can perform string operations on it, as described in the next section.

3.3   String Operations

GIS Python programming requires frequent string manipulation. To deal with file names, field names, and so forth, you'll need to be familiar with finding the length of a string, indexing into a string, concatenating, slicing, and checking for a sub-string in a string. Examples of each operation follow.

3.3.1   Find the Length of Strings

The built-in len function finds the length of a string literal:

```
>>> len('trees.shp')
9
```

Or the length of the value held by a string variable:

```
>>> data = 'trees.shp'
>>> len(data)
9
```

3.3.2   Indexing into Strings

Each character in a string has a numbered position called an *index*. The numbering starts with zero, in other words Python uses *zero-based indexing*. From left to right, the indices are 0, 1, 2, and so forth.

$$\text{fieldName} = \text{'COVER'}$$
$$\uparrow \uparrow \uparrow \uparrow \uparrow$$

Index number: 0 1 2 3 4

Indexing into a string means pointing to an individual character in a string using its index number. To index into a string variable, use square brackets after the variable name and place an index number inside the brackets. The general format for indexing looks like this:

```
variableName[index_number]
```

This example assigns a value to a string variable and then indexes the first character in the string value:

```
>>> fieldName = 'COVER'
>>> fieldName[0]
'C'
```

Since indexing is zero-based, the last valid index is the length of the string minus one. Attempting to using an invalid index number results in an IndexError:

```
>>> len(fieldName)
5
>>> fieldName[5]
Traceback (most recent call last):
File "<interactive input>", line 1, in <module>
IndexError: string index out of range
>>> fieldName[4]
'R'
```

Negative indices count one-based from the right. This can be useful for getting the last character without checking the string length.

```
>>> fieldName[-1]
'R'
```

It is not possible to change the value of an individual character of a string with indexing. Attempting to do so results in a TypeError, as shown in the following code:

```
>>> fieldName[0] = 'D'
Traceback (most recent call last):
File "<interactive input>", line 1, in <module>
TypeError: 'str' object does not support item assignment
```

This occurs because Python strings are *immutable*. The word 'immutable' is a synonym for unchangeable. Of course, we can change the value of a string variable by assigning an entirely new string to it, as in the example below, but in this context, immutability refers to this specific quality of not being able to change individual parts of an existing object. To change the value of fieldName from COVER to DOVER, you need to use a string literal:

```
>>> fieldName = 'DOVER'
```

Or, you can use the string replace method discussed in Section 3.4:

```
>>> fieldName = fieldName.replace('C','D')
```

3.3.3   Slice Strings

Indexing can be thought of as getting a substring which is only one character long. To get more than one character, use slicing instead of indexing. Slicing gets a substring (a slice) of a string. Slicing uses square brackets with a colon inside. A number on either side of the colon indicates the starting and ending index of the desired slice. The left number is inclusive, the right is exclusive.

```
>>> fieldName = 'COVER'
>>> fieldName[1:3]
'OV'
```

The letter O is the index 1 character in the string 'COVER' and the letter E is the index 3 character, so the slice starts at the letter O and ends just before the letter E.

If the left slice number is omitted, the slice starts at the beginning of the word. If the right slice number is omitted, the slice ends at the end of the word.

```
>>> fieldName[:3]
'COV'
>>> fieldName[1:]
'OVER'
```

We often want to use the *base name* of a file, the part without the extension, to build another file name, so that we can create output names based on input names. Counting from the right with a negative index can be used to find the base name of a file which has a three digit file extension. Omitting the left index starts from the beginning of the string. Using – 4 as the right index removes the last four characters— the three digit file extension and the period—leaving the base name of the file.

```
>>> inputData = 'trees.shp'
>>> baseName = inputData[:-4] # Remove the file extension.
>>> baseName
'trees'
```

This approach assumes you know the file extension length. A more general solution using the os module is discussed in Chapter 7.

3.3.4 Concatenate Strings

Concatenation glues together a pair of strings. You use the same sign for addition, but it acts differently for strings. The plus sign performs addition on numeric values and concatenation on strings.

```
>>> 5 + 6 # adding two numbers together
11
>>> '5' + '6' # concatenating two strings
'56'

>>> rasterName = 'NorthEast'
>>> route    = 'ATrain'
>>> output = rasterName + route
>>> output
'NorthEastATrain'br />
```

Both of the variables being concatenated must be string types or you'll get a TypeError. For example, using the plus sign between a numeric variable and a string variable causes an error. Python doesn't know whether to perform addition or concatenation.

```
>>> i = 1
>>> rasterName = 'NorthEast'
>>> output = rasterName + i
Traceback (most recent call last):
File "<interactive input>", line 1, in <module>
TypeError: cannot concatenate 'str' and 'int' objects
>>>
```

The `TypeError` says that a string cannot be concatenated with an integer object. Solve this problem by using the built-in `str` function which returns a string data type version of an input value. Type conversion is referred to as *casting*. To combine a string with a numeric value, cast the number to a string. Then the number is treated as a string data type for the concatenation operation:

```
>>> i = 145
>>> str(i)
'145'
>>> rasterName = 'NorthEast'
>>> output = rasterName + str(i)
>>> output
'NorthEast145'
```

We often use concatenation with slicing to create an output name based on an input file name.

```
>>> inputData = 'trees.shp'
>>> baseName = inputData[:-4]
>>> baseName
'trees'
>>> outputData = baseName + '_buffer.shp'
>>> outputData
'trees_buffer.shp'
```

3.3.5 Check for Substring Membership

The `in` keyword enables you to check if a string contains a substring. Python documentation refers to this as checking for 'membership' in a string. Suppose you want to check if a file is a buffer output or not, and you have named each buffer output file so that the name contains the string `buff`.

```
>>> substring = 'buff'
>>> substring in outputData
True
>>> substring in inputData
False
```

Table 3.2 Sequence operations on strings.

>>> exampleString = 'tuzigoot'		
Operation	**Sample code**	**Return value**
Length	`len(exampleString)`	8
Indexing	`exampleString[2]`	`'z'`
Slicing	`exampleString[:-4]`	`'tuzi'`
Concatenation	`exampleString+exampleString`	`'tuzigoottuzigoot'`
Membership	`'ample' in exampleString`	False

These string operations are powerful when combined with batch processing (discussed in Chapters 10 and 11). Table 3.2 summarizes these common string operations. As we'll discuss shortly, they can be applied to other data types as well. Strings and other data types that have a collection of items are referred to as *sequence* data types. The characters in the string are the individual items in the sequence. The operations in Table 3.2 can applied to any of the sequence data types in Python.

3.4   More Things with Strings (a.k.a. String Methods)

Along with the operations described above, processing GIS data often requires additional string manipulation. For example, you may need to replace the special characters in a field name, you may need to change a file name to all lower case, or you may need to check the ending of a file name. For operations of this sort, Python has built-in string functions called string methods. *String methods* are functions associated particularly with strings; they perform actions on strings. Calling a string method has a similar format to calling a built-in function, but in calling a string method, you also have to specify the string object—using dot notation. *Dot notation* for calling a method uses both the object name and the method name, separated by a dot. The general format for dot notation looks like this:

```
object.method(argument1, argument2, argument3,...)
```

'Object' and 'method' are object-oriented programming (OOP) terms and 'dot notation' is an OOP specialized syntax.

- Everything in Python is an *object*. For example, numbers, strings, functions, constants, and exceptions are all object*s*. Other programming languages use this term more narrowly to refer to data types which have associated functions and attributes. In Python, most objects have accompanying functions and attributes, which are also referred to as methods and properties. As soon as a variable is assigned a value, it is a string object which has string methods.
- A *method i*s a function that performs some action on the object. Methods are simply functions that are referred to as 'methods' because they are performed

on an object. The terms 'calling methods', 'passing arguments', and 'returning values' apply to methods in the same way they apply to functions. The example below calls the `replace` method. The variable, `line`, is assigned a string value. Then the dot notation is used to call the string method named `replace`. This example returns the string with commas replaced by semicolons:

```
>>> line = '238998,NPS,NERO,Northeast'
>>> line.replace(',' , ';')
'238998;NPS;NERO;Northeast'
```

- *Dot notation* is an object-oriented programming convention used on string objects (and other types of objects) to access methods and properties, that are specially designed to work with those objects.

The image below points out the components in the `replace` method example. The dot notation links the object method to the object. The object is the variable named `line`. The method (`replace`) takes two string arguments. These arguments specify the string to replace (a comma) and the replacement (a semicolon).

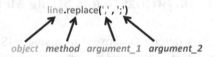

object method argument_1 argument_2

String methods need to be used in assignment statements. Strings are immutable, so string methods do not change the value of the string object they are called on; instead, they return the modified value.

For example, the `replace` call does not alter the value of the variable, `line`:

```
>>> line = '238998,NPS,NERO,Northeast'
>>> line.replace(',' , ';')
>>> line
'238998,NPS,NERO,Northeast'
```

An assignment statement can be used to store the return value of the method. Call the `replace` method on the right-hand side of an assignment statement like this:

Gets the return value

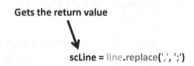

scLine = line.replace(',', ';')

```
>>> line = '238998,NPS,NERO,Northeast'
>>> semicolonLine = line.replace(',' , ';')
>>> semicolonLine
'238998;NPS;NERO;Northeast'
```

To alter the original variable called line, use it as the variable being assigned. In other words, you can put it on both sides of the equals sign:

```
>>> line = line.replace(',' , ';')
>>> line
'238998;NPS;NERO;Northeast'
```

Not all methods require arguments; However, even when you don't pass any arguments, you must use the parentheses. For example, to change the letters of a string to all uppercase letters, use the upper method like this:

```
>>> name = 'Delaware Water Gap'
>>> name = name.upper()
>>> print name
DELAWARE WATER GAP
```

The split and join methods are used in many GIS scripts. These methods involve another Python data type called a list (Python lists, the main topic of Chapter 4, are containers for holding sets of items). The split method returns a list of the words in the string, separated by the argument. In the example below, the split method looks for each occurrence of the forward slash (/) in the string object and splits the string in those positions. The resulting list has five items, since the string has four slashes.

```
>>> path = 'C:/gispy/data/ch03/xy1.txt'
>>> path.split('/')
['C:', 'gispy', 'data', 'ch03', 'xy1.txt']
```

The split and join methods have inverse functionality. The split method takes a single string and splits it into a list of strings, based on some delimiter. The join method takes a list of strings and joins them with into a single string. The string object value is placed between the items. For example, elephant is placed between 1, 2, and 3 here:

```
>>> numList = ['1', '2', '3']
>>> animal = 'elephant'
>>> animal.join(numList)
'1elephant2elephant3'
```

In the following example, the join method is performed on a string literal object, semicolon (;). The method inserts semicolons between the items and returns the resulting string:

```
>>> pathList = ['C:', 'gispy', 'data', 'ch03', 'xy1.txt']
>>> ';'.join(pathList)
'C:;gispy;data;ch03;xy1.txt'
```

IDEs make it easy to browse for methods of an object by bringing up a list of choices when you type the object name followed by a dot. For example, if you create a string variable named `path` and then type `path` in the PythonWin Interactive Window, a *context menu* appears with a list of string methods. This menu of choices is referred to as a context menu because the menu choices update dynamically based on the context in which it is generated. You can scroll and select a method by clicking on your choice.

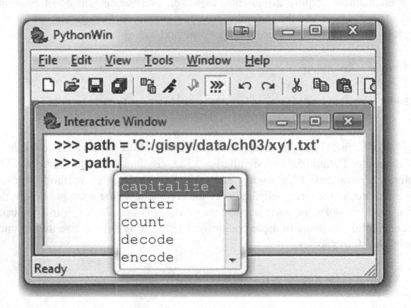

In PythonWin, context menus appear in the Interactive Window or after the code in a script has been executed (after the script has been run). In PyScripter, the context menus appear as soon as a variable has been defined within a script. If your choice doesn't appear in the context menu, check for spelling errors. Try using the context menu to bring up the `endswith` method for a string variable in the PythonWin Interactive Window. As you start typing the method name, the list box rolls to that name and you can use the 'Tab' key to complete the selection. The `endswith` method checks if the ending matches the argument and returns `True` or `False`:

```
>>> path = 'C:/gispy/data/ch03/xy1.txt'
>>> path.endswith('.shp')
False
>>> path.endswith('.txt')
True
```

The examples here demonstrated a few of the string methods. In fact there are many more, including `capitalize`, `center`, `count`, `decode`, `encode`, `endswith`, `expandtabs`, `find`, `index`, `isalnum`, `isalpha`, `isdigit`, `islower`, `isspace`, `istitle`, `isupper`, `join`, `ljust`, `lower`, `lstrip`, `partition`, `replace`, `rfind`, `rindex`, `rjust`, `rpartition`, `rsplit`, `restrip`, `split`, `splitlines`, `startswith`, `strip`, `swapcase`, `title`, `translate`, `upper`, and `zfill`. Help documentation and a comprehensive list of string methods is available online (search for 'Python String Methods'). String method names are often intuitive. Testing them in the Interactive Window helps to clarify their functionality.

3.5   File Paths and Raw Strings

When dealing with file paths, you will encounter strings literals containing escape sequences. *Escape sequences* are sequences of characters that have special meaning. In string literals, the backslash (\) is used as an *escape character* to encode special characters. The backslash acts as a line continuation character when placed at the end of a line in a string literal as described in Section 3.2. But when a backslash is followed immediately by a character in a string literal, the backslash along with the character that follows it are called an *escape sequence* and the backslash is interpreted as a signal that the next character is to be given a special interpretation. For example, the string literals `'\n'` and `'\t'` are escape sequences that encode 'new line' and 'tab'. New line and tab characters are used to control the space around the visible characters. They are referred to as *whitespace characters*, because the characters themselves are not visible when printed.

White space escape sequences	
\n	new line
\t	tab

When printed without the built-in `print` function in the Interactive Window, whitespace escape sequences are printed as entered:

```
>>> 'X\tY\tValue\n\n16\t255\t6.3'
'X\tY\tValue\n\n16\t255\t6.3'
```

When printed with the built-in `print` function, whitespace escape sequences are interpreted to modify the formatting:

```
>>> print 'X\tY\tValue\n\n16\t255\t6.3'
X    Y    Value

16    255    6.3
```

The strip method can be used to strip unwanted leading or trailing whitespace
from a string. This is often called for when processing files. The following example
prints a string before and after using the strip method:

```
>>> dataRecord = ' \n\t\tX\tY\tZ\tM\t'
>>> print dataRecord

            X    Y     Z    M
>>> dataRecord = dataRecord.strip()
>>> dataRecord
'X\tY\tZ\tM'
>>> print dataRecord
X    Y    Z    M
>>>
```

Escape sequences can lead to unintended consequences with file paths that contain backslashes. In this example, the t in terrain and the n in neuse are replaced
by whitespace when the string is printed:

```
>>> dataPath = 'C:\terrain\neuse_river'
>>> dataPath
'C:\terrain\neuse_river'
>>> print dataPath
C:      errain
euse_river
```

Here are three options for avoiding this problem:

1. Use a forward slash instead of a backward slash. The forward slash is not an
   escape character, but is interpreted correctly as a separator in file paths. This
   book uses forward slashes (southwest to northeast) instead of backward slashes,
   as in this example:

   ```
   >>> dataPath = 'C:/terrain/neuse_river'
   >>> print dataPath
   C:/terrain/neuse_river
   ```

2. Double the backslashes. This works because the first slash is an escape character
   that tells the code to interpret the second slash literally as a slash.

   ```
   >>> dataPath = 'C:\\terrain\\neuse_river'
   >>> print dataPath
   C:\terrain\neuse_river
   ```

3. Use *raw strings*. When you first come across a string literal preceded by a
   lowercase r, you might guess that it's a typo. But in fact, placing an r just before

a string literal creates a raw string. Python uses the raw value of a raw string. In other words, it disregards escape sequences, as in the following example:

```
>>> dataPath = r'C:\terrain\neuse_river'
>>> print dataPath
C:\terrain\neuse_river
```

3.6 Unicode Strings

When you start using ArcGIS functionality in Chapter 6, you will begin to see a lowercase u preceding strings that are returned by GIS methods. The u stands for unicode string. A *unicode string* is a specially encoded string that can represent thousands of characters, so that non-English characters, such as the Hindi alphabet can be represented. A unicode string is created by prepending a u to a string literal, as shown here:

```
>>> dataFile = u'counties.shp'
>>> dataFile
u'counties.shp'
>>> type(dataFile)
<type 'unicode'>
```

The difference between 'str' and 'unicode' string data types in Python lies in the way that the strings are encoded. The default encoding for Python 'str' strings is based on the American Standard Code for Information Interchange (ASCII). Because ASCII encodings were designed to encode English language characters, they can only represent hundreds of characters; whereas, the more recently developed unicode technique can encode thousands. Because of its capability to encode non-English languages, software programs, including the ArcGIS Python interface, have begun to use unicode encodings more often.

You don't need to know exactly how unicode or ASCII strings encoding works. You just need to know that in your GIS scripts, you can handle 'unicode' strings just like 'str' strings. They have the same string operations and methods. The following examples demonstrate using a few string methods and operations on the unicode variable dataFile:

```
>>> dataFile.endswith('.shp') # Does the string end with '.shp'?
True
>>> dataFile.startswith('co') # Does the string start with 'co'?
True
>>> dataFile.count('s') # How many times does 's' occur in the string?
2
```

The output, from methods and operations that return strings, is unicode when the input object is unicode:

```
>>> dataFile.upper() # Return an all caps. string.
u'COUNTIES.SHP'
>>> dataFile[5] # Index the 6th character in the string.
u'i'
>>> dataFile + dataFile # Concatenate two strings.
u'counties.shpcounties.shp'
```

Just as the quotation marks are not printed by the built-in `print` function, the unicode u is not printed when you use the built-in `print` function to print a unicode string:

```
>>> print dataFile
counties.shp
```

3.7   Printing Strings and Numbers

The built-in `print` function is used frequently in scripting, so we'll show a few examples of how it can be used. As mentioned earlier, the `print` function does not use parentheses around the arguments (though this changes in Python 3.0 in which the parentheses become required). The arguments are the expressions to be printed. We often want to print multiple expressions within the same print statement and these expressions need to be linked together so that the print statement uses them all. Here we demonstrate three approaches to linking expressions to be printed:

Commas. When commas are placed between variables in a `print` expression, the variable values are printed separated by a space. The print statement inserts a space where each comma occurs between printed items:

```
>>> dataFile = 'counties.shp'
>>> FID = 862
>>> print dataFile, FID
counties.shp 862
```

The expression can be a combination of comma separated string literals, numbers, and variables of assorted data types:

```
>>> print 'The first FID in', dataFile, 'is', FID, '!'
The first FID in counties.shp is 862 !
```

Concatenation. The spacing created by the commas may be undesirable in some situations. For example, we usually don't want a space before punctuation.

Concatenation can be used to avoid this problem, though concatenation introduces its own complications. Recall that concatenation uses a plus sign to join strings. If we replace each comma in the above expression with a plus sign, we get a `TypeError` because one piece of the expression is not a string:

```
>>> print 'The first FID in' + dataFile + 'is' + FID + '!'
Traceback (most recent call last):
File "<interactive input>", line 1, in ?
TypeError: cannot concatenate 'str' and 'int' objects
```

`FID` is an integer type, so it must be cast to string. This works, but the spacing isn't correct since the plus signs are not replaced by spaces:

```
>>> print 'The first FID in' + dataFile + 'is' + str(FID) + '!'
The first FID incounties.shpis862!
```

When using concatenation statements to print strings, you may have to tailor the spacing, by adding spaces in the string literals:

```
>>> print 'The first FID in ' + dataFile + ' is ' + str(FID) + '!'
The first FID in counties.shp is 862!
```

String formatting. The string `format` method provides an alternative that uses place-holders to allow you to lay out a string with the spacing you want and handle different variable types without casting. The `format` method is performed on a string literal object, which contains place-holders for variables and gets the variables as a list of arguments. Place-holders are inserted into string literals as numbers within curly brackets ({0}, {1}, {2}... and so forth). The numbers refer to the zero-based index of the arguments in the order they are listed. The string values of arguments are substituted for the place-holders. The method returns a string with these substitutions. This example calls the `format` method in a print statement:

```
>>> print 'The first FID in {0} is {1}!'.format(dataFile, FID)
The first FID in counties.shp is 862!
```

The `format` method uses the dot notation. Here's a detailed breakdown of our example:

- `'The first FID in {0} is {1}!'` is the object.
- `format` is the method.
- `dataFile` and `FID` are the arguments. The string value of `dataFile` is substituted for {0}. The string value of `FID` is substituted for {1} because dataFile appears first in the argument list and FID appears second.
- The return value is printed.

Triple quotes can be used in combination with the `format` method to create multi-line strings:

```
>>> print '''X    Y     Value
... -------------------
... {0}    {1}    {2}'''.format(16, 255, 6.3)
X    Y    Value
-------------------
16   255   6.3
```

3.8   Key Terms

`int` data type	Casting
`float` data type	The `in` keyword
Integer division	Dot notation
`str` data type	Objects
String literal	Methods
String variable	Context menus
Line continuation	Whitespace characters
Zero-based indexing	Escape sequences
Built-in `len` function	Raw strings
Slicing	Unicode strings
Concatenating	String formatting

3.9   Exercises

1. The Python statements on the left use string operations involving the variable `happyCow`. Match the Python statement with its output. All answers MUST BE one of the letters A through I. The string variable called `happyCow` is assigned as follows:

```
>>> happyCow = 'meadows.shp'
```

Python statement	Output (notice there are nine letters)
1. `happyCow[0]`	
2. `happyCow[0:5] + happyCow[-4:]`	A. `IndexError`
3. `len(happyCow)`	B. `'meado.shp'`
4. `happyCow[0:5]`	C. `'meado'`
5. `happyCow[-4:]`	D. `True`
6. `happyCow[11]`	E. `False`
7. `happyCow[:5]`	F. `11`
8. `happyCow in "5meadows.shp"`	G. `'w'`
9. `happyCow[5]`	H. `'.shp'`
10. `'W' in happyCow`	I. `'m'`

2. These Python statements use string methods and operations involving the variable `LCS_ID`. Determine if each Python statement is true or false. The double equals signs return true if the two sides are equal. The `!=` signs return true if the two sides are not equal. The string variable called `LCS_ID` is assigned as follows: `LCS_ID = '0017238'`

   (a) `'17' in LCS_ID`
   (b) `LCS_ID.isdigit()`
   (c) `LCS_ID.lstrip('0') == '17238'`
   (d) `LCS_ID.zfill(10) == '10101010'`
   (e) `LCS_ID + '10' == 17248`
   (f) `LCS_ID[6] == '3'`
   (g) `len(LCS_ID) == 7`
   (h) `LCS_ID[0:7] == '0017238'`
   (i) `int(LCS_ID) + 10 == 17248`
   (j) `LCS_ID != 17238`

3. The Python statements on the left use string methods and operations involving the variable `state`. Match the Python statement with its output. All answers MUST BE one of the letters A through L. The string variable called `state` is assigned as follows:

   `state = 'missiSSippi'`

Python statement	Output
1. `state.count('i')`	A. `'Mississippi'`
2. `state.capitalize()`	B. `'miRRiRRippi'`
3. `state.endswith('ippi')`	C. `['m', 'ss', 'SS', 'pp', '']`
4. `state.find('i')`	D. `'m'`
5. `';'.join([state,state])`	E. `'MISSISSIPPI'`
6. `state.lower().replace('ss','RR')`	F. `'missiSSippi;missiSSippi'`
7. `state.split('i')`	G. `True`
8. `state.upper()`	H. `'i'`
9. `state[7:]`	I. `4`
10. `state[1]`	J. `1`
11. `state[0:1]`	K. `False`
12. `'Miss' in state`	L. `'ippi'`

4. Test your understanding of this chapter's 'Key terms' by matching the Python statement with a term or phrase that describes it. The four variables used in the matching have been assigned as follows:

   ```
   >>> dataDir = 'C:/data'
   >>> data = 'bird_sightings'
   >>> count = 500
   >>> n = 3
   ```

Python statement	Output
1. `count/n`	A. Casting
2. `'nest'`	B. Indexing
3. `r'count\n'`	C. Slicing
4. `u'hatchling'`	D. Finding the length of a string
5. `len(data)`	E. Concatenating
6. `str(count)`	F. Line continuation
7. `dataDir + '/' + data`	G. String literal
8. `data[0:n]`	H. String method that is the inverse of the `join` method
9. `data[n]`	I. Escape sequence
10. `'bird' in data`	J. Integer division
11. `'{0} records'.format(count)`	K. Raw string
12. `data.split('_')`	L. Unicode string
13. `'Bird data \` `   Wing span'`	M. String formatting

5. **printPractice.py** Modify sample script 'printPractice.py', so that it prints the same statement four times, but using four distinct techniques—hard-coding, commas, concatenation, and string formatting. The hard-coding is already done. The other three techniques have been started but they contain mistakes. These techniques should each use all three of the provided variables. Once the mistakes have been corrected, run the script to make sure that it prints the statement identically four times:

```
Found 12 lights in the 5 mi. buffer and 20 intersections.
Found 12 lights in the 5 mi. buffer and 20 intersections.
Found 12 lights in the 5 mi. buffer and 20 intersections.
Found 12 lights in the 5 mi. buffer and 20 intersections.
```

6. This problem deals with date time stamps that have the form MM/DD/YYYY HH:MM:SSxM (where xM is AM for morning and PM otherwise). Part (a) creates a date-time variable which is used in all other parts. Solutions should work for other values of this variable. Write one or more lines of code to achieve the outcome described in each part.

   (a) Use an assignment statement to set a variable named `dt` to the following string literal: '07/28/2055 05:25:33PM'.
   (b) Use the `in` keyword in a code statement to print `True` if the value of `dt` is in the morning (and `False` otherwise).
   (c) Use slicing to extract the month from `dt`.
   (d) Use the `split` method twice and then use indexing to extract the hour from `dt`.
   (e) Use the `split` method twice and then use indexing to extract the year from `dt`.

7. Write one or more lines of code to achieve the outcome described in each part.

(a) Often geoprocessing involves sorting data files based on their names. For example, you may want to move all the files from Region 1 to a separate directory. Write a line of code to check if a variable named filename contains a substring that matches the value of a variable named code. The examples below show how the line of code could be used.

```
>>> # Sample input 1:
>>> filename = 'Forest361Region1_rec.shp'
>>> code = 'Region1'
>>> # Insert line of code here.
True
>>> # Sample input 2:
>>> filename = 'Marsh12Region4.shp'
>>> code = 'Region1'
>>> # Insert line of code here.
False
```

(b) Write a line of code that uses the rstrip method and an assignment statement to remove the trailing whitespace from the string variable named data. The variable contains a line of tab delimited North Carolina county forestry data. The example below shows how the line of code could be used.

```
>>> # Sample input:
>>> data = 'Site:\tNortheast Prong\tDARE\t01\t\n\n'
>>> data
'Site:\tNortheast Prong\tDARE\t01\t\n\n'
>>> # Insert line of code here.
>>> data
'Site:\tNortheast Prong\tDARE\t01'
```

(c) Suppose that we are editing data and saving the results in another file. We want to append the string '_edited' to the input base name to create the output name. For example, if the input file is named 'countiesNC.txt', the output file should be named 'countiesNC_edited.txt'. If the input file is named 'riversWVA.txt', the output file should be named 'riversWVA_edited.txt', and so forth. Write a line of code that uses slicing, concatenation, and a variable named input to assign a value to a variable named output. The example below shows how the line of code could be used.

```
>>> # Sample input:
>>> inputName = 'counties.shp'
>>> # Insert line of code here.
>>> print outputName
counties_edited.shp
```

(d) Write a line of code using the string `format` method to print the desired output. Pass the given variables as arguments. The output from the print statements should appear exactly as shown in the output below.

```
>>> eagleNests = 2
>>> forestAcreage = 10
>>> campsites = 5
>>> # Insert line of code here.
There are 2 nests and 5 campsites within 10 acres.
```

(e) Write a line of code using the string `format` method to print the desired output.

Pass the given variables as arguments. The output from the print statements should appear exactly as shown in the output below.

```
>>> bufferInput = 'crimeSites'
>>> num = 32
>>> # Insert line of code here.
crimeSites.shp buffered. Results saved in C:/buff32.shp
```

Chapter 4
Basic Data Types: Lists and Tuples

Abstract GIS scripting frequently involves manipulating collections of items, such as files or data records. For example, you might have multiple tabular data files with special characters in the fields preventing direct import into ArcGIS. A script to replace the characters can use lists to hold the file names and field names. Then the files and fields can be batch processed. Built-in Python data types, such as lists, are useful for solving this kind of GIS problem. This chapter presents the Python list data type, list operations and methods, the range function, mutability, and tuples. The chapter concludes with a debugging walk-through to demonstrate syntax checking and tracebacks.

Chapter Objectives

After reading this chapter, you'll be able to do the following:

- Create Python lists.
- Index into, slice, and concatenate lists.
- Find the length of a list and check if an item is in a list.
- Append items to a list.
- Locate online help for the list methods.
- Create a list of numbers automatically.
- Differentiate between in-place methods and methods that return a value.
- Create and index tuples.
- Check script syntax.
- Interpret traceback error messages.

4.1 Lists

A Python *list* is a data type that holds a collection of items. The items in a list are surrounded by square brackets and separated by commas. The syntax for a list assignment statement looks like this:

```
listVariable = [item1, item2, item3,...]
```

Here is an example in which `fields` is a Python list of field names from a comma separated value ('.csv') file containing wildfire data:

```
>>> fields = ['FireId', 'Org', 'Reg-State', 'FireType']
>>> fields
['FireId', 'Org', 'Reg-State', 'FireType']
>>> type(fields)
<type 'list'>
```

All of the items in the `fields` list are strings, but a list can hold items of varying data types. For example, a list can contain numbers, strings and other lists. The `stateData` list assigned here contains a string, a list, and an integer:

```
>>> stateData = ['Florida', ['Alabama', 'Georgia'], 18809888]
```

This list contains both numeric and string items:

```
>>> exampleList = [10000, 'a', 1.5, 'b', 'banana', 'c', 'cusp']
```

You can also create an empty list by assigning empty square brackets to a variable, as in this example:

```
>>> dataList = []
```

4.1.1 Sequence Operations on Lists

Lists, like strings are one of the sequence data types in Python. The sequence operations discussed in the context of string data types also apply to lists; the length of a list can be found, lists can be indexed, sliced, and concatenated, and you can check if an item is a member of a list. Table 4.1 shows sample code and output for each of these operations.

Table 4.1 Sequence operations on lists.

>>> exampleList = [10000, 'a', 1.5, 'b', 'banana', 'c', 'cusp']		
Operation	**Sample code**	**Return value**
Length	`len(exampleList)`	7
Indexing	`exampleList[6]`	`'cusp'`
Slicing	`exampleList[2:4]`	`[1.5, 'b']`
Concatenation	`exampleList + exampleList`	`[10000, 'a', 1.5, 'b', 'banana', 'c', 'cusp', 10000, 'a', 1.5, 'b', 'banana', 'c', 'cusp']`
Membership	`'prune' in exampleList`	`False`

The operations work on list items just like they work on string characters; however, lists are mutable. This means that indexing can also be used to *change* the value of an item in the list. Recall that strings are immutable and trying to change an indexed character in a string returns an error. In the following example, a list is created, then on the second line, an indexed list item is modified. The last line in the example shows that the first item in the list has been changed:

```
>>> exampleList = [10000, 'a', 1.5, 'b', 'banana', 'c', 'cusp']
>>> exampleList[0] = 'prune' # modifying the first item in the list.
>>> exampleList
['prune', 'a', 1.5, 'b', 'banana', 'c', 'cusp']
```

4.1.2 List Methods

Also like strings, list objects have a specific set of methods associated with them including `append`, `extend`, `insert`, `remove`, `pop`, `index`, `count`, `sort`, and `reverse`. For a complete description of each, search online for Guido Van Rossum's Python Tutorial for the 'More on lists' section. Like string methods, list methods use the dot notation:

```
object.method(arguments1, argument2, argument3,...)
```

There is one notable difference between list and string methods; it relates to mutability. Many list methods are in-place methods. *In-place methods* are those methods which change the object that calls them. Whereas, other methods do not alter the object that calls them; instead, they return a value. In-place list methods such as `append`, `extend`, `reverse`, and `sort` do not use an assignment statement to propagate the new list. Instead, they modify the existing list. The example below creates a `fireIDs` list and appends a new ID to the end of the list using the `append` method:

```
>>> # Initialize the list with 4 IDs.
>>> fireIDs = ['238998', '239131', '239135', '239400']
>>> newID = '239413'
>>> fireIDs.append(newID) # Changing the list in-place.
>>> fireIDs
['238998', '239131', '239135', '239400', '239413']
>>> # New value was appended to the end of the list.
```

Since `append` is an in-place method, the original list is modified, in-place. In contrast, the list method named 'count' is not an in-place method. It does not alter the

list, but returns a value instead. The following example creates a fireTypes list and
determines the number of code 11 fires that had occurred, using the count method:

```
>>> fireTypes = [16, 13, 16, 6, 17, 16, 6, 11, 11, 12, 14, 13, 11]
>>> countResults = fireTypes.count(11)
>>> print countResults
3 # The list contains three elevens.
>>> fireTypes # Note that the list contents are unchanged:
[16, 13, 16, 6, 17, 16, 6, 11, 11, 12, 14, 13, 11]
```

4.1.3   The Built-in range Function

The built-in range function is a convenient way to generate numeric lists, which
can be useful for batch processing tasks. The range function takes one to three
numeric arguments and returns a list of numbers. If you pass in one numeric argument, n, it returns a Python list containing the integers 0 through $n - 1$, as in the
following example:

```
>>> range(9)
[0, 1, 2, 3, 4, 5, 6, 7, 8]
```

By using a second argument, you can modify the lower bound:

```
>>> range(5,9)
[5, 6, 7, 8]
```

By using a third argument, you can change the step size:

```
>>> range(0,9,2)
[0, 2, 4, 6, 8]
```

The range function is used again in a later chapter to create numeric lists for
looping.

4.1.4   Copying a List

As we've just discussed, in-place methods like reverse and sort alter the order of
the list. For example, here the reverse method reverses the order of the fireIDs
list:

```
>>> fireIDs = ['238998', '239131', '239135', '239400']
>>> fireIDs.reverse()
```

```
>>> fireIDs
['239400', '239135', '239131', '238998']
```

In some cases, it may be necessary to keep a copy of the list in its original order. Python variables can be thought of as tags attached to objects. Because of this, there are two types of copy operations for objects: shallow copies and deep copies. A *shallow copy* attaches two tags to the same list object. A *deep copy* attaches each tag to a separate list object. The assignment statement shown in the following example doesn't work as you might intuitively expect. Python list a is copied to list b but both the lists are reversed when a is reversed. This is a shallow copy:

```
>>> a = range(1,11)
>>> a
[1, 2, 3, 4, 5, 6, 7, 8, 9, 10]
>>> b = a # "shallow copy" list a
>>> b
[1, 2, 3, 4, 5, 6, 7, 8, 9, 10]
>>> a.reverse() # reverse list a
>>> a
[10, 9, 8, 7, 6, 5, 4, 3, 2, 1]
>>> b
[10, 9, 8, 7, 6, 5, 4, 3, 2, 1] # list b is also reversed
```

Both tags a and b are pointing to the same object. Instead, we need to create a second object via a deep copy to retain a copy of the list in its original order. To do this we need to create a new list and attach the b tag to it. The built-in list function constructs a new list object based on a list given as an argument. This built-in function is used to create a deep copy in the next example. list(a) constructs a new list object and this is assigned to b. In the end, a is reversed, but b retains the original order:

```
>>> a = range(1,11)
>>> a
[1, 2, 3, 4, 5, 6, 7, 8, 9, 10]
>>> b = list(a) # "deep copy" list a
>>> b
[1, 2, 3, 4, 5, 6, 7, 8, 9, 10]
>>> a.reverse() # reverse list a
>>> a
```

```
[10, 9, 8, 7, 6, 5, 4, 3, 2, 1]
>>> b
[1, 2, 3, 4, 5, 6, 7, 8, 9, 10] # list b is not reversed
```

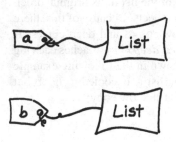

4.2   Tuples

A *tuple* is another Python data type for holding a collection of items. A tuple is a set of comma separated items, but round brackets (parentheses) are used instead of square brackets. When a tuple is printed, it is surrounded by parentheses. A tuple can be created with or without parentheses around the items:

```
>>> t = "abc", 456, 'wxyz'
>>> type(t)
<type 'tuple'>
>>> t
('abc', 456, 'wxyz')

>>> t2 = (4.5, 7, 0.3)
>>> type(t2)
<type 'tuple'>
>>> t2
(4.5, 7, 0.3)
```

Like strings, tuple items can be indexed but they are not mutable. It is not possible to assign individual items of a tuple with indexing:

```
>>> t[0]
'abc'
>>> t[1]
456
>>> t2[0]
4.5
>>> t2[0] = 5
Traceback (most recent call last):
File "<interactive input>", line 1, in <module>
TypeError: 'tuple' object does not support item assignment
```

Tuples are used in situations where immutability is desirable. For example, they are often used for (x, y, z) point coordinates or for holding the set of items in a data record.

4.3 Syntax Check and Tracebacks

Most of the examples in the book until this point have used the Interactive Window to explain the basic Python components. In upcoming chapters, we will begin to write complete scripts. Before we do so, it will be useful to become familiar with syntax checking in PythonWin. Chapter 2 discussed examples of errors, tracebacks, and exceptions thrown by interactive input. Now we'll compare interactive input error handling to script window error handling.

The Interactive Window automatically gives you immediate feedback in the form of a traceback message when you enter code that contains a syntax error, because it evaluates (runs) the code when you enter it. For example, a traceback message is printed, because of the following code, in which the `print` keyword is misspelled as `pirnt`:

```
>>> pirnt test
Traceback (File "<interactive input>", line 1
pirnt test
        ^
SyntaxError: invalid syntax
```

When you enter code in a PythonWin script window, you don't automatically receive immediate feedback as you're writing, but you can trigger syntax checking when you're ready by clicking a button. The PythonWin syntax checker is similar to a word processing spelling checker. The 'check' button, the one with the checkmark icon ( ), on the PythonWin standard toolbar checks the syntax without executing the code. It finds syntax errors, such as incorrect indentation, misspelling of keywords, and missing punctuation.

The feedback panel in the lower left corner of PythonWin shows the result of the check. If an error is detected, the feedback panel says "Failed to check—syntax error—invalid syntax" as shown in Figure 4.1. The cursor jumps to a position near to the first syntax error it detects in the script, usually the same line or the line following the erroneous line. To try this yourself,

1. Create a new script with one line of code that says the following: pirnt test

2. Initially, the 'Check' button is disabled (grayed out). You can't check the syntax until the script has been saved at least once. Save the script in 'C:\gispy\ scratch' as script 'test.py'.

3. Click in the 'test.py' script window to make sure your window focus is there and not in the Interactive Window.

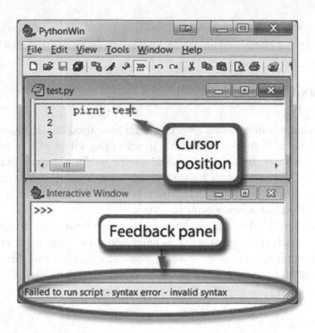

Figure 4.1 Syntax checker finds an error.

4. Now click the 'Check' button , which became enabled when the script was saved.
5. Observe the feedback panel and the cursor (see Figure 4.1). The feedback panel says `Failed to check-syntax error-invalid syntax` and the cursor has jumped to the erroneous line. It can't detect exactly what's wrong, so it gives this generic message and moves the cursor near to where it detects an error.
6. Next click the 'Run' button to see if it will run. The feedback shows a syntax error message—"Failed to *run script...*" (Figure 4.1). If you overlook the feedback panel, it may seem as if the program is not responding to the 'Run' button click. Check the feedback panel whenever you run a script.
7. The script will not run until the syntax errors are repaired; Repair the syntax error—change 'pirnt' to 'print'.
8. Click the 'Check' button again.
9. The feedback bar reports success: "Python and TabNanny successfully checked the file". The TabNanny, a script that checks for inconsistent indentation, is run as part of the syntax checking process.
10. A successful syntax check does not necessarily mean that the code is free of errors. Click the 'Run' button now that the syntax has been corrected. The feedback panel

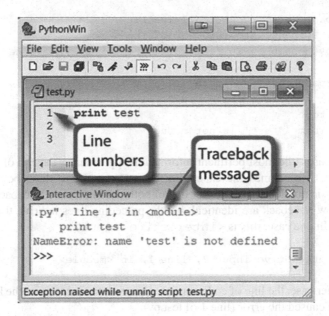

Figure 4.2 Exception raised.

reports that an exception was raised. This time a traceback error is printed in the Interactive Window (Figure 4.2) because the error was only detected when the script ran, not during the syntax checking phase.

The syntax check can only detect errors that violate certain syntax rules. Some errors can only be detected when Python attempts to run the code and then an error message appears in the feedback panel when it reaches a problem. When Python tries to print `test`, it throws an exception and prints a traceback message. A script traceback differs slightly from interactive input traceback because the message traces the origin of the error.

Entering the code from 'test.py' in the Interactive Window prints the following traceback:

```
>>> print test
Traceback (most recent call last):
File "<interactive input>", line 1, in <module>
NameError: name 'test' is not defined
```

Whereas, the traceback message in Figure 4.2 thrown by running 'test.py' reads as follows:

```
>>> Traceback (most recent call last):
  File "C:\Python27\ArcGIS10.3\Lib\site-
packages\pythonwin\pywin\framework\scriptutils.py", line 326, in
RunScript
    exec codeObject in __main__.__dict__
  File "C:\gispy\scratch\test.py", line 1, in <module>
    print test
NameError: name 'test' is not defined
```

In many cases, the most pertinent information appears near the end of the trace-back. The last line names the exception (NameError) and describes the error (name test is not defined). The last line of both the interactive traceback and the script window traceback are identical, but the preceding lines indicate the location of the error. In one case, this is <interactive input>.

```
  File "<interactive input>", line 1, in <module>
```

In the other case, the line of code is printed beneath the location of the line of the script which caused the error (line 1 of test.py).

```
  File "C:\gispy\scratch\test.py", line 1, in <module>
    print test
```

Python keeps track of a call stack. The call stack contains a list of files (and functions) that are called as Python runs. When 'text.py' is run from PythonWin, there are two items on the call stack, the 'scriptutils.py' and 'test.py' files. The 'scriptutils.py' file is called by PythonWin when you select 'Run'. This script, in turn, prompts the 'test.py' script to run. This is why the traceback lists 'scriptutils.py' and 'test.py' in this order. Can you tell which line of 'scriptutils.py' triggered 'test.py' to run? These lines show the line of code that was triggered in scriptutils.py:

```
  File "C:\Python27\ArcGIS10.3\Lib\site-
packages\pythonwin\pywin\framework\scriptutils.py", line 326, in
RunScript
    exec codeObject in __main__.__dict__
```

We can ignore this part of the traceback. The only part of the stack that is interesting for our purposes is the line of code which we wrote that caused the error. This is why we focus mainly on the last few lines of the message. Traceback messages print the script line number where the error occurred. The examples in the upcoming chapters of this book usually report only a portion of tracebacks. Turn on IDE line numbers to help with traceback interpretation.

To show line numbers in PythonWin:
Tools > Options...> Editor tab > Set "Line Numbers" Margin
Width >= 30 pixels.

To show line numbers in PyScripter:
Tools > Options > Editor Options > Display > Gutter:
check 'Visible' and 'Show line numbers'.

These examples show that the Interactive Window prints a traceback message immediately when you enter erroneous code because the code is evaluated immediately; whereas, script window code is only checked when you use the syntax check button or try to run the code. If a syntax error is found in a script, the feedback bar reports an error and the cursor moves near to the first detected error. Some errors are not detectable by the syntax check. These errors stop the script from running and print traceback messages in the Interactive Window. The traceback, read from bottom to top, names the exception, describes the error, and prints the erroneous line and the line number in the script where it was found. In summary, checking syntax, showing line numbers, and reading tracebacks will help you build working scripts as you progress through the upcoming chapters.

Note Always check the feedback panel and the Interactive Window when you run a script in PythonWin.

4.4 Key Terms

List data type in-place method
Mutable vs. immutable
Shallow copy vs. deep copy
Built-in list function
Sequence operations (length, indexing, concatenation, slicing, membership)

Range built-in function
List methods
tuple data type

4.5   Exercises

1. Practice interpreting exceptions and diagnosing errors as you follow the given steps. Then answer the follow-up questions.
   (a) Locate 'addWithErrors.py' in the Chapter 2 sample_scripts directory. Open it in an IDE and save it as 'addWithOutErrors.py'. This script is like 'add_version2. py' from Chapter 2, but several errors have been introduced.
   (b) Run the syntax check.
       The feedback panel says: "Failed to check—syntax error—invalid syntax" and the cursor jumps to line 7. In the steps below, you'll see that there are several errors, but the code is evaluated from top to bottom, so the cursor jumps to the first error Python detects. Do you see a problem on that line?
   (c) The keyword 'import' is misspelled. Replace 'iport' with 'import'.
   (d) Run the syntax check.
       The feedback panel says: "Failed to check - syntax error - invalid syntax" and now the cursor jumps to line 13. (Now line 13 is the first place an error is detected).
   (e) But there's nothing wrong with line 13! Can you find the error on the previous line? The closing parenthesis is missing on line 12. The cursor jumps to line 13 because Python detected an assignment statement before the parentheses were closed. Until it reached line 13, it didn't detect that there was anything wrong. Add ')' to the end of line 12. So that it looks like this:

   ```
   b = int(sys.argv[2])
   ```

   (f) Run the syntax check.
       The feedback panel says: "Failed to check—syntax error—invalid syntax" and the cursor jumps to line 14. Do you see the syntax error on line 14?
   (g) Line 14 is missing a quotation mark on the string. Add a quotation mark (") after the `print` keyword as follows (and note the changes in the color of the text): `"The sum is {0}.".format(c)`
   (h) Run the syntax check.
       The feedback panel says: "Python and TabNanny successfully checked the file". All the detectable syntax errors have been corrected, so the script can be run.
   (i) Run the script with no arguments by clicking the 'Run' button. The following traceback message is printed:

   ```
   File "C:\gispy\scratch\addWithOutErrors.py", line 10, in
   <module>
       a = sys.argv[1]
   IndexError: list index out of range
   ```

   (j) The traceback message says an `IndexError` occurred on line 10 of 'addWithOutErrors.py'. In this case, the traceback is generated because we didn't provide any user arguments. Index 1 is out of range, because no script arguments were used.

(k) Run the code with arguments 3 and 8. The following traceback message is printed:

```
File "C:\gispy\scratch\addWithOutErrors.py", line 13, in
<module>
c = a + b
   TypeError: cannot concatenate 'str' and 'int' objects
```

(l) The traceback message says a `TypeError` occurred on line 13 of the script. It says it cannot concatenate a string object with an integer object. When you get a type error, it's useful to check the type of the variables, using the built-in type function, to see if they are the data type that you expect them to be. Check the data types of 'a' and 'b' in the Interactive Window using the built-in `type` function:

```
>>> type(a)
<type 'str'>
>>> type(b)
<type 'int'>
```

(m) a is a string because arguments are strings unless they are cast to another type. b is cast to integer. Cast a to integer too by changing line 10 to look like this:

```
a = int(sys.argv[1])
```

(n) Run the script again. The feedback bar says "Script File 'C:\gispy\scratch\ addWithoutErrors.py' returned exit code 0" and the following prints in the Interactive Window.

```
>>> The sum is 11.
```

'Exit code 0' is good news. It means the script ran to completion without throwing any exceptions.

Follow-up questions:
  (i) What does the cursor do when you use the syntax checker and a syntax error occurs? Specifically, where does it go?
 (ii) How does PythonWin display a quotation mark that's missing its partner (the mark on the other end of the string literal)?
(iii) When you use your mouse to select a parenthesis that has a matching partner, how does PythonWin display the parentheses?
(iv) When you use your mouse to select a parenthesis that does not have a matching partner, how does PythonWin display the parenthesis?

2. The Python statements on the left use operations involving the list type variable
   places. Match the Python statement with its output. All answers MUST BE
   one of the letters A through I. If you think the answer is not there, you're on the
   wrong track! The list variable named places is assigned as follows:

   ```
   places = ['roads', 'cities', 'states']
   ```

Python statement	Output (notice there are nine letters)
1. places[0]	A. 3
2. places[0:2] + places[-3:]	B. 'roads'
3. places[0][0]	C. IndexError
4. len(places)	D. ['roads', 'cities']
5. places[0:5]	E. ['roads', 'cities', 'states']
6. places[-4:]	F. ['roads', 'cities', 'roads', 'cities', 'states']
7. places[11]	
8. places[:2]	G. False
9. 'towns' in places	H. 'r'
10. places[1]	I. 'cities'

3. The Python statements on the left use list methods and operations (and one string
   method) involving the list variable, census. Match the Python statement with the
   resulting value of the census variable. Consider these as individual statements,
   not cumulative. In other words, reset the value of census to its original value
   between each command. All answers MUST BE one of the letters A through G.
   **Notice that the question asks for matching the Python statement with the
   resulting value of the census variable and not matching the function with the
   value it returns. If you think the answer is not there, you're on the wrong
   track!**

   The list variable called census is assigned as follows:

   ```
   census = ['4', '3', '79', '1', '66', '9', '1']
   ```

Python statement	Resulting value of census (there are only 7 letters)
1. `len(census)`	A. `[2,'4','3','79','1','66','9','1']`
2. `census.insert(0, 2)`	B. `['1','1','3','4','66','79','9']`
3. `census.append(2)`	C. `['4','3','79','66','9','1']`
4. `census.remove('1')`	D. `['4','3','79','1','66','9','1']`
5. `census = '0'.join(census)`	E. `'403079010660901'`
6. `census.pop(3)`	F. `['1','9','66','1','79','3','4']`
7. `census.count('1')`	G. `['4','3','79','1','66','9','1',2]`
8. `census.sort()`	
9. `census.reverse()`	

4. Answer these three questions related to the difference between in-place methods and methods that return a value.

   (a) What is the value of the variable `census` after all of these statements are executed?

   ```
   >>> census = [4, 3, 79, 1, 66, 9, 1]
   >>> census.sort()
   >>> census.pop()
   >>> census.reverse()
   ```

   ```
   (1) [4, 3, 79, 1, 66, 9, 1]
   (2) [79, 66, 9, 4, 3, 1, 1]
   (3) [79, 66, 9, 4, 3, 1]
   (4) [66, 9, 4, 3, 1, 1]
   ```

   (b) What is the value of the variable `happySheep` after ALL of these statements are executed? Hint: if you think the correct answer is not there, you're on the wrong track.

   ```
   >>> happySheep = 'knoll.shp'
   >>> happySheep.replace('ll', 'ck')
   >>> happySheep.upper()
   >>> happySheep[:4]
   ```

    (1) 'Knock.shp'
    (2) 'knoll.shp'
    (3) 'KNO'
    (4) 'KNOCK.SHP'

(c) Use an example from parts a and b to explain what an in-place method does, as opposed to a method that only returns a value.

5. Select the line or lines of code that set listA to [0, 10, 20, 30, 40].

(a) listA = range(10, 50, 10)
(b) listA = range(0, 40, 10)
(c) listA = range(0, 50, 10)
(d) listA = range(0, 40)
(e) listA = range(40, 0, -10)
(f) listA = range(0, 41, 10)

6. The Northeastern region fire data is a text file containing a set of fields followed by records containing entries for those fields. The following lines of code use Python lists to store the field names and the first data record:

```
>>> fields = ['FireId', 'Org', 'Reg-State', 'FireType']
>>> record = ['238998', 'NPS', 'Northeast', '11']
```

Select the line(s) of code that use indexing to print only the value of the Reg-State field in the given record.

(a) **print** fields[3]
(b) **print** record[1:3]
(c) **print** record[2]
(d) **print** fields + records
(e) **print** record[-2]

7. The following code uses list methods and operations to find the minimum value of a list. Write a line of code that does the same thing but uses a Python built-in function (with no dot notation).

```
>>> fireTypes = [8, 4, 2, 5, 7]
>>> fireTypes.sort() # Sort the numeric list
>>> print fireTypes[0] # Use indexing to print the minimum
2
```

8. The following Interactive Window code samples use square braces in three different ways. Use chapter key terms from Section 4.4 to describe what each code statement is doing.

(a) >>> theList = ['spam', 'eggs', 'sausage', 'bacon', 'spam']
(b) >>> theList[1]
    'eggs'

(c) >>> theList[:3]
    ['spam', 'eggs', 'sausage']

9. Answer the following questions about the traceback message shown below:

   (a) What is the name of the exception?
   (b) What is the name of the script containing the error?
   (c) Which line of the script contains the error?
   (d) Explain why this error might have occurred.

```
Traceback (most recent call last):
  File "C:\Python27\ArcGIS10.2\Lib\site-
packages\pythonwin\pywin\framework\scriptutils.py",
line 326, in RunScript
    exec codeObject in __main__.__dict__
  File "C:\gispy\sample_scripts\ch04\County.py",
    line 23, in <module> District = sys.argv[5]
IndexError: list index out of range
```

10. **noMoreErrors.py** The sample script named 'noMoreErrors.py' currently contains five errors. Modify the script to remove all errors. As you remove errors, note whether the error resulted in a feedback bar message or a traceback message and note the message. Modify the last five lines of the script to report the error messages you encountered. The first two of these have been completed for you already. When the script is repaired, the first portion of the output will look like this:

```
NoMoreErrors.py
b
Mucca.gdb
Dyer20
DYER
#1. FEEDBACK BAR: Failed to check - syntax error - EOL while
    scanning
string literal
#2. TRACEBACK: IndexError: string index out of range
```

Error messages for #3–5 will also be printed, when the script is complete, but they are omitted here.

Chapter 5
ArcGIS and Python

Abstract ArcGIS provides a palette of sophisticated tools for processing and analyzing geographic data. There are several ways in which these tools can be used. For example, ArcToolbox tools can be run from ArcToolbox by clicking on the tool and filling out a form to specify the parameters; they can be run from ModelBuilder Models, and they can be run from Python scripts using the arcpy package. This chapter discusses ArcToolbox, ModelBuilder, ArcCatalog, Python's ArcGIS capabilities, the arcpy package, arcpy functions, and environment settings.

Chapter Objectives
After reading this chapter, you'll be able to do the following:

- Describe the ArcToolbox hierarchy.
- Search for tools in ArcCatalog.
- Locate tool help on ArcGIS Resources online.
- Export a script from a visual workflow model.
- Modify and run exported scripts.
- Preview geoprocessing output.
- Release locks on data.
- Explain, in general terms, the capabilities of the `arcpy` package.
- Define the Python terms *module* and *package*.
- Set geoprocessing environment variables.

5.1 ArcToolbox

ArcGIS provides a large suite of tools for processing data. The ArcToolbox panel in ArcCatalog (and ArcMap) lists the toolboxes—3D Analyst, Analysis, Cartography, and so forth (Figure 5.1). The tools are grouped into toolboxes by the type of actions they perform and each toolbox contains toolsets that further group the tools by their functionality. Each toolbox (🧰) and toolset (🔧) can be expanded to show the contents. Figure 5.2 shows the Extract, Overlay, Proximity, and Statistics toolsets in the Analysis toolbox. In this figure, the Proximity toolset is also expanded.

Figure 5.1 The
ArcToolbox application
contains numerous
toolboxes (only a few
shown are here).

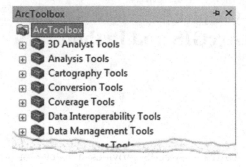

Figure 5.2 The Analysis
toolbox and the Proximity
toolset.

This toolset contains six tools, Buffer, Create Thiessen Polygons, etc. Though the icons vary on the third level, these are all tools. A tool can be run from ArcToolbox, by clicking on the tool, filling out the form that it launches, and clicking 'OK'.

The availability of tools depends on the license level of ArcGIS desktop installed (basic, standard, or advanced). Some tools, such as Spatial Analyst tools, are not available at the basic and standard levels. When these tools are available, an extension needs to be checked out for the tools to become functional. For example, to use Spatial Analyst tools with an advanced install, the Spatial Analyst extension must be checked out (ArcCatalog > Customize > Extensions > Check Spatial Analyst). Scripts need to check out these extensions too, as shown in the upcoming chapter.

The 'Search' panel available in ArcCatalog (and ArcMap) is useful for navigating the tools. Click the 'Search' button 🔍 to open this panel, select the 'Tools' option, and type a tool name. Figure 5.3 shows the results of a search. Click on a tool in the results (e.g., Buffer (Analysis) Tool) and it opens the GUI to run the tool. The form has a 'Tool Help' button which will launch the local ArcGIS Desktop help for that tool. As you work through the examples in this book, you can locate tools in this way and become familiar with their functionality by reading the help and running them using the GUI interface.

Figure 5.3 The search
panel.

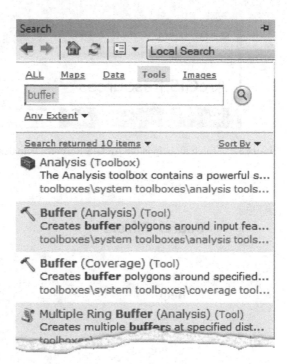

5.2   ArcGIS Python Resources

The 'ArcGIS Resources' site (resources.arcgis.com) is the foremost reference you
will need for working with Python in ArcGIS. The site provides full documentation
for ArcGIS Python functionality. The 'Search ArcGIS Resources' box is indispens-
able for working with Python in ArcGIS. Use the search box to get a list of pages
matching your query and use the search filters to narrow the search (Figure 5.4). For
example, enter 'buffer' in the search box. This returns thousands of results including blogs, bug reports, web mapping help, and so forth. Narrow the search by using
the 'Help' and 'Desktop' filters as shown Figure 5.5.

Notice that the results are different from the ArcCatalog search results for the
same term (Figure 5.3); the ArcGIS Desktop help is organized differently than
the online help. The online help provides the most current, comprehensive documentation. A set of descriptive identifiers is provided for each link in the online
help. The Buffer (Analysis) link has the identifiers 'Tool Reference' and 'ANALYSIS'
(Figure 5.6). The last identifier shows the date the content was last modified. Each
ArcGIS tool has a 'Tool Reference' page that corresponds to the built-in help for
that tool. Chapter 6 discusses components in 'Tool Reference' pages.

This site is referred to as the 'ArcGIS Resources' site in this book. Key search
terms will be provided to direct you to specific help topics within this site.

Figure 5.4 Search for help in the ArcGIS Resource help box.

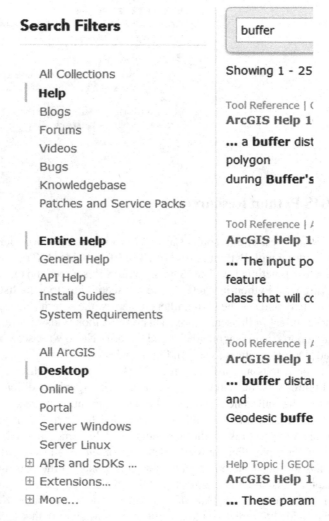

Figure 5.5 Filter the search for relevant results. The search shown here only returns 'Help' for ArcGIS 'Desktop' topics.

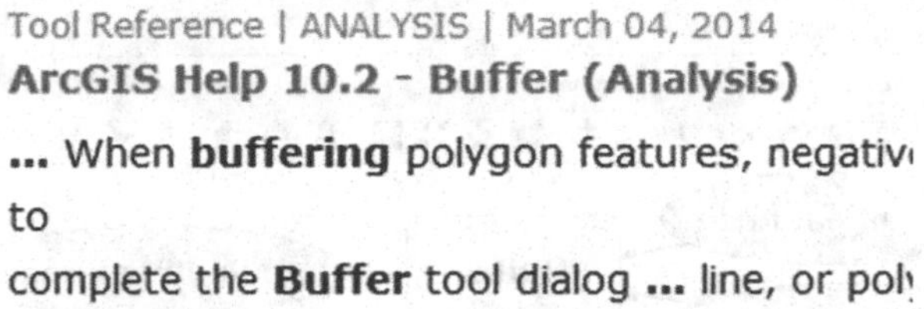

Tool Reference | ANALYSIS | March 04, 2014
ArcGIS Help 10.2 - Buffer (Analysis)

... When **buffering** polygon features, negativ₁

to

complete the **Buffer** tool dialog **...** line, or pol₁

Figure 5.6 Descriptive identifiers such as 'Tool reference' and 'ANALYSIS' appear above each search result.

5.3 Exporting Models

Python can call almost all the tools in ArcToolbox and, in this way, repetitive processes can be automated. Before we begin writing scripts from scratch, we'll start with an example automatically generated by the ArcGIS ModelBuilder application. ModelBuilder is an application built into ArcCatalog (and ArcMap) that allows users to create a workflow visualization, called a *model*. Models not only visualize the workflow, but can also be can be run to execute the workflow. ArcGIS Toolbox tools can also be run via ModelBuilder; Tools can be dragged into the model panel and connected to create the workflow. When a model runs, it executes tools and the underlying code statements that correspond to pieces of the model. The underlying code can be exported to a Python script and we can compare the workflow visualization with the code. Follow steps 1–3 to create and export a simple model.

1. In ArcCatalog, to create a model like the one in Figure 5.7:

   - Launch ModelBuilder from the button ⬚ on the Standard toolbar.
   - Open ArcToolbox with the ArcToolbox button ⬚ on the Standard toolbar.
   - Locate the Buffer (Analysis) tool in ArcToolbox (ArcToolbox > Analysis Tools > Proximity > Buffer)
   - Click the Buffer tool and drag it into ModelBuilder from its toolbox in ArcToolbox. A rectangle labeled 'Buffer' will appear.
   - Right-click on the new Buffer rectangle > Make variable > From Parameter > Input Features.
   - Right-click on the Buffer rectangle (again) > Make variable > From Parameter > Distance.
   - Double click on the Input Features oval and browse to: 'C:/gispy/data/ch05/park.shp'.
   - Double click on the Distance oval and set it to 100 feet.
   - Right-click on the Input Features oval > rename 'inputFeatures'.
   - Right-click on the Distance oval > rename 'distance'.
   - Right-click on the Output Feature Class oval > rename 'outputFeatures'.
   - Check that the result looks like Figure 5.7.

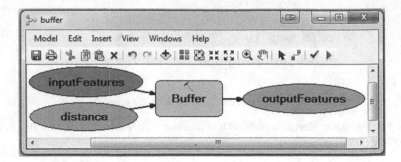

Figure 5.7 ModelBuilder Model to buffer input features.

```
1   # -*- coding: utf-8 -*-
2   # --------------------------------------------------------------------------
3   # buffer.py
4   # Created on: 20xx-05-22 12:17:03.00000
5   #    (generated by ArcGIS/ModelBuilder)
6   # Description:
7   # --------------------------------------------------------------------------
8
9   # Import arcpy module
10  import arcpy
11
12
13  # Local variables:
14  inputFeatures = "C:\\gispy\\data\\ch05\\park.shp"
15  distance = "100 Feet"
16  outputFeatures = "C:\\gispy\\data\\ch05\\park_Buffer.shp"
17
18  # Process: Buffer
19  arcpy.Buffer_analysis(inputFeatures, outputFeatures, distance, "FULL", "ROUND", "NONE", "")
```

Figure 5.8 A script exported from the model shown above.

2. Run the model to confirm that it works (Model menu > Run).
3. Export the model as a script (Model > Export > Pythons Script). This should generate a script like Figure 5.8, shown with line numbers to the left of the code. Open the script in PythonWin to view the code.

To compare the model and script, we'll look at each line of code.

- Lines 1–7, 9, 13, and 18 are comments.
- Line 10 imports `arcpy`. This line of code enables the script to use ArcGIS commands. We'll talk more about this in a moment.
- Lines 14–16 are assignment statements for string variables, `inputFeatures`, `distance`, and `outputFeatures`. These correspond to the model variables, the ovals that specify input and output for the tool. The string literal assigned to the script variables depends on the value assigned in ModelBuilder. For example, on line 13, `inputFeatures` is being assigned the value 'C:\\gispy\\data\\ ch05\\park.shp' because the model variable was given this value before the model was exported.

- Line 19 calls the Buffer (Analysis) tool. Running a tool in Python is like calling a function. We call the tool and pass arguments into it. The tool does our bidding and creates output or returns values. The variables and string literals in the parentheses are passing information to the tool. The Buffer tool requires three input parameters (the other parameters are optional). These required parameters are represented by the ovals in our model. The first three arguments in the Python correspond to these parameters. When the code was exported, it filled in default values for the rest of the parameters. In summary, line 19 acts like the rectangle in the model; it creates a buffer around the input features by the given distance and saves the results in the output features.

With these observations, it's possible to get a feel for the connection between the model components and the lines of code in the script. In theory, exported models could be used as a starting point for scripts, but this approach can be cumbersome for several reasons. First, scripting enables a more flexible, reusable complex workflow, including functionality beyond ArcGIS geoprocessing. Second, exported scripts usually require modification to perform as desired, making it more efficient to modify existing code samples than to build and export models.

The model/script comparison above provides some intuition for how the Python code is working, though more detailed explanation is needed. The next sections address this need with discussions on importing `arcpy`, using dot notation with `arcpy`, and calling tools.

5.4 Working with GIS Data

Each time you run a Python script that generates geoprocessing output, you will want to check the results. Esri geographic data features and the tables associated with them can be viewed in the ArcCatalog 'Preview' tab. Browse to a shapefile in the ArcCatalog 'Catalog Tree' and select the 'Preview' tab. Select 'Geography' preview from the bar at the bottom of the Preview pane to view the geographic features. Then select 'Table' to see the associated attribute table. Figures 5.9 and 5.10 show the geography and table views of 'park.shp'. Only seven rows of the table are shown in Figure 5.10. There are 426 rows in total, one data record for each polygon.

When the data is being viewed in ArcCatalog, it is locked, so that other programs can't modify it simultaneously. A file with an 'sr.lock' extension appears in Windows Explorer when the data is locked. For example, the file could be named something like 'park.shp.HAL.5532.5620.sr.lock' on a computer named 'Hal'. When you perform processing on the file in a Python script, you need to make sure that the data is not locked. To unlock the data after previewing it in ArcCatalog, select the parent workspace ('C:/gispy/data/ch05' in Figures 5.9 and 5.10) and refresh ArcCatalog (press F5). Selecting another file within the same workspace and refreshing the Catalog Tree will not release the lock; the parent workspace must be refreshed.

Note Unlocking the data before performing Python geoprocessing is critical, else the script may give unforeseen errors.

To see the geographic view of more than one file at a time, you need to use ArcMap. To view your data in ArcMap, the simplest approach is to browse to the data in the ArcCatalog tree embedded in ArcMap and drag/drop it onto a blank map. As long as ArcMap is still running, data used in this way will be locked. Even if the map document is closed, the locks may not be released until the program itself is exited.

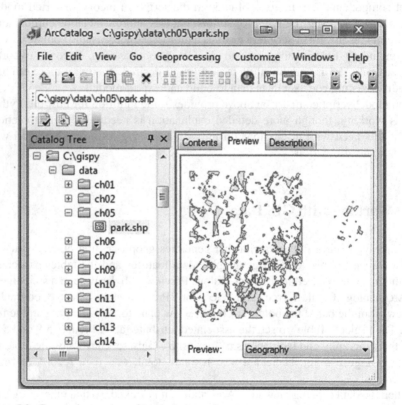

Figure 5.9 Geography preview of 'park.shp'.

5.5   ArcGIS + Python = arcpy

Now that you're familiar with ArcToolbox, ModelBuilder, and ArcCatalog data previews, it is time to introduce the arcpy Package. Geoprocessing scripts begin by importing arcpy (e.g., Figure 5.8); The arcpy package is Python's

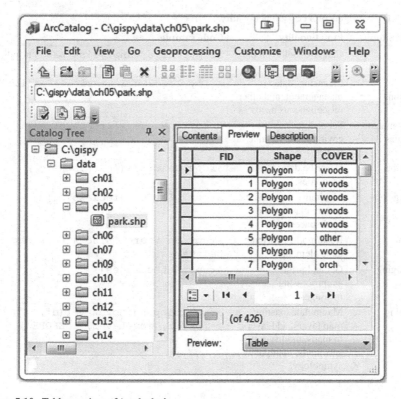

Figure 5.10 Table preview of 'park.shp'.

means for accessing ArcGIS from Python. To use ArcGIS functionality in Python, a script needs to import `arcpy`. The terms `import`, package, and `arcpy` are explained here:

- The keyword `import` is used to reference a module or a package. This provides access to functionality beyond the built-in Python functionality. As discussed in Chapter 2, the term following the `import` keyword is a module or a package.
- Recall that a *module* is a single Python script ('.py' file) containing tightly related definitions and statements. A *package* is a special way of structuring a set of related Python modules. A package is a directory containing modules and sometimes subpackages, which also contain modules. A module named '__init__.py' tells Python that the directory contains a package. Modules structured within a package and items in those modules can be accessed using the dot notation.
- `arcpy` is a package installed with ArcGIS. `arcpy` can also be thought of as a Python object that has numerous methods, including geoprocessing tools. Search under the ArcGIS install and you will find an `arcpy` directory. This is the package being imported. `arcpy` provides a variety of functionality that we'll be using throughout the remainder of the book. Table 5.1 lists the main `arcpy` topics covered in this book.

Table 5.1 Highlights of `arcpy` functionality.

ArcGIS topic	Functionality	Code sample
Tools	Call ArcToolbox tools	`arcpy.Buffer_analysis('park.shp', 'output.shp', '1 Mile')`
Other functions	Licensing, data management, and other miscellaneous needs	`arcpy.CheckOutExtension('Spatial')`
Environment variables	Set and get environment variable values	`arcpy.env.overwriteOutput = True`
Describe	Describe data properties	`arcpy.Describe('park.shp')`
Listing data	List items (e.g., rasters, feature classes, and workspaces)	`arcpy.ListFeatureClasses()`
Cursor	Read/modify attribute table elements	`arcpy.da.SearchCursor('park.shp', fieldNames)`
Messaging	Get and print geoprocessing messages	`arcpy.GetMessages()`
Mapping	Manipulate existing map layers, add layers to maps, modifying surrounds, manipulate symbology	`arcpy.mapping.MoveLayer(df, refLayer, moveLayer, 'BEFORE')`

The `arcpy` package has an object-oriented design which can be defined using object-oriented terms, some of which were already used in Chapters 3 and 4:

- Everything in Python is an *object*, including modules and packages. `arcpy` is an object. `arcpy` also uses objects, such as a `ValueTable` object, a `Describe` object, and a `Result` object which are discussed in upcoming sections.
- Objects have methods associated with them. A *method* is a function that performs some action on the object. Methods are simply a specific type of functions. The terms 'calling methods', 'passing arguments', and 'returning values' apply to methods just as they apply to functions (see Section 2.4.4).

Dot notation generates context menus for `arcpy`, showing a list of available `arcpy` methods and properties. When you use dot notation with strings and lists, context menus automatically appear, but these are built-in data types. In order to view context menus for `arcpy`, you must first import `arcpy` to give Python access to the information it uses to populate the context menu. In other words, it recognizes the word `arcpy` as a package. Once you have imported `arcpy`, within a PythonWin session, the context menus are available for the entire session. Try the code shown in the screen shot below to see the context menu for `arcpy` properties are case-sensitive and must be spelled correctly. Selecting choices from the context menu ensures accurate spelling and capitalization. To use the context menu, you can start typing a name and the menu will scroll to the closest choice. Pressing the 'Tab' key insert the current selection into the code. If you don't see your choice, you know your spelling or capitalization is incorrect. The `arcpy` menu contains a list of `arcpy` functions, environment settings, tools, and modules.

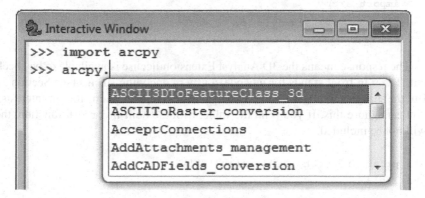

With each software release, `arcpy` functionality grows. It would be difficult to learn about every `arcpy` object, method, and property. In fact, you don't need to know all these details; they can be referenced in the help documentation. Instead, we aim for a general exposure to available capabilities. Figure 5.11 provides an overview of `arcpy` functionality. Symbols are used for properties, methods, and objects as shown in the key (bottom right). In this abbreviated object model diagram, the boxes enclose functionality categories and only a few examples are shown for each category. The contents are not exhaustive, but Figure 5.11 will provide a reference for the `arcpy` discussion. Packages, modules, and classes are Python constructs for organizing code. These constructs can contain functions (or methods) and properties. To see a complete list of functions/methods and properties for any of these constructs, search the online ArcGIS Resources site. The diagram is available as arcpyCheatSheet.pptx in the Chapter 5 data directory (C:/gispy/data/ch05). Chapters 6, 9, 11, 17, and 24 reference this document. If possible, print this document and keep it handy for upcoming chapters.

5.6 arcpy Functions

The `arcpy` functions (top left box in Figure 5.11) provide support for geoprocessing workflows. For example, functions can be used to list datasets, retrieve a dataset's properties, check for existence of data, validate a table name before adding it to a geodatabase, or perform many other useful scripting tasks. The syntax for calling functions in the `arcpy` package uses the dot notation with `arcpy` before the dot and the function name after the dot:

```
arcpy.functionName(argument1, argument2, argument3,...)
```

In the following example, the function name is 'CheckExtension' and it takes one argument, the extension code, '3D', for the 3D Analyst toolbox:

```
>>> import arcpy
>>> arcpy.CheckExtension('3D')
u'Available'
```

The response means the 3D Analyst Extension license is available to be checked out. The u in front stands for unicode, a way of encoding strings (see Section 3.6). But the string encoding doesn't have any practical repercussions for our interests, so you can ignore this. If you print the value, with the built-in print function, the u will not be included.

```
>>> print arcpy.CheckExtension('3D')
Available
```

Notice, the example above did not import arcpy. For a given PythonWin session, once arcpy is imported, it doesn't need to be imported again. But if PythonWin is closed and reopened, arcpy needs to be imported again.

Functions that don't require arguments still need to use parentheses. For example, the ListPrinterNames function lists the printers available to the caller and takes no arguments:

```
>>> arcpy.ListPrinterNames()
[u'Use Mac Printer', u'Send To OneNote 2010', u'Microsoft
XPS Document Writer', u'Fax', u'Adobe PDF']
```

Many arcpy functions return values. An assignment statement can capture the return value in a variable. The following code creates an arcpy Point object. The CreateObject function returns a Point object and it is stored in the variable named pt. The second line prints the value of this variable, a 2-dimensional Point object located at (0,0):

```
>>> pt = arcpy.CreateObject('Point')
>>> pt
<Point (0.0, 0.0, #, #)>
```

The arcpy functions serve a variety of scripting needs, some dealing with administrative concerns such as licensing (e.g., CheckExtension), others dealing with data management. For example, the Exists function takes one argument, a dataset and checks if it exists:

```
>>> arcpy.Exists('C:/gispy/data/ch05/park.shp')
True
```

Other functions deal with topics such as geodatabase management, messaging, fields, tools, parameters, and cursors. Search for an 'alphabetical list of `arcpy` functions', on the ArcGIS Resources site to see a complete list of functions. Technically, the ArcToolbox tool functions are `arcpy` functions, but these are listed separately elsewhere on the site. The syntax for calling tools is similar to the syntax for calling other `arcpy` functions. Before learning to call tools you need to know a little about managing environment settings. The syntax for environment settings uses dot notation as well.

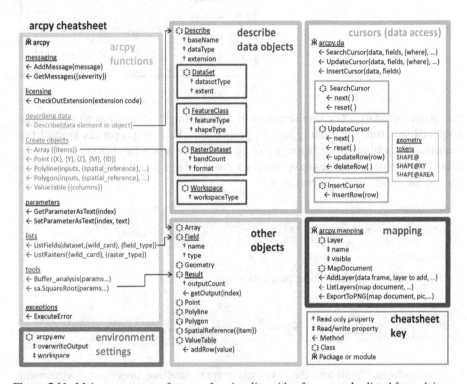

Figure 5.11 Main components of `arcpy` functionality with a few examples listed for each item.

5.7 Environment Settings

Each tool has settings it uses to execute an operation, such as a tolerance or output location. The *environment settings* are conditions that can be applied to all tools within an application (e.g., all operations in ArcCatalog can be limited to an extent set in the environment settings). ArcGIS has default values for these settings and users can modify the values via a dialog box in ArcGIS (Geoprocessing > Environments launches the dialog box shown in Figure 5.12). Python commands

Figure 5.12 Interface for
manually modifying
ArcMap environment
settings.

can also be used to get or set the values of these settings. arcpy has an env class, a special structure for holding related properties. The env properties control the environment settings. env belongs to arcpy and the properties belong to the env class, so the property names have two dots, one after arcpy and one after env. The format for setting these properties is:

```
arcpy.env.property = value
```

The format for getting these properties is:

```
variable = arcpy.env.property
```

The workspace path and the overwrite output status are important environment properties. The workspace path specifies a structure such as a directory or file geodatabase that contains pertinent data. Tools look for input data in the workspace and place the output in the workspace. If the input data resides in the workspace, only the name of the file needs to be used. arcpy will automatically search in the workspace. Similarly, the workspace is the default location for output, unless otherwise specified.

```
>>> # Setting the current workspace path
>>> arcpy.env.workspace = 'C:/Data/Forestry'
>>> # Getting the current workspace path
>>> mydir = arcpy.env.workspace
>>> mydir
u'C:/Data/Forestry'
```

Setting the workspace is simply a string assignment. No checking is done at this point to ensure that the workspace exists. This only occurs when a tool attempts to use the workspace to read or write data. The overwrite output status, either `True` or `False`, controls whether or not existing files are allowed to be overwritten by output from tools. The built-in constants `True` or `False` must be capitalized. The overwrite output status can also be set to 1 or 0 (1 for `True` and 0 for `False`). The default value for the `overwriteOutput` properties is `False`.

```
>>> arcpy.env.overwriteOutput
False
```

This protects the user from unintentionally overwriting a file, but is inconvenient during software development when scripts need to be run more than once for testing. For complex scripts, it's useful to set overwriteOutput to `True` placing this statement near the beginning of the script, after `arcpy` is imported, but before any tool calls are made:

```
>>> arcpy.env.overwriteOutput = True
```

Since Python is case-sensitive, be sure to use lower camel case for `overwriteOutput`. For environment settings, no error will appear if the capitalization is incorrect; it will simply not work as expected. In the following example, we use the wrong capitalization and no error is reported, but the value of the `overwriteOutput` property is not changed to `False`:

```
>>> arcpy.env.overwriteoutput = False
>>> arcpy.env.overwriteOutput
True
```

Other environment variables may be useful for specific problems. For example, when you are creating raster data sets, you may want to set the tile size property which specifies the height and width of data stored in blocks. The default size is 128 by 128 pixels:

```
>>> arcpy.env.tileSize
u'128 128'
```

Type the following line of code to print a complete list of available environment properties:

```
>>> arcpy.ListEnvironments()
```

5.8   Key Terms

ModelBuilder models   ArcCatalog tool search
import keyword   ArcGIS Resources site
arcpy package   ModelBuilder
Python package   Model parameter
Python module   ArcCatalog geography and table previews
Environment settings   Data locks
ArcToolbox

5.9   Exercises

1. Use the given steps to create and test 'aggregate.py'. The steps walk through
   creating a model, exporting it, and running it as a script. The script will call the
   Aggregate Polygons (Cartography) tool, a data summary technique which groups
   features that are very close together. If input features are within a specified
   aggregation distance, this script will combine them into a single unit. The figures
   below show an original shapefile in Figure 5.13a and an aggregated version in
   Figure 5.13b.
   (a) Step 1: *Create a new toolbox.* Browse to C:\gispy\sample_scripts\ch05 in
       ArcCatalog and create the new toolbox there.
   (b) Step 2: *Create the model.* The model should use the tool named 'Aggregate
       Polygons'. Browse to the tool (ArcToolbox > Cartography Tools > Generaliz
       ation > Aggregate Polygons), then drag and drop it onto the model. Give
       the tool two input parameters: input features and aggregation distance.

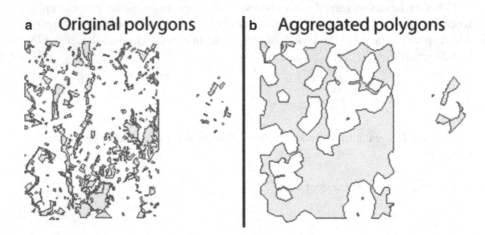

a   **Original polygons**   b   **Aggregated polygons**

Figure 5.13   park.shp with no aggregation (**a**) and after 1500 foot aggregation (**b**).

Double-click on the input variables and give them default values of 'C:\\gispy\\data\\ch05\\park.shp' and 100 feet. The model will become colored once these values are set. Rename the variables to meaningful, succinct names: inputFeatures, aggDistance, outputFeatures, and outputTable. Make all four variables model parameters (right-click on the oval>Model Parameter and a 'P' appears by the oval). Set the `inputFeatures` 'P' first and `aggDistance` 'P' second so that your script has the input features as the first parameter and the aggregation distance as the second parameter.

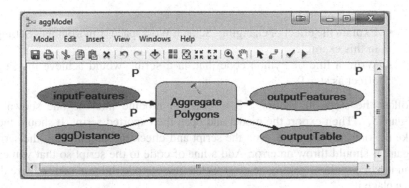

(c) Step 3: *Export the model*. Call it aggregateExport.py and save it in 'C:\gispy\sample_scripts\ch05'.

(d) Step 4: *Run the model in PythonWin*. Open the script in PythonWin and run the script with arguments: In the 'Run Script' window 'Arguments' text box, insert the arguments the script needs, a shapefile and an aggregation distance. Try the following input example:

  C:/gispy/data/ch05/park.shp "1500 feet"

Be sure to separate the two arguments by a space. Click 'OK' to run the script. Next, in the script, change the names of the output features and table to "C:\\gispy\\data\\ch05\\park_Agg500.shp" and "C:\\gispy\\data\\ch05\\park_Agg500_Tbl", respectively, and run the script again with an aggregation distance of 500 feet, by modifying the aggregation distance value in the 'Arguments' text box.

(e) Step 5: *Check the output*. View the two output shapefiles in ArcCatalog by selecting each one in turn in the 'Catalog Tree' and selecting the 'Preview' tab. Select the 'Table' view from the drop-down menu at the bottom of the 'Preview' tab and observe the number of records in the output files as compared to the input file.

2. Try the following steps to experiment with environment settings and answer the questions that follow:

   1. Launch ArcCatalog.

   2. Search for the Buffer (Analysis) tool in the search window ◻.

   3. Double-click on the tool link to launch the tool dialog.

   4. Drag C:/gispy/data/ch05/park.shp into the 'Input Features' parameter box.

   5. Observe the 'Output Feature Class' path.

   6. Cancel the tool.

   7. Open the environment settings dialog (Geoprocessing>Environments…)

   8. Click on Workspaces to expand this category.

   9. Set 'Scratch Workspace' to C:/gispy/scratch

 10. Click OK to accept the changes and close the dialog box.

 11. Repeat Steps 2–5

     Follow-up questions:
- Explain the effect of changing 'Scratch Workspace' Environment Settings in this example.
- Write a line of Python code that, in a script, would achieve the same affect as step 9.

3. Follow the steps in Section 5.3 to create a module like the one shown in Figure 5.7. Then export the model and open the exported script. It should look like the one in Figure 5.8. Run the script and check the output. Run the script again. It should throw an error. Add a line of code to the script so that you can run it twice without getting an error. What line of code did you add? Where did you place it?

Chapter 6
Calling Tools with Arcpy

Abstract ArcGIS offers a large array of sophisticated geoprocessing tools which can be accessed from Python using the arcpy package. Chapter 5 introduced the arcpy package and used it to set environment variables. This chapter focuses on how to call ArcGIS geoprocessing tools with arcpy and how to use the tool help documentation and get sample code. We explain the general format for calling tools and how to handle linear unit parameters, multi-value parameters, optional parameters, and return values. We also discuss tools which require specialized approaches, including Calculate Field, Raster Calculator, Make XY Event Layer tools, and tools in the Spatial Analyst toolbox. Then we show how to call tools in a sequential work-flow and how to call custom tools.

Chapter Objectives
After reading this chapter, you'll be able to do the following:

* Invoke geoprocessing tools with Python.
* Interpret ArcGIS Resources scripting help for geoprocessing tools.
* Copy code snippets from hand-run tools.
* Calculate field values with Python expressions.
* Format linear units, multi-value input, and optional input for GIS tools.
* Consume values returned by GIS tools.
* Call Spatial Analyst tools and perform map algebra.
* Save temporary raster and feature layer data.
* Use output from a GIS tool as input to another GIS tool.
* Import custom toolboxes and call custom tools.

6.1 Calling Tools

ArcGIS users know how to run tools from ArcToolbox. ArcToolbox tools, like other `arcpy` functions, use dot notation with `arcpy` before the dot. The tool name and toolbox alias separated by an underscore are placed after the dot. All ArcGIS tools with the exception of tools in the Spatial Analyst Toolbox can be called with the following syntax:

L. Tateosian, *Python For ArcGIS*, DOI 10.1007/978-3-319-18398-5_6

```
arcpy.toolName_toolboxAlias(arg1, argt2, arg3,...)
```

For example, the RasterToASCII tool is in the Conversion Toolbox. The following example takes two arguments, the input raster name ('C:/gispy/data/ch06/getty_rast') and the output ASCII text file name ('C:/gispy/data/ch06/output.txt'). The first line sets the workspace, so we don't need to specify the full path file names:

```
>>> arcpy.env.workspace = 'C:/gispy/data/ch06/'
>>> inputRaster = 'getty_rast'
>>> outputFile = 'output.txt'
>>> arcpy.RasterToASCII_conversion(inputRaster, outputFile)
```

The toolbox alias needs to be part of the tool call because some tools have the same name. For example, there is a Buffer tool in both the Analysis and Coverage Toolboxes. Table 6.1 lists a few toolbox alias examples. Search the ArcGIS Resources site with the phrase 'toolbox alias' for a complete list. The following example finds a one quarter mile buffer around the polygons in the shapefile. If the workspace is not set and the full file path for the output is not specified, the output is placed in the same directory as the input. The first line sets `overwriteOutput` to `True`, in case 'parkBuffer.shp' already exists in 'C:/gispy/data/ch06'.

```
>>> arcpy.env.overwriteOutput = True
>>> arcpy.Buffer_analysis('C:/gispy/data/ch06/park.shp',
                          'parkBuffer.shp', 0.25 miles')
```

The name of the input includes the path where it is stored. We call this the *full path file name* or the *absolute path*. If a full path file name is specified in a tool call, the full path file location is used instead of the workspace. In the following example, the tool looks for the input file within the `arcpy` workspace, but the output is placed in a different directory because the full path file name is used for the output:

```
>>> arcpy.env.workspace = 'C:/gispy/data/ch06/'
>>> arcpy.Buffer_analysis('park.shp',
                          'C:/gispy/scratch/parkBuffer.shp',
                          '0.25 miles')
```

Table 6.1 Selected toolbox aliases.

Toolbox	Alias
3D Analyst Tools	3d
Analysis Tools	analysis
Cartography Tools	cartography
Conversion Tools	conversion
Data Interoperability Tools	interop
Data Management Tools	management
Spatial Analyst Tools	sa
Spatial Statistics Tools	stats

We know that the Buffer (Analysis) tool requires three arguments: the input data, the output file name, and a buffer distance and that there are several more optional arguments. The next section explains the ArcGIS Resources site tool help information, such as the required/optional parameters for a tool.

6.2 Help Resources

The general format is consistent for most tools, but each tool has unique parameter requirements. To call tools from scripts, you need to know how to interpret the documentation and find code samples. The next sections walk through the tool help documentation and explain how to generate custom code samples.

6.2.1 Tool Help

As mentioned in Chapter 5, the ArcGIS Resources site hosts a 'Tool reference' page for each tool. Search on the name of the tool and filter by 'Help' and 'Desktop' to find a tool's page. Tool reference pages begin with an explanation of how the tool works (usually consisting of Summary, Illustration, and Usage sections). The next sections, the 'Syntax' and 'Code sample' sections are guides for Python scripting.

As an example, search for the Buffer (Analysis) tool. Locate the 'Syntax' section for the Buffer tool. At the very beginning of this section, tool syntax is represented by a *code signature*. This is a blueprint for how to call the tool via Python. The code signature for the Buffer (Analysis) tool looks like this:

```
Buffer_analysis (in_features, out_feature_class,
buffer_distance_or_field, {line_side}, {line_end_type},
{dissolve_option}, {dissolve_field})
```

It shows the toolName_toolboxAlias followed by a list of parameters (A 'parameter' is specified in the tool's signature; whereas, an 'argument' is a specific value passed in for a tool call). Python scripts need to list the parameters in the same order shown in the code signature. Optional parameters are surrounded by curly brackets in the code signature. For example, the last four buffer parameters are optional.

The table below the code signature provides more information about these parameters. Figure 6.1 shows the first few rows of the Buffer (Analysis) tool parameter table. The table has three columns:

- The 'Parameter' column contains the parameter name as given in the code signature. If the parameter is optional, the word 'optional' appears behind the name. The Python code often uses a string to pass in the parameter value. In some

Syntax
Buffer_analysis (in_features, out_feature_class, buffer_distance_or_field, {line_side}, {line_end_type}, {dissolve_o{
{dissolve_field})

Parameter	Explanation	Data Type
in_features	The input point, line, or polygon features to be buffered.	Feature Layer
out_feature_class	The feature class containing the output buffers.	Feature Class
buffer_distance_or_field	The distance ar~ ~ith ~	Linear unit ~

Figure 6.1 The code signature and top portion of the parameter table in the 'Syntax' section of the Buffer (Analysis) tool help page.

cases, Python lists are used to supply values for *multivalue inputs*, for parameters which accept more than one value, as discussed in Section 6.3.3.

• The 'Explanation' column describes the purpose of the parameter, gives a list of acceptable values (where applicable), and indicates the default behavior for optional parameters.

• The 'Data Type' column lists the Esri data type the tool needs for each parameter. These data types are specialized structures related to ArcGIS data. The data types in this column do not refer to Python data types. For example, the in_features and out_feature_class are Feature Layer and Feature Class data types, respectively. 'Feature Layer' and 'Feature Class' refer to Esri data files such as shapefiles, but we use a Python string to specify the names of these files. The 'Linear Unit' is a distance, but we use a Python string to specify that too. In the following example, Python string literals are used to specify a Feature Layer, a Feature Class, and a Linear Unit:

```
>>> arcpy.Buffer_analysis('C:/gispy/data/ch06/park.shp',
                          'parkBuffer.shp', '0.25 miles')
```

The 'Code Sample' section following the 'Syntax' section provides sample scripts that can be a good starting point for beginners. Sometimes they can be used with only slight adjustments to accommodate the differences in input data.

6.2.2   Code Snippets

Code snippet captures provide an additional source for code samples. ArcGIS enables you to run a tool 'by hand' using the dialog box and then copy the Python code it creates from that tool run. Here are the steps to follow:

1. Locate the tool in ArcToolbox.
2. Execute the tool using its dialog box.
3. Open the 'Results' window (Geoprocessing menu > Results).

4. In the Results window, right-click on the tool name in the results listed there and select 'Copy as Python snippet'.
5. Paste the snippet into PythonWin and examine the syntax.

This technique is particularly useful when the structure of the required input is complex. For example, the Weighted Overlay (Spatial Analyst) tool requires an arcpy weighted overlay table parameter that can be complicated to compose by hand. Running the tool with your desired input and copying the snippet will provide an easier starting point.

6.3 Tool Parameters

Many geoprocessing tools require a linear unit parameter (e.g., a buffer distance for buffer analysis, an aggregation distance for aggregation tools, a simplification tolerance for simplification tools, and so forth). Other tools require a code expression parameter (e.g., an expression to calculate a value for the field calculation tool). Still other tools accept one or multiple values for a single parameter (e.g., a single file or a list of files). Meanwhile, most tools have both required and optional parameters. The next sections discuss how Python handles linear unit, code expression, multivalue, and optional parameters.

6.3.1 Linear Units

A linear unit is an Esri data type for Euclidean distances. When we use a tool's GUI interface, there's a text box and a combo box for typing a number and selecting a unit of measure (Figure 6.2). In Python, linear units are specified as a string, with a number and a unit of measure, separated by a space (e.g.,`'5 miles'`). A list of recognized units of measure is given in Table 6.2. Singular and plural forms of these units are both accepted. Also, linear unit names are not case sensitive. If only a numeric value is specified and a distance unit is not specified (e.g., `'5'`), the tool uses the units of the input feature, unless the Output Coordinate System environment property has been set.

Table 6.2 Linear unit of measure keywords (not case sensitive).

centimeters	decimal degrees	decimeters	feet	inches	kilometers	meters	miles
millimeters	nautical miles	points	unknown	yards			

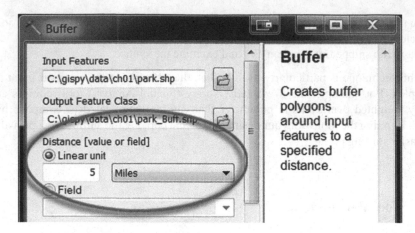

Figure 6.2 Tool GUIs collect units of measure (e.g., distance, radius, area, etc.) with a text box for the number and a combo box for the unit. Here the number 5 has been entered in the text box and Miles has been selected as the unit.

6.3.2   Python Expressions as Inputs

Some tools, such as the Calculate Field and Calculate Value (Data Management) tools, take parameters which are themselves Python expressions. The expression can be very simple. For example, the following code sets every entry in the `'result'` field of the 'data1.shp' shapefile to 5:

```
data = 'C:/gispy/data/ch06/data1.shp'
fieldName = 'result'
expr = 5
arcpy.CalculateField_management(data, fieldName, expr, 'PYTHON')
```

The last parameter in this Calculate Field call indicates the type of the expression (The choices are `'PYTHON'`, `'PYTHON_9.3'`, or `'VB'`. The latter, which stands for Visual Basic, is the default). The Calculate Field expressions can also use the values in other fields and combine them with mathematical operators. In these expressions, field names need to be surrounded by exclamation points. The following code calculates the `'result'` field again using an expression which multiplies the value in the `'measure'` field by two and subtracts the value in the `'coverage'` field:

```
expr = '2*!measure! - !coverage!'
arcpy.CalculateField_management(data, fieldName, expr,'PYTHON')
```

Note that the expression must be compatible with the field type. The 'result' field is a 'float' type field, so mathematical values can be used. Feature class field calculation expressions can also use the 'shape' field (Only feature classes have a 'shape' field). The shape field contains a set of arcpy Geometry objects (listed on the arcpy cheat sheet under 'Other objects'). Geometry objects have properties such as 'area' and 'length'. These properties can be used with dot notation on the 'shape' field in the expressions, still surrounded by exclamation points. The following code uses the 'area' geometry property to calculate a field named 'PolyArea' for the 'special_regions.shp' polygon shapefile:

```
data = 'C:/gispy/data/ch06/special_regions.shp'
fieldName = 'PolyArea'
expr = '!shape.area!'
arcpy.CalculateField_management(data, fieldName, expr,'PYTHON')
```

6.3.3 Multivalue Inputs

Some tools accept multiple values as input for a single parameter. If this is the case, a Python list appears behind the variable name in the 'Parameter' column. For example, the Merge (Data Management) tool takes a list of input datasets. The table entry for this parameter is shown in Figure 6.3. Multivalue input tools are usually ones that combine the input in some manner (e.g., merge, intersect, or union the data). A Python list may be the simplest way to input these values, especially if you already have the input in a list. However, there are two other ways to provide input to these tools. They also accept a semicolon delimited string or a ValueTable object. In the following code a Python list is used to specify the three input files to merge:

```
>>> inputFiles = ['park.shp', 'special_regions.shp', 'workzones.shp']
>>> arcpy.Merge_management(inputFiles, 'mergedData.shp')
```

In some cases, you may have values in a *multivalue string*, a string where the values are separated (or delimited) by semicolons. Multivalue parameters can be specified in this way too. Here the three inputs are given in a multivalue string:

```
>>> inputFiles = 'park.shp;special_regions.shp;workzones.shp'
>>> arcpy.Merge_management(inputFiles, 'mergedData2.shp')
```

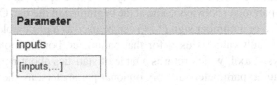

Parameter
inputs
[inputs,...]

Figure 6.3 The [inputs, ...] notation indicates that you can pass in a list of input files for this parameter.

A `ValueTable` is an `arcpy` object for storing rows and columns of information. To use it, you create a `ValueTable` object using the `ValueTable` function. This function returns a `ValueTable` object, which has methods and properties. One of the `ValueTable` methods is `addRow`. This example shows how to use the value table to merge three files:

```
>>> vt = arcpy.ValueTable()
>>> vt.addRow('park.shp')
>>> vt.addRow('special_regions.shp')
>>> vt.addRow('workzones.shp')
>>> arcpy.Merge_management(vt, 'mergedData3.shp')
```

The merge example is a one-dimensional example. That is, it only has one item in each row. The true advantage of the `ValueTable` object is for dealing with higher dimensions of data. The `ValueTable` approach provides a convenient way to organize data when the input is a list of lists. As an example, the 'in_features' argument for the Intersect tool can be a list of input file names or alternatively, it can be a list of file names and priority rankings, as in the following example:

```
>>> inputFiles = [['park.shp', 2], ['special_regions.shp', 2],
                  ['workzones.shp',1]]
>>> arcpy.Intersect_analysis(inputFiles, 'intersectData.shp')
```

Instead of using a list of lists, you could use a `ValueTable` for the intersection example as follows:

```
>>> vt = arcpy.ValueTable()
>>> vt.addRow('park.shp 2')
>>> vt.addRow('special_regions.shp 2')
>>> vt.addRow('workzones.shp 1')
>>> arcpy.Intersect_analysis(vt, 'intersectData.shp')
```

6.3.4 Optional Parameters

Tool parameters must be used in the order they are listed in the tool code signature. The required tool parameters always come at the beginning of the list. Optional parameters can be omitted or a number sign ('#') can be used as a place holder. In either of these cases, the default value is used for that parameter. For example, the Polygon Neighbors (Analysis) tool, which returns a table of statistics about neighboring polygons has two required parameters and six optional parameters, as shown in the code signature:

```
PolygonNeighbors_analysis (in_features, out_table, {in_fields},
{area_overlap}, {both_sides}, {cluster_tolerance}, {out_linear_units},
{out_area_units})
```

In the following example, the Polygon Neighbors tool is called two different ways, but these lines of code are two equivalent ways of using the default values for the six optional parameters in the tool:

```
>>> arcpy.env.workspace = 'C:/gispy/data/ch06/'
>>> arcpy.env.overwriteOutput = True

>>> # Use default values for the last 6 args.
>>> arcpy.PolygonNeighbors_analysis('park.shp', 'PN.dbf')

>>> # Another way to use default values for the last 6 args.
>>> arcpy.PolygonNeighbors_analysis('park.shp', 'PN.dbf', '#',
'#','#', '#', '#', '#')
```

If you want to set some but not all optional parameters, you must use number signs as place holders for interior optional arguments. In the following example, no place holder is needed, because we're using the first optional argument and we can simply omit the last five parameters:

```
>>> # Use default values for the last 5 parameters.
>>> arcpy.PolygonNeighbors_analysis('park.shp', 'PN.dbf', 'COVER')
```

However, if we want to set the 'area_overlap' parameter, but use the default value for the 'in_fields' parameter, the place holder is needed. The 'in_fields' parameter precedes 'area_overlap' in the parameters list, so if we failed to use '#' as a place holder, the tool would assume 'AREA_OVERLAP' was a field name.

```
>>> # Use default value for in_fields, but set the value for area_overlap.
>>> arcpy.PolygonNeighbors_analysis('park.shp','PN.dbf','#',
'AREA_OVERLAP')
```

6.4   Return Values and Result Objects

When a geoprocessing tool is run, the tool returns an `arcpy Result` object. The `Result` object contains information about the tool run, such as input values for the parameters and whether or not it was successful. It also contains a list of one or more output values. The returned object can be stored in a variable by using an assignment statement:

```
variable = arcpy.toolName_toolboxAlias(arg1, arg2, arg3,...)
```

In the following example, the `Result` object returned by the Polygon Neighbors tool is stored in a variable named 'pnResult':

```
>>> pnResult = arcpy.PolygonNeighbors_analysis('park.shp', 'PN.dbf')
```

The built-in `type` function shows that `pnResult` is an `arcpy Result` object:

```
>>> type(pnResult)
<class 'arcpy.arcobjects.arcobjects.Result'>
```

When printed, the `Result` object prints the first item in the tool output list. In the case of the Polygon Neighbors tool, the first output is the full path file name for the output:

```
>>> print pnResult
C:/gispy/data/ch06\PN.dbf
```

`Result` objects have a set of methods and properties. Figure 6.4 expands on the list given in the arcpyCheatSheet.pptx diagram. The `outputCount` property and

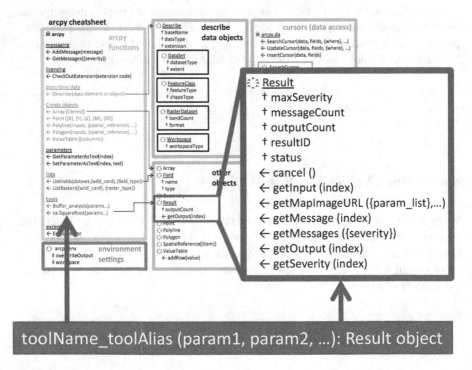

Figure 6.4 `Result` object methods and properties.

the `getOutput` method provide access to the output information. The following statement shows that the Polygon Neighbor tool returned one output:

```
>>> pnResult.outputCount
1
```

To get the output, use the `getOutput` method with an index. It is zero-based indexing, so to get the first output, use index zero. The following statement gets the first (and only) output from the `pnResult` object:

```
>>> pnResult.getOutput(0)
u'C:/gispy/data/ch06\PN.dbf'
```

The output depends on the nature of the tool. The main purpose of some tools, such as Buffer and Polygon Neighbors, is to create an output file. While the main purpose of other tools, such as Get Count and Average Nearest Neighbors is to compute numeric values. For tools like Buffer and Polygon Neighbors, there is one output value, the name of the output file.

```
>>> res = arcpy.Buffer_analysis('park.shp', 'outBuff.shp', '4 miles')
>>> print res
C:/gispy/data/ch06\outBuff.shp
```

Whereas, for tools like Get Count and Average Nearest Neighbor, the tool call returns the calculated results.

```
>>> resGC = arcpy.GetCount_management('park.shp')
>>> print resGC
426
```

Though the print statement is convenient for printing the first result, the `getOutput` function must be used to retrieve further results or to use numeric results in calculations. The following code throws an exception because a `Result` object cannot be added to a number:

```
>>> resGC + 25
Traceback (most recent call last):
  File "<interactive input>", line 1, in <module>
TypeError: unsupported operand type(s) for +: 'Result' and 'int'
```

Instead, the value must be retrieved from the `Result` object using the `getOutput` method:

```
>>> count = resGC.getOutput(0)
>>> count
u'426'
```

But `getOutput` returns a string, so this can't be used directly:

```
>>> count + 25
Traceback (most recent call last):
  File "<interactive input>", line 1, in <module>
TypeError: coercing to Unicode: need string or buffer, int found
```

Instead, it needs to be cast to an integer:

```
>>> int(count) + 25
451
```

The `getOutput` method also needs to be used when a tool returns more than one result. For example, the Average Nearest Neighbor (Spatial Statistics) tool computes a nearest neighbor ratio and a z-score which tell us something about the clustering of our points. Suppose we have points representing acts of kindness, we might be interested in whether the points exhibit clustering or if they resemble a random distribution. Example 6.1 calls the Average Nearest Neighbor tool and prints the results using `getOutput`. According to the results, the points do exhibit clustering. The tool help explains the meaning of these output parameters.

Example 6.1

```
# avgNearNeighbor.py
# Purpose: Analyze crime data to determine if spatial
#          patterns are statistically significant.
import arcpy

arcpy.env.workspace = 'C:/gispy/data/ch06'

annResult = arcpy.AverageNearestNeighbor_stats('points.shp',
  'Euclidean Distance')

print 'Average Nearest Neighbor Output'
print 'Nearest neighbor ratio: {0}'.format(annResult.getOutput(0))
print 'z-score: {0}'.format(annResult.getOutput(1))
print 'p-value: {0}'.format(annResult.getOutput(2))
print 'Expected Mean Distance: {0}'.format(annResult.getOutput(3))
print 'Observed Mean Distance: {0}'.format(annResult.getOutput(4))
```

Printed output:

```
>>> Average Nearest Neighbor Output
Nearest neighbor ratio: 0.853047
z-score: -5.802495
p-value: 0
Expected Mean Distance: 753.825013
Observed Mean Distance: 643.048013
```

6.5 Spatial Analyst Toolbox

The spatial analyst toolbox contains a variety of tools for calculating spatial measures such as density, distance, and neighborhood properties as well as map algebra, logical, and mathematics tools, extraction, generalization, and interpolation tools, and other specialized tools for hydrology, groundwater, and solar calculations, to name just a few. This section contrasts the Python approach for the Spatial Analyst toolbox with the approach for other toolboxes. Specifically, this section discusses the Python format for calling these tools, the import statements in the tool help, and access to raster calculation.

6.5.1 Calling Spatial Analyst tools

The syntax for calling Spatial Analyst tools is slightly different from other tools. Double dot notation must be used instead of an underscore to specify the toolbox.

```
arcpy.toolboxAlias.toolName(arg1, arg2, arg3,...)
```

There's another difference in the way Spatial Analyst tools work. Instead of returning a `Result` object, Spatial Analyst tools return `Raster` objects. The `Raster` object temporarily holds the output raster in memory. Unless explicitly saved, it will be deleted when the IDE session is closed. To store the raster permanently, we have to save the raster in a separate step. The following code assigns the environment settings, checks out the Spatial Analyst extension and then calls the Square Root tool on a raster named `'getty_rast'`. The `Raster` object being returned is assigned to a variable named `outputRast`.

```
>>> import arcpy
>>> arcpy.env.workspace = 'C:/gispy/data/ch06/'
>>> arcpy.env.overwriteOutput = True
>>> inRast = 'getty_rast'
>>> arcpy.CheckOutExtension('Spatial') u'CheckedOut'
>>> outputRast = arcpy.sa.SquareRoot(inRast)
```

The next step is to save the output raster. The `Raster` objects have a `save` method which takes one optional parameter, a name. When the Square Root tool was run, it automatically generated a name. If no parameter is used, the raster is saved by that name. The following line of code saves the output raster with the default name:

```
>>> outputRast.save()
>>> outputRast
C:\gispy\data\ch06\squar_ras
```

In this case, the automatically generated name is `'squar_ras'`. The following line of code saves the raster and names it `'gettySqRoot'`:

```
>>> outputRast.save('gettySqRoot')
```

Raster names have some constraints. When you don't specify an extension, the Esri GRID raster format is used. The names of GRID rasters must be no longer than 13 characters, though the error says something more generic than that, as shown here:

```
>>> outputRast.save('gettySquareRoot')
Traceback (most recent call last):
File "<interactive input>", line 1, in <module>
RuntimeError: ERROR 010240: Could not save raster dataset to
C:\gispy\data\ch06\gettySquareRoot with output format GRID.
>>> len('gettySquareRoot')
15
```

Other toolboxes automatically save the output to a long-term storage location such as a shapefile on the computer hard disk drive when the tool is called. The Spatial Analyst tools keep the raster data in-memory (a sort of short term memory storage which is fast to access). The raster `save` method commits the raster to long-term memory. This extra step of saving the raster is designed for optimization. Raster calculations might involve a chain of sequential calls that create intermediate rasters. If only the final output is needed (and not the intermediate rasters), the script can opt to only save the last one. This can save significantly on processing time for large rasters.

6.5.2 Importing spatial analyst

The ArcGIS Resources Spatial Analyst tool reference pages sometimes use a variation of the import statement in the code samples as in the following code:

```
>>> from arcpy.sa import *
```

This statement creates a direct reference to the Spatial Analyst tool; It provides a shortcut so Spatial Analyst tools can be called without prepending `arcpy.sa`. For example, it allows you to replace this statement:

```
>>> outputRast = arcpy.sa.SquareRoot(inRast)
```

with this statement:

```
>>> outputRast = SquareRoot(inRast)
```

This type of import saves some typing, but it can lead to confusion. For example, the Spatial Analyst has an `Int` tool which converts each cell raster value to an integer; At the same time, Python has a built-in integer function named `int`. Using the special import shortcut could lead to mystifying code such as the following:

```
>>> outputRast = Int(inRast)
>>> outputNum = int(myNum)
```

These statements look almost identical, but one takes a raster and returns a `Raster` object; The other takes a number and returns a number. Using the standard import and full tool call (e.g., `arcpy.sa.Int`) avoids ambiguities.

6.5.3 Raster Calculator

Python can call any tool in ArcToolbox with the exception of the Spatial Analyst Raster Calculator tool. The Raster Calculator performs what is known as 'map algebra'; it performs mathematical operations on each cell of a raster grid. For example, multiplying a raster by two using the Raster Calculator doubles the value stored in each cell. The Raster Calculator tool is not designed to be called from a stand-alone script. Instead, the script can call tools from the Spatial Analyst Math Toolset. For example, multiplying the cells in 'getty_rast' by 5 and subtracting 2 from each cell, in Raster Calculator would look like this: `5*'getty_rast'-2`. The values in 'getty_rast' are 1 and 2, so the resulting values are 3 and 8. To do this in a Python script, you could use two tool calls. Output from the first statement would provide input for the second as shown in Example 6.2. Since the initial values are 1 and 2, outRast1 has the values 5 and 10. These values are used in the Minus tool call, so the values in outRast2 are 3 and 8. The second to last line of code saves this raster. The last line of code deletes the `Raster` objects using the `del` keyword. This avoids locking issues.

Example 6.2

```
# computeRastEquation.py
# Purpose: Calculate 5 * 'getty_rast' - 2

import arcpy

arcpy.env.overwriteOutput = True
arcpy.env.workspace = 'C:/gispy/data/ch06/rastSmall.gdb'
arcpy.CheckOutExtension('Spatial')

outRast1 = arcpy.sa.Times(5, 'dataRast')
outRast2 = arcpy.sa.Minus(outRast1, 2)
outRast2.save('equationRast')
del outRast1, outRast2
```

Another way to handle raster calculations is to create `arcpy Raster` objects using the `Raster` method and then construct expressions using mathematical operators in combination with the `Raster` objects. The following code is equivalent to the code used in Example 6.2, producing values in `equationRast2` identical to those in `equationRast`:

```
>>> rastObj = arcpy.Raster('dataRast')
>>> outRast = 5*rastObj - 2
>>> outRast.save('equationRast2')
>>> del outRast
```

One advantage of this approach is that pairwise mathematical operators ($*$, $/$, $+$, $-$, etc.) can be used more than once, so that more than one pair of items can be multiplied in one statement, whereas, the Times (Spatial Analyst) tool can only handle pairs. The following code uses multiple `Raster` objects and numbers and mathematical operators in a single expression and saves the results in the `'output'` raster:

```
>>> r1 = arcpy.Raster('out1')
>>> r2 = arcpy.Raster('out2')
>>> r3 = arcpy.Raster('out3')
>>> outRast = (5*r1*r2*r3)/2
>>> outRast.save('output')
>>> del outRast
```

6.6   Temporary Feature Layers

Unlike the Spatial Analyst raster computations, most ArcGIS tools that compute vector data output create a long term file (such as a shapefile or geodatabase feature class) without calling a `save` function. Some exceptions can be found in the Data Management toolbox. Some of these tools create a *temporary feature layer*, a collection of features in temporary memory that will not persist after the IDE session ends. For example, the Make XY Event Layer (Data Management) tool creates a temporary feature layer. This tool takes a tabular text file and creates a new point feature layer based on the x and y coordinate fields in the file. The following code makes a temporary layer file (`'tmpLayer'`) from the 'xyData.txt' file which contains 'x', 'y', and 'butterfly' fields.

```
>>> myData = 'C:/gispy/data/ch06/xyData.txt'
>>> arcpy.MakeXYEventLayer_management(myData, 'x','y','tmpLayer')
```

You can perform additional tool calls on the temporary layer within the same script (or within the same Interactive Window session). The following code gets the record count:

```
>>> countRes = arcpy.GetCount_management('tmpLayer')
>>> countRes.getOutput(0)
u'8'
```

No file named `'tmpLayer'` ever appears in the 'C:/gispy/data/ch06' directory. To save this file in long-term memory, you'll need to call another `arcpy` tool. There are a number of Data Management tools that can be used for this. The following code uses the 'CopyFeatures' tool to save the layer to a shapefile:

```
>>> arcpy.CopyFeatures_management('tmpLayer', 'butterflies.shp')
<Result 'C:/gispy/data/ch06\\butterflies.shp'>
```

The output from the Make XY Event Layer (Data Management) tool is used as input for the 'Copy Features' tool and a 'butterflies.shp' file is saved in the Chapter 6 data directory. When the current IDE session ends (for example, when PythonWin is closed), this file will persist, but `'tmpLayer'` will be destroyed.

6.7   Using Variables for Multiple Tool Calls

Calling more than one tool in a script often involves using the output from one tool as the input for the next tool. In this case, it becomes useful to employ variables to store tool arguments so that string literals are not repeated. The temporary feature layer example in the previous section repeats the string literal 'tmpLayer' three times. Using a variable name would make it easier to change this name if needed (since it would only need to be changed in one place). Additionally, output file names can easily be created based on input file names. For example, the following code slices the file extension and appends 'Buffer.shp' to the name:

```
>>> fireDamage = 'special_regions.shp'
>>> fireBuffer = fireDamage[:-4] + 'Buffer.shp'
>>> fireBuffer
'special_regionsBuffer.shp'
```

In Example 6.3, 'special_regions.shp' represents regions damaged by a fire. To find the portion of the park (in 'park.shp') which lies within 1 mile of the fire damage, we buffer the fire damaged regions and then clip the park polygons on the buffer zone. The output from the Buffer tool is input for the Clip tool. Figure 6.5 shows the input data, the intermediate output, and the final output.

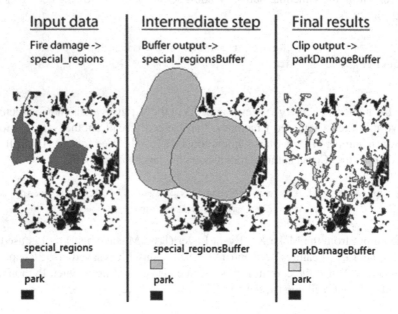

Figure 6.5 The input (left), intermediate output (center), and final output (right) from Example 6.3. The script finds the portion of the park which likes within one mile of the fire damage by first buffering the fire damage regions, then clipping the park data on this buffer. The output from the intermediate step is the input to the next step.

Example 6.3

```
# buffer_clip.py (hard-coded version)
# Purpose: Buffer a zone and use it to clip another file
# Input: No arguments needed.

import arcpy

arcpy.env.overwriteOutput = True
arcpy.env.workspace = 'C:/gispy/data/ch06'
outDir = 'C:/gispy/scratch/'
# Set buffer params
fireDamage = 'special_regions.shp'
fireBuffer = outDir + fireDamage[:-4] + 'Buffer.shp'
bufferDist = '1 mile'
# Set clip params
park = 'park.shp'
clipOutput = outDir + park[:-4] + 'DamageBuffer.shp'

arcpy.Buffer_analysis(fireDamage, fireBuffer,bufferDist)
print '{0} created.'.format(fireBuffer)
arcpy.Clip_analysis(park, fireBuffer, clipOutput)
print '{0} created.'.format(clipOutput)
```

Example 6.4 finds the lengths of the fire damage boundaries in two steps. First a field is added to store this information. Second, shape lengths are calculated in the new field. Both of these tools use the same input dataset and field name, and these variable values are set before the tool calls. The other two Calculate Field parameters are set as string literals and only used once in this example. The length is calculated using the '!shape.length!' expression and 'PYTHON' indicates that it is a Python type expression.

Example 6.4

```
# addLengthField.py
# Purpose: Add a field containing polygon lengths to the shapefile.
# Usage: No arguments needed.
import arcpy
arcpy.env.workspace = 'C:/gispy/data/ch06'

inputFile = 'special_regions.shp'
fieldName = 'Length'
```

```
arcpy.AddField_management(inputFile, fieldName, 'FLOAT')
arcpy.CalculateField_management(inputFile, fieldName,
'!shape.length!', 'PYTHON')
```

6.8   Calling Custom Tools

Many useful tools can be found by searching online. You may find models or Script
Tools that you want to call from within your code. To use custom tools (those that
are not built-in to ArcToolbox) use the `arcpy ImportToolbox` function and
then call the tool using the tool name and the toolbox alias. Tool names can differ
from the labels displayed in ArcCatalog. Labels are allowed to contain spaces. To
determine the tool name, right-click on the tool and view the 'Name' property on
the 'General' tab. To determine the toolbox alias, right-click on the toolbox and
view the 'Alias' property on the general tab. If it doesn't already have one, you can
set the alias here.

Example 6.5 pertains to 'customTools.tbx' which has a Script Tool named
'Inventory'. The toolbox alias is 'custom'. The tool points to a script named 'inventory.py'. The header comments in this script say the following:

```
# Usage: workspace, level_of_detail (SUMMARY_ONLY or DETAILED_INVENTORY)
# Example: C:/gispy/data/ch06/ SUMMARY_ONLY summaryFile.txt
```

We call the tool requesting a summary of the 'C:/gispy/data/ch06/' directory.

Example 6.5

```
# callInventory.py
# Purpose: Call the inventory tool.
import arcpy

arcpy.env.workspace = 'C:/gispy/sample_scripts/ch06'
inputDir = 'C:/gispy/data/ch06/'
arcpy.ImportToolbox('customTools.tbx')
arcpy.Inventory_custom(inputDir, 'SUMMARY_ONLY', 'summaryFile.txt')
```

The output text file from this tool, 'summaryFile.txt', looks something like this:

```
Summary of C:\gispy\data\ch06
-------------------------------------

27 Files
10 ShapeFiles
1 Workspaces
(1) Folders
(2) TextFiles
1 RasterDatasets
```

6.9   A Word About Old Scripts

Python programming for ArcGIS has changed a great deal since it was first introduced. ArcGIS Python is backwards compatible. In other words, Python code written for previous versions of ArcGIS will still run in the newer versions. For example,
an ArcGIS 9.2 Python script will work in ArcGIS 10.1. The `arcpy` package is how
ArcGIS 10.x gives Python access to geoprocessing. In 9.x versions, the `arcgis-`
`scripting` package was used. `arcpy` encompasses and enhances arcgisscript-
ing. Table 6.3 shows prior version equivalents to importing `arcpy`.

6.10   Discussion

Calling tools in ArcGIS follows a basic pattern with some variations. The parameters required by the tools vary based on the kinds of input the tool needs. To use
Python for ArcGIS, you'll need to become familiar with the basic pattern and use
the tool help for more detailed information. Context menus help with spelling and
parameter requirements. The ArcGIS Resources tool help parameter tables and
script examples provide more detailed information. This chapter presented the basic
pattern for calling tools and highlighted the most common parameter usage questions. Spatial Analyst toolbox objects such as the `arcpy.sa.RemapValue` and
`arcpy.sa.WOTable` are somewhat complex and highly specialized. For tools
that require these arguments, run the tool via ArcToolbox and 'Copy the code snippet' to get an example with sample input data. For the vast majority of tool calls
though, you won't need to copy code snippets, because the syntax will become
routine after a little practice.

The `ValueTable`, `Result`, and `Raster` objects introduced in this chapter
are just three of the many `arcpy` objects. The `arcpy` cheat sheet detail diagrams
in this book, like the one in Figure 6.3, provide selected method and property names
for several `arcpy` objects. These can be used as a quick reference, but the ArcGIS
Resource site contains full property and method documentation for each `arcpy`
object.

Table 6.3 Geoprocessing in Python.

ArcGIS version	Python code statements
9.1	`from win32com.client import Dispatch`
	`gp = Dispatch('esriGeoprocessing.GPDispatch.1')`
9.2	`import arcgisscripting`
	`gp = arcgisscripting.create()`
9.3	`import arcgisscripting`
	`gp = arcgisscripting.create(9.3)`
10.x	`import arcpy`

When searching, for an `arcpy` object, use the word `arcpy` along with the name of the object to improve the search results. For example, instead of searching for 'result', search for 'result arcpy', to bring the `Result` object help page to the top.

> **Note** When you search for `arcpy` objects on the ArcGIS Resources site, add the word `arcpy` to the search to bring the `arcpy`-related results to the top of the search.

6.11 Key Terms

Full path file name   Multivalue string
Python expression parameters   ValueTable object
Optional parameters   Temporary feature layers
Result object   Map Algebra
Multivalue input   Raster object

6.12 Exercises

Write a Python script to solve each problem and test the script on sample data in 'C:/gispy/data/ch06/'. Hard-code the workspace to this directory. Hard-coding means specifying it with a string literal (`arcpy.env.workspace = 'C:/gispy/data/ch06'`). To achieve a deeper understanding of the code syntax, write the scripts without using ModelBuilder. Name the script as specified in bold print at the beginning of each question.

1. **erase.py** 'special_regions.shp' represents areas where a fire damaged a park, 'park.shp'. Use the Erase (Analysis) tool to create a shapefile in 'C:/gispy/scratch' called 'no_damage.shp' which shows all the park regions that did not sustain damage.

2. **freq.py** The land cover types in 'park.shp' include 'woods', 'orchards', and 'other'. Park rangers would like a table including the frequencies of polygons with each cover type. Write a script to perform a Frequency (Analysis) on the 'COVER' field that will generate this table. Create the output in 'C:/gispy/scratch' and name it 'COVER_freq.dbf'.

3. **split.py** 'park.shp' is a large park. Maintenance has outlined work zones for which different teams will be responsible. 'workzones.shp' specifies the zones. The manager would like to give each team leader a map that only includes her own work zone. Write a script to call the Split (Analysis) tool to generate the

desired shapefiles. Using the Zone field to split the park. Have the script place the output shapefiles in 'C:/gispy/scratch'.

4. **square.py** Compute the square of the cell values in Gettysburg battlefield Esri GRID raster, 'getty_rast', using the Square (Spatial Analyst) tool. Create the output raster in 'C:/gispy/scratch' and name it 'squareGetty'.

5. **getCount.py** Use the Get Count (Data Management) tool to determine the number of polygons in 'park.shp'. Get the count using the arcpy Result object. Output should look like this, though you may get a different count if 'park.shp' has been modified:

```
>>> There are 425 polygons in park.shp
```

6. **distribution.py** Practice performing sequential geoprocessing steps. To see how the land cover polygons are distributed within 'C:/gispy/data/ch06/park.shp', perform two tool calls in this script. Use the Feature to Point (Data Management) tool to find the centroid of each polygon. Name this result 'C:/gispy/scratch/centroid.shp'. Then use the Create Thiessen Polygons (Analysis) tool to find the Voronoi regions of these centroids and name the result 'C:/gispy/scratch/voronoi.shp'.

7. **hullPoints.py** Practice performing sequential geoprocessing steps. To find four corner points for 'C:/gispy/data/ch06/data1.shp', perform two tool calls in this script: Use the Minimum Bounding Geometry (Data Management) tool to find the hull of the points. Name this result 'C:/gispy/scratch/boundingPoly.shp'. Then call the Feature Vertices to Points (Data Management) tool on this bounding polygon and name this result 'C:/gispy/scratch/outerPoints.shp'.

8. **avgRast.py** Write a script that uses Raster objects and Spatial Analyst to calculate an average raster value from three rasters 'out1', 'out2', and 'out3' in the sample database, in rastSmall.gdb. Save the output raster in rastSmall.gdb as 'C:/gispy/scratch/avgRast'. Use a few simple mathematical operations instead of using the Cell Statistics tool. Be sure to delete the output Raster object using the del keyword.

9. **reclassify.py** To explore using the RemapValue object, run the Reclassify (Spatial Analyst) tool 'by hand' as follows: Reclassify the Gettysburg battlefield Esri GRID raster (getty_rast) 'VALUE' field so that all raster areas with a value of 1 have a new value of 100 and all areas with a value of 2 have a new value of 200. Name the output raster 'C:/gispy/scratch/reclassGetty'. Next, copy a code snippet from the results to see the syntax for the RemapValue object, and write a script that calls the Reclassify tool with the same input and output. In the script, be sure to delete the output Raster object using the del keyword.

Chapter 7
Getting User Input

Abstract Scripts that are designed to be flexible can be reused with varying parameter values. Relying on hard-coded parameters, specified with string literals, numbers and so forth, means users can't change the values without opening the script and modifying the code. A flexible script that accepts arguments for the parameters can be easier to reuse. The geoprocessing examples in Chapter 6 specified the tool parameters inside the script. These processes can be made dynamic by accepting user input. This chapter covers the `arcpy` approach to receiving arguments and contrasts this with using `sys.argv`. Then it discusses argument spacing and `os` module methods for handling file path arguments and getting the script path.

Chapter Objectives
After reading this chapter, you'll be able to do the following:

- Capture user input with two different techniques.
- Predict script behavior in case of too few arguments.
- Group characters in arguments appropriately.
- Explain why single quotations cannot be used to group arguments.
- Extract the file name, directory, and file extension from a full path file name.
- Join base file names with paths.
- Get the size of a file.
- Get the path of the current script.

7.1 Hard-coding versus Soft-coding

Hard-coded tool parameters are encoded specifically within the script, so that the code needs to be altered in order to change this values. More flexible scripts get input from the user—they are *soft-coded*. The input and output features ("park.shp" and "boundingBoxes.shp") in Example 7.1 are hard-coded.

© Springer International Publishing Switzerland 2015
L. Tateosian, *Python For ArcGIS*, DOI 10.1007/978-3-319-18398-5_7

Example 7.1

```
# boundingGeom.py
# Purpose: Find the minimum bounding geometry of a set of features.

import arcpy

arcpy.env.overwriteOutput = True
arcpy.env.workspace = 'C:/gispy/data/ch07'

inputFeatures = 'park.shp'
outputFeatures = 'boundingBoxes.shp'

arcpy.MinimumBoundingGeometry_management(inputFeatures,
                                         outputFeatures)
```

In this chapter, we compare two techniques for accessing user arguments:

1. An `arcpy` function named `GetParameterAsText`.
2. The `sys` Python standard module variable named `sys.argv`.

7.2   Using GetParameterAsText

The `arcpy` package has a `GetParameterAsText` function that gets input from
the user. To use this `arcpy` function:

1. Use `arcpy.GetParameterAsText(index)`, starting with `index = 0`,
   to replace the hard-coded parameters.
2. Add usage and example comments in the header to tell the user how to run the
   script.

Example 7.2 shows the script modified with these changes. Indices start at zero
and correspond to the order in which the arguments are passed. To run the script

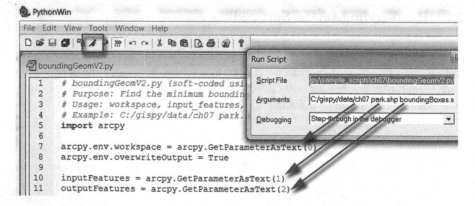

Figure 7.1 User arguments are consumed by the `GetParameterAsText` function.

with user input in PythonWin, place arguments, separated by spaces, in the 'Arguments' list in the 'Run Script' dialog. In PyScripter, go to Run>Command Line Parameters... and place the arguments in the text box (space separated). Make sure that 'Use Command Line Parameters?' is checked. Figure 7.1 shows 'boundingGeomV2.py' being run in PythonWin with these arguments: "C:/gispy/data/ch07" "park.shp" "boundingBoxes.shp".

Example 7.2

```
# boundingGeomV2.py (soft-coded using arcpy)
# Purpose: Find the minimum bounding geometry of a set of features.
# Usage: workspace, input_features, output_features
# Example: C:/gispy/data/ch07 park.shp boundingBoxes.shp

import arcpy

arcpy.env.overwriteOutput = True
arcpy.env.workspace = arcpy.GetParameterAsText(0)

inputFeatures = arcpy.GetParameterAsText(1)
outputFeatures = arcpy.GetParameterAsText(2)

arcpy.MinimumBoundingGeometry_management(inputFeatures,
                                         outputFeatures)
```

7.3 Using `sys.argv`

If the script imports the `arcpy` package, `GetParameterAsText` is a good approach to use. However, if the script is designed to be used on a machine that might not have ArcGIS installed, the built-in system module should be used instead. This `sys` module has a property named `argv` (short for argument vector). The `sys.argv` variable stores:

1. The name of the script, including the path where it is stored (the full path file name of the script).
2. The arguments passed into the script by the user.

`sys.argv` is a zero-based Python list that holds the script name and the arguments. *The first item in the list is the full path file name of the script itself.* The next item (indexed 1) is the first user argument. To soft-code a script using the `sys` module, you need to do these three things:

1. Import sys.
2. Replace hard-coded parameters with sys.argv[index], starting with index = 1 (since the zero index is used for the script path).
3. Add usage and example comments in the header to tell the user how to run the script.

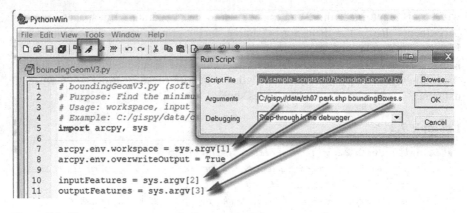

Figure 7.2 User arguments automatically populate the sys.argv list.

To get three arguments passed by the user, 'boundingGeomV3.py' in Example 7.3 uses sys.argv[1], sys.argv[2], and sys.argv[3].

Example 7.3

```
# boundingGeomV3.py (soft-coded using sys)
# Purpose: Find the minimum bounding geometry of a set of features.
# Usage: workspace, input_features, output_features
# Example: C:/gispy/data/ch07 park.shp boundingBoxes.shp

import arcpy, sys

arcpy.env.overwriteOutput = True
arcpy.env.workspace = sys.argv[1]

inputFeatures = sys.argv[2]
outputFeatures = sys.argv[3]

arcpy.MinimumBoundingGeometry_management(inputFeatures,
                                         outputFeatures)
```

Figure 7.2 shows 'boundingGeomV3.py' being run in PythonWin with these arguments: "C:/gispy/data/ch07" "park.shp" "boundingBoxes.shp".

7.4   Missing Arguments

A script can fail if the user does not provide enough arguments. When an argument is missing, the GetParameterAsText method itself does not throw an error. Instead, it returns an empty string. If the script in Example 7.2 is run without any arguments, the script assigns an empty string to the input and output features. Then when the tool is called, the following error is thrown, since the tool is being run on empty strings (only the last three lines of the error are shown here):

```
ERROR 000735: Input Features: Value is required
ERROR 000735: Output Feature Class: Value is required
Failed to execute (MinimumBoundingGeometry).
```

When an argument is missing in a script that instead uses the `sys.argv` method, the script throws an `IndexError` because `sys.argv` is a Python list and the code is trying to use a list index that is beyond the end of the list. If Example 7.3 is run without arguments, the following error is thrown:

```
arcpy.env.workspace = sys.argv[1]
IndexError: list index out of range
```

7.5 Argument Spacing

Arguments are space delimited in the 'Arguments' list. Consequently, if an argument itself has embedded spaces, it must be surrounded by quotations else it will be interpreted as more than one argument. For example, if the intended argument is a file path such as "C:/African Elephant/rasters", Figure 7.3a will produce an incorrect result, but Figure 7.3b will produce the desired results.

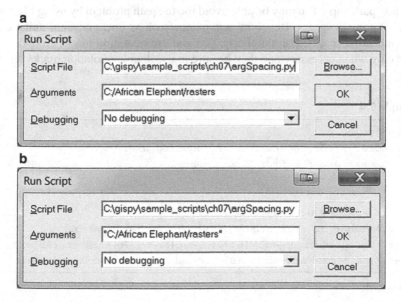

Figure 7.3 The argument is a file path with an embedded space between 'African' and 'Elephants'. Run script (**a**) will interpret it as two arguments ('C:/African' and 'Elephant/rasters'). Run script (**b**) will interpret it as one argument ('C:/African Elephant/rasters').

'argSpacing.py' in Example 7.4 illustrates this problem. Running it with the input shown in Figure 7.3a produces this output:

```
>>> Number of user arguments: 2
The first argument: C:/African
The second argument: Elephant/rasters
```

Running it with the input shown in Figure 7.3b produces this output:

```
>>> Number of user arguments: 1
The first argument: C:/African Elephant/rasters
The second argument:
```

Usually in Python, single and double quotes are interchangeable. Arguments are an exception. The quotation marks around arguments passed in through an IDE must be double, not single. PythonWin and PyScripter do not interpret the single quotations as an indication of grouping, so passing 'C:/My data/rasters' into 'arg-Spacing.py' returns two arguments:

```
>>> Number of user arguments: 2
The first argument: 'C:/African
The second argument: Elephant/rasters'
```

The single quotation marks appear in the printed output too. Python IDE's do not strip off the single quotes, so even if you do not have spaces within an argument, using single quotes can cause failure. For example, use "park.shp" as a script argument not 'park.shp'. You may be able avoid the file path problem by using file paths that don't contain spaces. But some arguments, such as linear or areal units, must have embedded spaces. For simplicity, you can use quotation marks around arguments only when necessary and then be sure to use double quotation marks instead of single.

Example 7.4

```python
# argSpacing.py
# Purpose: Print the number of incoming user
# arguments and the first 2 arguments.

import arcpy

numArgs = arcpy.GetArgumentCount()
print 'Number of user arguments: {0}'.format(numArgs)
print 'The first argument: {0}'.format(arcpy.GetParameterAsText(0))
print 'The second argument: {0}'.format(arcpy.GetParameterAsText(1))
```

> **Note: Use Double Quotes for IDE Arguments**
>
> No quotation marks are needed for arguments that don't have embedded spaces. However, if you do need a space within a script argument, you must use double quotes, not single quotes. For example, use "5 miles", not '5 miles'.

7.6 Handling File Names and Paths with os Module Functions

GIS scripts often need to separate or join file names and file paths. The standard os (operating system) module introduced briefly in Chapter 2, has a path sub-module for file path name manipulation, including the functions dirname, basename, join, and getsize. Since path is a submodule of os, double dot notation (os.path.functionName) is used to access these functions. The following code uses the basename method to extract the file's base name from the full path file name:

```
>>> import os
>>> inFile = 'C:/gispy/data/ch07/park.shp'
>>> # Get only the file name.
>>> fileName = os.path.basename(inFile)
>>> fileName
'park.shp'
```

The following code uses the dirname method to extract the directory path from the full path file name:

```
>>> # Get only the path.
>>> filePath = os.path.dirname(inFile)
>>> filePath
'C:/gispy/data/ch07'
```

The following code uses join to create a full path file name from the two parts:

```
>>> # Join the arguments into a valid file path.
>>> fullPath = os.path.join(filePath, fileName)
>>> fullPath
'C:/gispy/data/ch07\\park.shp'
```

This operation could be performed with concatenation too, but the join automatically takes care of slashes. In Example 7.5, 'copyFile.py' uses os.path. basename to get the input file name so that it can be copied to a backup directory. This script also uses os.path.join to create the new full path name.

Example 7.5

```
# copyFile.py
# Purpose: Copy a file.
# Usage: source_full_path_file_name, destination_directory
# Example: C:/gispy/data/ch07/park.shp C:/gispy/scratch/
# The C:/gispy/scratch directory must already exist
# for this example to work.

import arcpy, os

inputFile = arcpy.GetParameterAsText(0)
outputDir = arcpy.GetParameterAsText(1)

baseName = os.path.basename(inputFile)
outputFile = os.path.join(outputDir, baseName)

arcpy.Copy_management(inputFile, outputFile)

print 'inputFile =', inputFile
print 'outputDir =', outputDir
print
print 'baseName =', baseName
print 'outputFile = ', outputFile
```

Printed output:

```
inputFile = C:/gispy/data/ch07/park.shp
outputDir = C:/gispy/scratch

baseName = park.shp
outputFile = C:/gispy/scratch\park.shp
```

Other os.path methods are useful as well. In Example 7.6, 'compact.py' uses the getsize method to compare the size in bytes of a Microsoft database file before and after it has been compacted. The output from running this script with C:/gispy/data/ch07/cities.mdb looks something like this:

```
>>> cities.mdb file size before compact: 1552384 bytes.
>>> cities.mdb file size AFTER compact: 397312 bytes.
```

Example 7.6

```
# compact.py
# Purpose: Compact a file
# Usage: Full path file name of an mdb file.
# Example: C:/gispy/data/ch07/cities.mdb

import arcpy, os

# Get user input and extract base file name.
fileName = arcpy.GetParameterAsText(0)
baseName = os.path.basename(fileName)

# Check size
size = os.path.getsize(fileName)
print '{0} file size before compact: {1} bytes.'.format(baseName,
                                                         size)

# Compact the file
arcpy.Compact_management(fileName)

# Check size
size = os.path.getsize(fileName)
print '{0} file size AFTER compact: {1} bytes'.format(baseName,
                                                       size)
```

Chapter 6 demonstrated how to use slicing to separate a file path from its extension when the length of the extension is known. For example, the following code removes a three character extension (and the dot):

```
>>> myShapefile = 'parks.shp'
>>> rootName = myShapefile[:-4]
>>> rootName
'parks'
```

If the extension length is unknown, an os.path method can be used to split a file extension from its name. The os.path.splitext splits the file name at the dot in the name (if there is more than one dot, it uses the last one). It returns a tuple containing the two parts, the root name and the extension:

```
>>> os.path.splitext(myShapefile)
('parks', '.shp')
```

If the name has no extension, the first item is the name and the second is an empty string:

```
fc = 'farms'
>>> os.path.splitext(fc)
('farms', '')
```

Indexing the first item retrieves the root name:

```
>>> os.path.splitext(myShapefile)[0]
'parks'
>>> os.path.splitext(fc)[0]
'farms'
```

Slicing, on the other hand, may not work as expected, if the file extension length is unknown:

```
>>> fc[:-4]
'f'
```

7.6.1   Getting the Script Path

A simple way to share a script is to provide the script and the data it needs within a single common folder. The relative path to the data remains the same as long as the recipient leaves the data in the same relative position. In this case, the script can use its own location as the workspace. The command os.path.abspath(__file__) can be used inside a script to get the full path file name of the script being run by PythonWin. The __file__ variable returns the name of the script. Depending on how the script is run, this may or may not be the full path file name. But calling the abspath method on the __file__ constant returns the full path file name. The dirname method can then be used to get the script's directory as shown in Example 7.7. This script also demonstrates another useful os command method mentioned in a previous chapter. The listdir method returns a list of file names for every file in the directory that's passed into it as an argument. The output from running this script in a directory containing 9 files looks like this:

```
['argSpacing.py', 'boundingGeom.py', 'boundingGeomV2.py',
'boundingGeomV3.py', 'buffer_clipv2.py', 'compact.py', 'copyFile.py',
'near.py', 'scriptPath.py']
```

Example 7.7

```
# scriptPath.py
# Purpose: List the files in the current directory.
# Usage: No user arguments needed.
import os

# Get the script location
scriptPath = os.path.abspath(__file__)
scriptDir = os.path.dirname(scriptPath)

print '{0} contains the following files:'.format(scriptDir)
print os.listdir(scriptDir)
```

Avoid NameErrors!

PythonWin sometimes causes us to overlook missing import statements or assignment statements because variables keep their values throughout a PythonWin session. Try the following:

1. Run 'scriptPath.py' and keep PythonWin open.
2. Delete the import statements and save the script.
3. Run it again. It runs successfully. No need to import `sys` and `os`, right?
4. Close PythonWin.
5. Reopen PythonWin and run 'scriptPath.py' a third time. Now it throws a `NameError` exception.

If you run a script that imports a module, other scripts run within the same PythonWin session won't need to import that module to use it. A `NameError` won't occur unless you close PythonWin, then reopen it and run the script which is missing the import.

Always perform a final test within a fresh PythonWin session before sharing code or, alternatively, run scripts with PyScripter, which checks for imports with each script run.

7.7 Key Terms

Hard-coding vs. Soft-coding
`arcpy.GetParameterAsText(index)`
`sys.argv`
`os.path.dirname(fileName)`
`os.path.basename(fileName)`
`os.path.splitext(fileName)`
`os.path.abspath(path)`
`__file__`
File base name
Full path file name

7.8 Exercises

1. The code in 'near.py' uses the Near (Analysis) tool, which determines the distance from each feature in one dataset to the nearest feature in another dataset, within a given search radius. (e.g., a hiker can find historic landmarks close to the trails). The script default values point to trail and landmarks shapefiles. Inspect and run 'near.py' as described below to answer the following questions:

   (a) Run with arguments: # # #
       Which two shapefiles did the script use for the Near (Analysis) tool? How many nearby features were found?

(b) Run with arguments: # #
    Why didn't the script run successfully?
(c) Run with arguments: # # "100 kilometers"
    How many nearby features were found? Why does this give a higher value than part a?
(d) Run with arguments: # # 100 kilometers
    How many nearby features were found? Why does this give a lower value than part c?
(e) Run with arguments:
    'C:/gispy/data/ch07/data/landmarks.shp' 'C:/gispy/data/ch07/data/trails.shp' '50 miles'
    Why does this throw an error?
(f) Run with arguments:
    C:/gispy/data/ch07/My data/landmarks.shp C:/gispy/data/ch07/My data/trails.shp "50 miles"
    Why does this throw an error?
(g) Run with arguments:
    "C:/gispy/data/ch07/My data/landmarks.shp" "C:/gispy/data/ch07/My data/trails.shp" "50 miles"
    Why do you need all of these quotation marks?

2. Sample script 'buffer_clipv2.py' takes an input workspace, an output directory, a fire damage shapefile, a buffer distance, and an input polygon shapefile as arguments (It is just like sample script 'buffer_clipv2.py' except for the soft-coding). Which of the following arguments for sample script 'buffer_clipv2.py' is invalid input and why?

   C:/gispy/data/ch07/ C:/gispy/scratch/ special_regions.shp "1 mile" park.shp
   "C:/gispy/data/ch07/" "C:/gispy/scratch/" "special_regions.shp" "1 mile" "park.shp"
   C:/gispy/data/ch07/ C:/gispy/scratch/ special_regions.shp 1 mile park.shp
   C:/gispy/data/ch07/ C:/gispy/scratch/ special_regions.shp '1 mile' park.shp

3. **buffer_clipv3.py** Revise 'buffer_clipv2.py', replacing the `sys.argv` approach for gathering user input with `GetParameterAsText` methods, and save it as 'buffer_clipv3.py'. Test 'buffer_clipv3.py' with the input example in the header comments of the script. Then test it with no arguments. Run sample script 'buffer_clipv2.py' with no arguments and explain what error occurs and why there is a difference in the errors.

4. **handlePaths.py** Practice using `os` module commands by writing a script that takes two arguments, a full path file name and a base file name of another file. Then use `os` commands in the script to print each of the following:

   (a) The base name of the first file name.
   (b) The extension of the first file name (make sure this works for files without extensions).
   (c) The full path file name of the second file, assuming it's in the same workspace as the first.
   (d) The size of the second file.
   (e) A list of files in the script's directory.

Example input: C:/gispy/data/ch07/park.shp xy1.txt

Example output:
```
The first file is: park.shp
The first file extension is: .shp
The full name of the second file is: C:/gispy/data/ch07\xy1.txt
The size of the second file is: 42 bytes
C:\gispy\sample_scripts\ch07 contains the following files:
['argSpacing.py', 'boundingGeom.py', 'boundingGeomV2.py',
'boundingGeomV3.py', 'buffer_clipv2.py', 'compact.py',
'copyFile.py', 'handlePaths.py', 'near.py', 'scriptPath.py']
```

5. **shape2kml.py** It takes two steps to convert shapefiles to KML using Esri tools:

   (a) Make a feature layer from the input shapefile with the Make Feature Layer (Data Management) tool.
   (b) Use the Layer to KML (Conversion) tool to convert the layer to KML.

   Write a script that converts a shapefile to a KML file. It should take four arguments: the workspace, the shapefile, the full path output KML file name, and the scale. Notice, the Layer to KML tool requires that the output name has a ".kmz" extension. So, for example, to convert the shapefile 'C:/gispy/data/ch07/park.shp' to a file named 'C:/gispy/scratch/park.kmz' at a scale of 1:24000, the script arguments should be as follows:

   Example input:
   C:/gispy/data/ch07/ park.shp C:/gispy/scratch/park.kmz 24000

   Example output:
   ```
   Temporary layer tempLayer created.
   C:/gispy/scratch/park.kmz created.
   ```

6. **calcGeom.py** Write a script that adds a new float field to a given polyline or polygon shapefile and uses the Calculate Field tool to calculate the length or area of each feature in the given shapefile and populate the new field with the result. The script should also allow the user to specify the unit of measure (e.g., acres, miles, meters, ...). Use a Python expression (see Section 6.3.2) to specify the field calculation. Program this without using conditional constructs (discussed in an upcoming chapter). Instead, use the string `format` method to construct the Python expression based on the given geometry and unit of measure. The script should take four arguments ordered as follows: the full path file name of the input data, a field name, the geometric property (area or length), and a unit of measure. As an example, say you want to calculate the length of the trails in kilometers in the 'C:/gispy/data/ch07/trails.shp' file and you want to place the results in a new field named 'kmLength'. The input arguments for this scenario would appear as follows:

   C:/gispy/data/ch07/trails.shp kmLength length kilometers

7. **fileCompare.py** Write a script that compares two ASCII text files to determine if there are any differences. Use the `arcpy` File Compare (Data Management) tool which returns a `Result` object and writes a report on the differences between the files. Use the default setting for `file_type`. Set the `continue_compare` parameter to `'CONTINUE_COMPARE'`. The script should take three arguments, the full path file names of both input files, and the full path file name of the output file. It should create the output file and print feedback as shown in the example output. The file comparison verdict (true/false) is the second piece of information in the `Result` object. So, to print the true or false verdict, you'll need to use something like this:

```
resultObj.getOutput(1).
```

Example input:

C:/gispy/data/ch07/xy1.txt C:/gispy/data/ch07/xy_current.txt C:/gispy/scratch/diffOutput.txt

Example output:

```
>>> Comparison results have been written to:
C:/gispy/scratch/diffOutput.txt
Are the input files (xy1.txt and xy_current.txt) the same? false
```

Chapter 8
Controlling Flow

Abstract As you study the capabilities of programming for GIS, applications to your own work may come to mind. For example, you may need to run a geoprocessing tool on distributed batches of datasets or tweak multiple tables before they can be imported into ArcGIS. Before writing scripts for tasks like these, it's helpful to outline the main steps without belaboring syntax details, but instead using *pseudocode*, a generic format for outlining workflow. You also need to be familiar with the building blocks of workflow. How will the code make decisions based on the input? How will it repeat a process for each file? Chapters 9 and 10 describe the Python syntax for 'branching' and 'looping', but that part will seem easy, compared to the logic involved in designing the workflow. This chapter uses pseudocode examples to introduce 'branching' and 'looping'.

Chapter Objectives
After reading this chapter, you'll be able to do the following:

- Define the terms algorithm and pseudocode.
- Translate a GIS problem into a set of succinct steps.
- Use the terms looping, and branching to a describe workflow.
- Identify the three key components in a looping structure.
- Use indentation to group blocks of pseudocode.
- Interpret pseudocode.

8.1 Outlining Workflow

From a high-level perspective, all workflows are made up of three basic structures: *sequential* steps, *decision-making*, and *repetition*. Directions from point A to point B are given in sequential steps. Checking if data is zipped, tarred, or uncompressed is done with decision-making. Clipping every file in a geodatabase is performed with repetition. Decision-making and repetition are also referred to, respectively, as *branching* and *looping*.

Table 8.1 shows simple examples of these structures expressed with pseudocode. The examples include two common types of repetition. Workflows usually involve some combination of sequence, repetition, and decision-making steps.

© Springer International Publishing Switzerland 2015   133
L. Tateosian, *Python For ArcGIS*, DOI 10.1007/978-3-319-18398-5_8

Table 8.1 Pseudocode examples of the three basic flow structures.

Flow structure	Example
Sequence	Hitch horse to sleigh. Go over the river. Go through the woods. Arrive at grandmother's house on the right.
Repetition (looping)	FOR each file in state layers geodatabase Clip the file on the county boundary. ENDFOR

	SET buffer distance to 1 mile.
	WHILE buffer distance < 6 miles
	    Buffer the shapefile with the buffer distance.
	    INCREMENT buffer distance by 1 mile.
	ENDWHILE

Decision-making (branching)	IF data is compressed THEN
	    Uncompress data.
	ENDIF

Examples 8.1–8.3 demonstrate these structures with pseudocode for three applications (clipping batches of data, tweaking tabular data, and downloading data). The pseudocode in Example 8.1 outlines steps for finding the weighted sum of a set of rasters to analyze dumping patterns for sediment dredged from canals. To create the rasters, the workflow repeats a sequence of steps (steps 3–5) for each input table. These steps convert tabular data to a spatial format, clip the spatial data using the disposal region, and convert the vector results to raster format.

Example 8.1: Pseudocode for processing dredging sediment data repeats steps 3–5 for each table.

```
1  GET a list of the weekly data base tables.
2  FOR each table in the list
3      Make a spatial layer from the table.
4      Clip the new layer on the disposal site extent.
5      Convert clipped vector data to raster format.
6  ENDFOR
7  Find weighted sum of output rasters.
```

Pseudocode uses capitalized keywords such as GET and FOR...ENDFOR to denote workflow structures and actions. Table 8.2 has a list of common pseudocode keywords. Keywords with openings and closings such as FOR...ENDFOR, WHILE...ENDWHILE, and IF...ELSE...ENDIF, are *block structures*. Items within a block structure are indented the same amount to indicate that they are related. Since steps 3–5 are surrounded by the FOR...ENDFOR structure, this block of pseudocode is indented, reinforcing that these three steps are repeated for each table. This also emphasizes that step 7 is not repeated for each table. Step 7 needs to find the

Table 8.2   A short list of common pseudocode keywords.

Action	Pseudocode keywords
Input	READ, GET
Output	PRINT, SHOW
Initialize	SET
Add one or subtract one	INCREMENT or DECREMENT
Repetition (looping)	WHILE...ENDWHILE, FOR...ENDFOR
Decision-making (branching)	IF, THEN, ELSE IF, ELSE, ... ENDIF
Function (procedure or subroutine)	FUNC, ENDFUNC, CALL, RETURN

weighted sum of all the output rasters, so this must only occur one time—after all the output rasters have been created. The FOR...ENDFOR structure is usually referred to as a FOR-loop because the workflow *loops* back to repeat steps. For example, if there are three tables in the list, the line number workflow will run 1-2-3-4-5--2-3-4-5--2-3-4-5--2-3-4-5--6-7.

Though each line in the pseudocode corresponds to a step in the process, the steps are defined broadly enough so that details do not obscure the overall flow. For example, making a spatial layer from a table might involve finding the latitude and longitude field names and choosing a projection but the pseudocode doesn't need to include these details.

Example 8.2 outlines steps for automatically reading data links from a data hosting Web page, downloading the data from those links, and uncompressing the data. The workflow contains decision-making structures as well as a repetition structure because each data file is only uncompressed if needed. The paired steps 8–9 and steps 11–12 handle two distinct kinds of compression files. These lines are indented three levels because the IF...ELSE IF...ENDIF structure is inside the FOR... ENDFOR structure. Since it is inside of the FOR...ENDFOR structure, decision-making occurs for each dataset linked to the page.

Example 8.2: Pseudocode for automatically fetching and uncompressing data files listed on a web site.

```
 1   GET the data center web page URL.
 2   READ the data center web page contents.
 3   GET a list of links to data in the page contents.
 4   FOR each data link
 5       Fetch the data from the specified link.
 6       Save the data.
 7       IF the data is zip compressed THEN
 8           Unzip data.
 9           Delete zip file.
10       ELSE IF the data is tar compressed THEN
11           Untar data.
12           Delete tar file.
13       ENDIF
14   ENDFOR
```

Example 8.3 specifies the steps to prepare some tabular data with spatial fields to be imported into ArcGIS. In this example, the input tabular spatial data needs several modifications to fulfill GIS file import formatting requirements: the field names are not in the first row, the table has unnecessary trailing columns, and the field names have unacceptable special characters. This example uses both WHILE-loops and FOR-loops for repetition.

- FOR-loops are used to repeat steps *for each* item in a list of items such as, tables, data links, or field names. In this kind of FOR-loop, a *looping variable* successively gets the value of (or *iterates* through) each item in the list.
- WHILE-loops are used to repeat steps for a set of numbers (for example, row numbers 1–5 or buffer distances of 2–10 miles). A looping variable iterates through the numerical values. WHILE-loops can also be used to stop after an unknown number of steps based on the fulfillment of some condition (e.g., loop while the desired field header has not been found).
- There are three basic components of the WHILE-loop, initializing, checking, and updating the looping variable. In Example 8.3, these WHILE-loop components are as follows:

  1. Initialize looping variable: SET row number to 1
  2. Test looping variable: row number < field name row number
  3. Update looping variable: INCREMENT row number

Initializing the WHILE-loop variable sets the starting value, so that it can be tested with a condition which is either true or false. This controls the flow of the steps (if the condition is true, go to step 5; if not, go to step 8). In other words, when this condition is false, the loop terminates. If the condition is true, the indented steps, such as 'delete row' and 'increment row number' in Example 8.3 are performed. In most cases, the looping variable should be updated at the very end of the loop, after all of the other indented steps. Omitting the looping variable update results in never-ending repetition, since the loop condition never becomes false.

When a FOR-loop is used on a list of items, the three loop components are usually implied with the FOR each statement. For example, for lines 14–16 in Example 8.3, these components could be described as follows:

1. Start at the beginning of the list: Initialize the looping variable to the first field name in the list.
2. Check if there are still more items in the list: If the looping variable is still set to a field name, go to step 15. If not, go to step 17.
3. When the indented steps have been performed, point the looping variable to the next field name in the list or an 'end of list' flag.

Notice that in all of the FOR-loop and WHILE-loop examples, the looping variable is used inside the loop. In Example 8.1, the table is made into an X Y layer. In Example 8.2, the data is fetched from the specified link. In Example 8.3, the row is deleted (line 5), the column is deleted (line 10), and the field name is inspected for special characters. In general, the looping variable should be explicitly used inside the loop in some step other than the update. Otherwise, an important step may be missing.

Example 8.3: Pseudocode for preparing a table for ArcGIS import by deleting unneeded rows and columns and repairing field names.

```
1   FUNC preprocessTable (data table, field name row number, first
    invalid column number)
2       OPEN data table.
3       SET row number to 1.
4       WHILE row number < field name row number
5           Delete row.
6           INCREMENT row number.
7       ENDWHILE
8       SET column number to first invalid column number.
9       WHILE column number <= number of columns
10          Delete column.
11          INCREMENT column number.
12      ENDWHILE
13      GET a list of field names in the table.
14      FOR each field name in the list
15          Replace special characters in field name with underscore.
16      ENDFOR
17      Save the data table.
18      Close the data table.
19  ENDFUNC
```

In addition to sequences, decision-making, and repetition structures, code reuse structures, such as functions are often represented in pseudocode. Functions (also known as procedures), group a set of related steps so that they can be re-used by referring to them by name. The first line of Example 8.3 uses the keyword FUNC, short for function. This indicates that lines 2–18 make up a function named preprocessTable. If your workflow involves modifying multiple tables, you can CALL the preprocessTable function for each one as shown in Example 8.4.

Example 8.4: Pseudocode for preparing a set of tables for ArcGIS import; For each table, it deletes the first 5 rows and the columns beyond column 50, and repairs the field names.

```
1   GET a list of table names.
2   FOR each table name in the list
3       CALL preprocessTable(table name, 6, 51).
4   ENDFOR
```

Indentation is important in pseudocode to show the relationship of steps to each other and reinforce the block structures. It's also important in Python. Examples 8.1 and 8.4 only have two levels of indentation. The second level is due to their

FOR-loop blocks. Example 8.2 has three levels because the IF...ELSE IF...ENDIF decision-making blocks are nested inside the FOR-loop block. Example 8.3 has three levels of indentation as well; Code is indented inside the FUNC block and indented again inside the loop blocks. A *nested structure* is a block structure that occurs inside of another block structure and requires an additional level of indentation.

Example 8.5 has four levels of indentation. This pseudocode is designed to handle a set of tables each containing the same misspelled field name ('vlue' should be 'value'). The tables are distributed throughout a set of subdirectories, so we need to look in each directory, at each table, and each field name. There are three FOR-loops: one for subdirectories, one for tables, and one for field names. The contents of each loop is indented one more time than the previous one. These three loops are nested because the entities have a hierarchical relationship: each subdirectory has a set of tables and each table has a set of field names. The fourth level of indentation is due to the decision-making block (IF...ENDIF) to check the name of each field and replace it if necessary.

Note Any block structure nested within another block structure needs to be indented. Steps that are outside of a block structure need to be correctly dedented (moved back a notch).

Example 8.5: Pseudocode for correcting a misspelled field name (vlue to value) in each of a set of tables that are distributed through a set of subdirectories.

```
 1  FOR each subdirectory of the current directory
 2     FOR each table in this subdirectory
 3        FOR each field in this table
 4           GET the field name.
 5           IF the field name is 'vlue' THEN
 6              SET field name to 'value'.
 7           ENDIF
 8        ENDFOR
 9     ENDFOR
10  ENDFOR
```

Summing up, here are some rules-of-thumb for describing workflow with pseudocode:

- Decide carefully about what to put inside, outside, before, and after the repetition structure. Should it happen many times or just once?
- Use WHILE-loops for numerical looping variables or when the number of repetitions is unknown ahead of time vs. FOR-loops for a list of items.

- Update the looping variable in the last step of a WHILE-loop.
- Use the looping variable at least once inside a FOR-loop.
- Use the looping variable at least twice inside a WHILE-loop, once before the update.
- Indent steps within a single block structure the same amount to indicate that the steps are related.
- Indent steps within a nested block structure one more time than the previous level of indentation.

Pseudocode provides a convenient tool for sketching the outline of a workflow. Considering the sequential, repetition, decision-making structures provides a frame of reference for breaking the workflow into steps. Exposure to these concepts in general terms should make it easier to pick up the corresponding Python components discussed in the upcoming chapters.

8.2 Key Terms

Workflow structures: sequential steps, repetition, decision-making
Looping
Branching
Pseudocode
Block structures

WHILE-loop
FOR-loop
Looping variable
Iterate
Functions (procedures, subroutines)
Nested structures

8.3 Exercises

1. Rewrite the workflows described below in terms of pseudocode. Use appropriate **structural components** (sequences, decision-making, repetition), **keywords** and **indentation**. Answers will vary.

   (a) Get from your living room to your bathroom.
   (b) Check the geometry type of a given shapefile. If the geometry type is polygon, find the area of each polygon and print it. If the geometry type is line, find the length of each line and print it. Otherwise do nothing.
   (c) For all the rasters in a database, perform reclassification.
   (d) The ArcGIS polygon aggregation tool combines polygons that are within the specified distance of each other. For a given shapefile, perform polygon aggregation for aggregation distances of 100 feet, 200 feet, 300 feet, and so forth, up to and including 1000 feet. Use a WHILE-loop.
   (e) A KMZ compression format file contains multiple KML files. Unzip a KMZ file and then convert each KML file into a shapefile. Next, find the intersection

of *all* the shapefiles. The final output will be one shapefile containing the intersections of all the shapefiles.

(f) Some of the data in a set of dBASE tables was entered inaccurately. That is, in some records, the park ID is missing. You want to delete these corrupted records. All the tables are sitting together in a single directory. Loop through each record of each table in the directory. Use a FOR-loop inside a FOR-loop for this nested looping and use two levels of indentation.

2. Step through this pseudocode by hand. Hypothesize what it will print for the input x values of 100, 5, 0, and −5. (Zero is an even number, but it is neither positive nor negative.)

```
GET x
WHILE x is less than 100
    IF x is even THEN
        SET x to x + 50
        PRINT x
    ELSE
        IF x is positive THEN
            PRINT x
            SET x to x + 1
        ELSE
            PRINT x
        ENDIF
    ENDIF
ENDWHILE
```

Input	Hypothesized output	Script output
100		
5		
0		
−5		

Next test your answers by running the sample script 'numGames.py', which encodes the pseudocode as Python. How did you do? Report if your answers were correct or not and explain any mistakes in your predictions.

Chapter 9
Decision-Making and Describing Data

Abstract

Yellowstone Hiker's Flowchart

Scripts routinely need to perform different operations based on some deciding criteria. The decision may be very simple, ('if the field doesn't exist, add the field') or it may be more complex ('if the shapefile has point geometry, compute a buffer, but if it has polygon geometry, perform an intersection'). Chapter 8 introduced 'decision-making' structures. These are often referred to as 'branching' structures, because the workflow diagram branches where the decision occurs. For example, the 'Yellowstone Hiker's Flowchart' branches in three places: 'Do I see an animal?', 'Is it a bear?' and 'Is it a killer rabbit?'. Pseudocode uses IF, THEN, ELSE IF, ELSE, and ENDIF key words to express decision-making workflows. This chapter presents the Python syntax, conditional expressions, ArcGIS tools that make selections, the `arcpy Describe` method, handling optional input, and creating directories.

Chapter Objectives

After reading this chapter, you'll be able to do the following:

- Implement IF, ELSE IF, and ELSE structures in Python.
- Explain when to use only an IF block, when to use an ELSE IF block, and when to use an ELSE block.
- Specify decision-making conditions with comparison, logical, and membership operators.
- Select data within a table using SQL comparison and logical operators.
- Design syntactically and logically sound compound conditional expressions.
- Identify code testing cases for branching.
- Use data properties to make decisions.
- Handle optional user input.
- Safely create output directories.

"That one's past its 'Sell-By' date."

Decision-making is expressed with conditional constructs. If some condition is true, some action is taken. E.g., if the animal is a bear, ring a bell. In Python, conditional constructs begin with the if keyword. This keyword, followed by a condition, followed by a block of indented code, make up the simplest conditional construct. The syntax looks like this:

```
if condition:
    code statement(s)
```

If the condition is true, the indented block of code is executed. Otherwise, it is skipped. A condition is checked and an action is taken (or not), based on that condition. The semicolon and the indentation are required. At least one indented code statement must follow the line that ends in a colon (:). The conditional construct is the first instance of a compound code block discussed so far in this book. We'll introduce a number of additional compound code blocks in upcoming chapters. All compound code blocks in Python require a colon at the end of the first line and indentation in the line that follows.

Example 9.1

Python	Pseudocode

`if speciesCount < 500:`
`    print 'Endangered species'`
`print speciesFID`
	`IF species count < 500 THEN`
	`    PRINT 'Endangered species'`
	`ENDIF`
	`PRINT species FID`

The code in Example 9.1 only prints 'Endangered species' if the count is low, but prints the species ID for *every* species. The species ID print statement is outside of the `IF` block since it is dedented. Examples 9.1–9.4 show Python on the left and Pseudocode on the right. Can you spot the four differences between the Python code and pseucode? Python uses a colon at the end of the first line as opposed to a THEN keyword. Python keywords are lower case, pseudocode keywords are uppercase. Python uses the `elif` keyword as opposed to the pseudocode `ELSE IF` keywords. Pseudocode uses `ENDIF` to enclose the `IF` block statements; Python uses indentation instead.

Python IDEs automatically move the cursor to an indented position when you start a compound code block. If you notice that the code is not automatically indenting, it means you forgot to type the colon. When using the Interactive (or Interpreter) Window for compound code blocks, IDEs print ellipses (…) to indicate that you're inside a compound code block. To end a compound code block in the Interactive Window, you need to press the 'Enter' key twice.

```
>>> speciesCount = 250
>>> if speciesCount < 500:
...     print 'Endangered species'
...
Endangered species
```

Example 9.1 has no contingency plan. If the species count is 500 or higher, no special action is taken. It uses only one decision-making code block. When you want to execute two different actions depending on if the condition is true or false, add an `ELSE` block after the IF block. Example 9.2 has two decision-making code blocks. It uses `if` and `else` to handle the two possible outcomes (valid or not). Either way it prints a phrase, either confirming that the area is valid or warning that it's not.

Example 9.2: Check for valid polygon areas.

`if area > 0:`
    `print 'The area is valid.'`
`else:`
    `print 'The area is invalid.'`
	`IF polygon area > 0 THEN`
	    `PRINT The area is valid`
	`ELSE`
	    `PRINT The area is invalid`
	`ENDIF`

To handle more complex decisions, where you want to look at multiple conditions, you can use multiple decision-making blocks. Suppose you're making plans for the weekend. If the weather is sunny and calm, you'll swim in the ocean; if the weather is not sunny and calm, but sunny and windy, you'll fly a kite instead. If the weather is cold and clear, you'll ice-skate, and in all other conditions, you'll stay home and write some Python geoprocessing code. The pseudocode for your weekend plans would be as follows:

```
Check the weather forecast
IF weather is sunny and calm THEN
    Swim in ocean.
ELSE IF weather is sunny and windy THEN
    Fly a kite.
ELSE IF weather is cold and clear THEN
    Go ice-skating.
ELSE
    Write GIS Python code.
ENDIF
```

To make the decision like a computer, you would read from top to bottom and check the weather against each of these weather conditions until you reach one that matches. Then you would do the activity that was planned for that weather. You wouldn't even consider any activities in branches below where you found the match. If none of the weather conditional match, you still have an activity planned. In rain or snowy conditions, for example, you will write Python code! The pseudocode uses ELSE IF to check the alternative weather conditions and ELSE to catch all other weather. Python keywords `elif` and `else` can optionally follow an `if` code block, to specify more than one action. A colon and indentation are required for `elif` and `else` since these are block code statements (see Examples 9.3 and 9.4). A decision-making code block can use any number of `elif` keywords, but it can use at most one `else` keyword. Using one or more `elif` allows you to specify multiple conditions. An ELSE block is used for default actions and must come last, if it is used. Since `else` is used as a catch-all, it is not followed by a condition, only a colon.

9.1 Conditional Expressions

The condition following `if` or `elif` keywords is a Boolean expression. Expressions that are being used for their true or false value are referred to as *Boolean expressions*. In this book, the keywords `if` or `elif` followed by a Boolean expression is referred to as a *conditional expression*. A Boolean expression can be evaluated for its truth value using the built-in `bool` function, which returns a value of `True` or `False`. Any object in Python can be evaluated in this way. For example, 0 is evaluated as `False`, but 5 is evaluated as `True`:

```
>>> bool(0)
False
>>> bool(5)
True
```

The majority of object values evaluate as true. Table 9.1 shows seven false Python objects. Two built-in constants (`False` and `None`) evaluate as false. The number zero evaluates as false, but all other numbers evaluate as true. Empty strings, tuples, and lists evaluate as false, but all other values of these data types evaluate as true. An empty string literal value has two quotation marks (single or double) with nothing to separate them (not even a space). An empty list is just a pair of square brackets (spacing doesn't matter). Knowing this short list of false objects will help you build well-structured conditional expressions.

The conditional expressions in Example 9.1 (`speciesCount < 500`) and Example 9.2 (`area > 0`) compare variables to numeric values, using less than and greater than signs. These signs are examples of *comparison operators*. Operators like these are often needed to construct conditional expressions. Comparison, logical, and membership operators are discussed next.

Comparison, logical, and membership operators are collectively referred to as Boolean operators. Table 9.2 shows examples of these types of Boolean operators.

Table 9.1 False Python constants and data values.

Python value	Description
`False`	Built-in Python constant
`None`	Built-in Python constant
`0`	The number zero
`' '`	An empty string. (No spaces between the quotes)
`()`	An empty tuple
`[]`	An empty list
`{}`	An empty dictionary

Table 9.2 Boolean operators.

Python comparison operators	
(x and y are objects such as numbers or strings)	
`x < y`	`# x is less than y`
`x > y`	`# x is greater than y`
`x <= y`	`# x is less than or equal to y`
`x >= y`	`# x is greater than or equal to y`
`x == y`	`# x is equal to y`
`x != y`	`# x is not equal to y`
Python logical operators	
(x and y are objects, often Boolean expressions)	
x **and** y	True if both x and y are true.
x **or** y	True if either x or y or both are true.
not x	True if x is false.
Membership operators	
(x is any object and y is a sequence type object)	
x **in** y	True if x is in y.
x **not in** y	True if x is not in y.

9.1.1 Comparison Operators

Comparison operators use the familiar < and > symbols from mathematics for 'less than' and 'greater than'. 'Less than or equal to' and 'greater than or equal to' are represented by the <= and >= operators respectively. x and y being compared in Table 9.2 can be any type of object. These operators reference an ordering assigned to objects. Numbers are less than characters. Capital string characters are less than lower case and letters are ordered based on position in the alphabet.

```
>>> 100 < 'A' < 'a' < 'b'
True
```

Comparison operators compare individual characters in strings from left to right until a smaller character is reached.

```
>>> 'aa' < 'ab'
True
```

Numerical ordering applies to numbers, but not to strings of numeric characters. Comparison operators compare numeric characters in strings from left to right until a smaller numerical character is reached. The following example shows that the number 128 is evaluated as greater than 15, but the string '128' is evaluated as less than the string `'15'` because the second character of `'128'` is lower than the second character `'15'` (i.e., `'2' < '5'`):

```
>>> 128 > 15
True
>>> '128' > '15'
False
```

Equality is tested with the double equals sign (the == operator). Inequality ('not equal to') is tested with the ! = operator. Example 9.3 checks the equality of a pair of strings.

Example 9.3: Print the ID numbers of highways and rivers.

`if classType == 'major highway':`
    `print 'Highway--', FID`
`elif classType == 'river':`
    `print 'River--', FID`
	`IF class type is highway THEN`
	    `PRINT Highway ID number`
	`ELSE IF class type is river THEN`
	    `PRINT River ID number`
	`ENDIF`

Example 9.3 only prints an `FID` if the class type is highway or river, but Example 9.4 will always print an `FID`, since an `ELSE` block was added.

Example 9.4: Print the ID number for *all* class types.

`if classType == 'highway':`
    `print 'Highway--', FID`
`elif classType == 'river':`
    `print 'River--', FID`
`else:`
    `print 'Other--', FID`
	`IF class type is highway THEN`
	    `PRINT Highway ID number`
	`ELSE IF class type is river THEN`
	    `PRINT River ID number`
	`ELSE`
	    `PRINT other class type ID number`
	`ENDIF`

Note Cast numerical script arguments to numbers when using them in numerical comparisons. Without casting, comparisons may not work as expected:

```
>>> 25 < '20'
True
```

9.1.2 Equality vs. Assignment

One common mistake is using = when == is intended. The single equals sign (=) is an assignment operator, not a comparison operator. To set x equal to five, use x = 5. But to compare x to five, use x == 5. The double equals tests for equality. Attempting to perform an assignment in a conditional expression results in an error:

```
>>> if x = 5:
Traceback ( File "<interactive input>", line 1
if x = 5:
        ^
SyntaxError: invalid syntax
```

9.1.3 Logical Operators

Logical operator keywords and, or, and not are used to encode logical conditions. For example, they can check if a condition is not true, or if two conditions are true, or if one condition is true and another isn't, or if either of two conditions are true, and so forth. The following code prints the file name if it has an '.shp' or '.txt' extension:

```
>>> fileName = 'park.shp'
>>> if fileName.endswith('.shp') or fileName.endswith('.txt'):
...        print fileName
...
park.shp
```

The following code prints file names that do not have a '.csv' extension:

```
>>> if not fileName.endswith('.csv'):
...        print fileName
...
park.shp
```

Expressions that use and and or are called *compound conditional expressions*, because they combine two or more conditions. The expressions on either side of the keywords and and or are evaluated independently. Sometimes this doesn't correspond to how we would normally formulate a condition in natural language. For example, suppose we want to print the species name only when the species is salmon or tuna. This might erroneously get translated to Python like this:

```
>>> species = trout'
>>> if species == 'salmon' or 'catfish':
...        print species
...
trout
```

This code fails because of how it evaluates the conditions: It checks if the species is 'salmon'. Then it checks if 'catfish' is true—which it is, since it's not an empty string. No value of species will make this statement false, since the string literal 'catfish' is always true. A complete comparison expression is needed on both sides, as in the following code:

```
>>> if species == 'salmon' or species == 'catfish:
...     print species
...
```

9.1.4   Membership Operators

Membership operators can also be harnessed to form conditional expressions. Logical operators can be stringed together to test more than two conditions, as in the following example:

```
if classType == 'major highway' or classType == 'river' or \
   classType == 'stream' or classType == 'bridge':
    print classType, '--',FID
else:
    print 'Other--', FID
```

However, this is an ideal situation for using the in keyword instead. This example shows a more elegant solution to check if the class is in one of the special categories:

```
specialTypes = ['highway', 'river', 'stream', 'bridge']
if classType in specialTypes:
    print classType, '--',FID
else:
    print 'Other--', FID
```

Conversely, it can be used to identify only those items that are not in the list:

```
specialTypes = ['highway', 'river', 'stream', 'bridge']
if classType not in specialTypes:
    print classType, '--', FID
```

9.2   ArcGIS Tools That Make Selections

Many ArcGIS tools and `arcpy` functions have a `where_clause` parameter that can be used to make a selection of features, such as selecting the features in 'park.shp' which have a cover type of 'orch' (for orchard). For example, the third parameter of the Select (Analysis) tool shown here is named `where_clause`.

Syntax

Select_analysis (in_features, out_feature_class, {where_clause})

Parameter	Explanation	Data Type
in_features	The input feature class or layer from which features are selected.	Feature Layer
out_feature_class	The output feature class to be created. If no expression is used, it contains all input features.	Feature Class
where_clause (Optional)	An SQL expression used to select a subset of features. The syntax for the expression differs slightly depending on the data source. For	SQL Expression

These expressions should not be confused with the Python expressions used for calculating values in some tools, such as the Calculate Field (Data Management) tool (discussed in Section 6.3.2). `where_clause` parameters are expressions that are either true of false and they must be specified as SQL expressions, as opposed to Python expressions. SQL is a specialized programming language for managing databases. The term *where-clause* refers to the SQL keyword 'WHERE' which is used to make selections in SQL syntax. The syntax for SQL in ArcGIS may differ slightly from other versions of SQL with which you are familiar. Like Python conditional expressions, SQL where-clauses use comparison and logical operators. <, <=, >, and >= are used in the same way, but SQL uses a single equals sign (=) to check equality, unlike Python which reserves the single equals sign for assignment. Inequality is represented by <> in SQL, whereas Python uses ! =. The same logical operators can be used in SQL but they are usually uppercased in SQL (AND, OR, and NOT); Whereas, in Python, they must be lowercase. Table 9.3 can be used as a quick reference for these operators.

The syntax for `where_clause` ArcGIS tool parameters is slightly tricky because this is not stand alone SQL, but rather, SQL embedded in Python. The `where_clause` tool parameter is specified as a Python string, so it has to be surrounded by quotation marks like this:

```
>>> whereClause1 = 'RECNO >= 400'
>>> type(whereClause1)
<type 'str'>
```

Table 9.3 Python vs. SQL operators.

Python	SQL
Comparison operators	
<	<
>	>
<=	<=
>=	>=
==	=
!=	<>
Logical operators	
and	AND
or	OR
not	NOT

For these examples, consider the shapefile 'park.shp' which has a numeric field named 'RECNO' and a text field named 'COVER'. whereClause1 specifies the values of the 'RECNO' field greater than or equal to 400. The field name is specified on the left of the comparison operator (conventionally uppercased, though this is not case sensitive). To the right of the comparison operator the field value is specified. The entire phrase is surrounded by quotation marks to make it a Python string. Example 9.5 shows this clause being used in a tool call. It is passed as the third parameter into the Select (Analysis) tool to specify the selection to be made. The output file will contain only those records with 'RECNO' field values greater than or equal to 400.

Example 9.5: Using a hard-coded where-clause.

```
# where_clause1.py
# Purpose: Select features with high reclassification numbers.
# Usage: No arguments needed.

import arcpy
arcpy.env.workspace = 'C:/gispy/data/ch09'
inputFile = 'park.shp'
whereClause = 'RECNO >= 400'
arcpy.Select_analysis(inputFile, 'C:/gispy/scratch/out.shp', whereClause)
```

Compound clauses can be made with logical operators. Just like in Python, a complete expression must be used on either side of the logical operator. For example, this clause specifies the 'RECNO' field values between 400 and 410:

```
>>> whereClause2 = 'RECNO > 400 AND RECNO < 410'
```

In contrast, 'RECNO > 400 AND < 410' is not a valid SQL expression and will cause an error.

For text fields, an extra set of quotation marks need to be used around the field value. Two distinct types of quotation marks are used (single for the interior quotes, double for the exterior quotes), so that Python matches them correctly. For example, the following clause could be used to select the records where the `'COVER'` field has the value `'orch'`:

```
>>> whereClause3 = "COVER = 'orch'"
```

If single quotation marks are used throughout the phrase, a syntax error occurs:

```
>>> whereClauseBogus = 'COVER = 'orch''
Traceback ( File "<interactive input>", line 1
whereClauseBogus = 'COVER = 'orch''
                                 ^
SyntaxError: invalid syntax
>>>
```

Python, reads the quotation marks from left to right and assumes that the second quotation mark it reaches (the one just before `orch`) matches the first one, leaving the remainder of the characters outside the string.

In summary, the format for simple `where_clause` expressions is one of the following:

```
"NUMERIC_FIELD_NAME comparison_operator numeric_value"
"TEXT_FIELD_NAME comparison_operator 'text_value'"
```

In some applications the field name or the field value to be used in the expression is not determined beforehand, so it can't be hard-coded. The `where_clause` parameter is a Python string, so we can use the string `format` method to build the clause with variables. The following code uses a variable field value to create `whereClause4` (whose value is identical to `whereClause3`):

```
>>> fieldValue = 'orch'
>>> whereClause4 = "COVER = '{0}'".format(fieldValue)
>>> whereClause4
"COVER = 'orch'"
```

The `format` method substitutes the arguments into the string for the placeholders. Notice the single quotation marks are still used around the placeholder `{0}` so that they are in place when the replacement is made. The same method could be used for a variable field name or for both a variable field name and value, as in the following example:

```
>>> fieldName = 'COVER'
>>> fieldValue = 'orch'
>>> whereClause5 = "{0} = '{1}'".format(fieldName, fieldValue)
>>> whereClause5
"COVER = 'orch'"
```

Example 9.6 uses a string formatted clause to extract values from a raster. The field name and value are passed into the script by the user. If the script is run with the example arguments, 'COUNT' and '6000', the output raster only contains cells where the value of the 'COUNT' field is greater than 6000. In Example 9.6, the selected values are saved in a new raster; where_clause parameters can also be used to select records for operations on a subset of the records as discussed next.

Example 9.6: Using a where clause with variables.

```
# where_clause2.py
# Purpose: Extract raster features by attributes based on
#          user input.
# Usage: fieldName fieldValue
# Example input: COUNT 6000

import arcpy, sys

arcpy.env.workspace = 'C:/gispy/data/ch09'
arcpy.CheckOutExtension('Spatial')

inputRast = 'getty_rast'

fieldName = sys.argv[1]
fieldValue = sys.argv[2]

whereClause = '{0} > {1}'.format(fieldName, fieldValue)

outputRast = arcpy.sa.ExtractByAttributes(inputRast, whereClause)
outputRast.save('C:/gispy/scratch/attextract')
del outputRast
```

9.2.1   Select by Attributes and Temporary Feature Layers

The Select Layer By Attribute (Data Management) tool is another tool that uses a where_clause parameter. This tool requires a little bit of explanation for use in Python because both the input and output are temporary layers not feature classes such as shapefiles or geodatabase feature classes (Section 6.6 discussed the temporary layers in the context of the Make XY Event Layer (Data Management) tool). Performing a selection on a feature class and saving the selection as a feature class requires making a feature layer, selecting, and saving the selection, as in the following steps:

Step 1. The 'Make Feature Layer' tool can be used to create a temporary layer from a feature class. The following code creates a temporary feature layer:

```
>>> arcpy.env.workspace = 'C:/gispy/data/ch09/tester.gdb'
>>> tmp = 'tmpLayer'
>>> arcpy.MakeFeatureLayer_management('park', tmp)
```

Step 2.   The following code prepares a compound `where_clause` variable to find the small wooded plots (those with `area < 20000`) with 'woods' land cover and performs the selection:

```
>>> maxArea = 20000
>>> coverType = 'woods'
>>> whereClause = "Shape_area < {0} AND \
COVER ='{1}'".format(maxArea, coverType)
>>> arcpy.SelectLayerByAttribute_management(tmp,
                                            'NEW_SELECTION',
                                            whereClause)
<Result 'tmpLayer'>
```

Step 3.   Finally, the following code saves the temporary layer to a feature class:

```
>>> output = 'smallWoods'
>>> arcpy.CopyFeatures_management(tmp, output)
<Result 'C:/gispy/data/ch09\\smallWoods'>
```

9.3   Getting a Description of the Data

For geoprocessing, we often make decisions based on data properties. The `arcpy` function named `Describe` allows you to access properties of a data object. It takes one argument, the name of an Esri data element or geoprocessing object. Examples of valid arguments include a Feature Class name, a Layer name, a Raster Dataset name, and so forth. The following code calls the `Describe` function with the Raster Dataset named `'getty_rast'` as an argument:

```
>>> import arcpy
>>> arcpy.env.workspace = 'C:/gispy/data/ch09'
>>> rastFile = 'getty_rast'
>>> desc = arcpy.Describe(rastFile)
```

The `Describe` function returns a `Describe` object, which is stored in the variable named `desc`:

```
>>> desc
<geoprocessing describe data object object at 0x0092D2D0>
```

9.3.1 *Describe Object Properties*

The `Describe` object has a set of properties with information about the data. These properties fall into two categories:

- Universal properties—any valid input has all of these properties.
- Specialized properties—these properties depend on the data type of the input.

Whenever you create a `Describe` object, it has values for the set of universal properties that are common to any object being described. These are things like `baseName`, `dataType`, and `extension`, to name just a few. To access these properties, use dot notation with the `Describe` object. In this example, we print a few of the universal properties:

```
>>> desc.dataType
u'RasterDataset'
>>> desc.baseName
u'getty_rast'
>>> desc.extension
u''
>>> desc.catalogPath
u'C:/gispy/data/ch09/getty_rast'
```

The additional, specialized properties depend on the data type of the argument. For example, when the argument is a `RasterDataset` type, like 'getty_rast', there are a set of `RasterDataset` Properties, such as `bandCount`, `compressionType`, and `format`. Again, access these with dot notation:

```
>>> desc.CatalogPath
u'C:/gispy/data/ch09/getty_rast'
>>> desc.bandCount
1
>>> desc.compressionType
u'RLE'
>>> desc.format
u'GRID'
```

The `Describe` object `format` property should not be confused with the string method by the same name. The `format` property refers to the image format (GRID, JPEG, PNG, etc.)

Many of the properties return string values, but some return `arcpy` objects. These objects may have their own set of methods and properties, which can be accessed with dot notation. For example, the `extent` property returns an `Extent` object with minimum and maximum values for x, y, z, and m, as well as other properties. The following code assigns an `Extent` object to a variable named bounds, then uses the variable to find the maximum x value:

```
>>> bounds = desc.extent
>>> bounds
<Extent object at 0x3338c50[0x125b4d70]>
>>> bounds.XMax
2167608.390378157
```

9.3.2 Lists of Properties

The ArcGIS Resources site lists the universal and specialized properties. The 'Describe object Properties' page lists the universal properties; any valid input has all of these properties. The 'Describe (arcpy)' page has a list of the specialized data types, each linked to a list of properties.

Many of the special data types have access to a few general types. The `Dataset` properties and `Table` properties are available to many types of `Describe` objects. For example, the 'Raster Dataset Properties' page says 'Dataset Properties also supported.' This means Raster Dataset type `Describe` objects also have all the `Dataset` properties, such as, `extent` and `spatialReference`.

Figure 9.1 depicts a portion of the `Describe` functionality. The `Describe` function is one of the `arcpy` functions. It returns a `Describe` object. The box highlighted in Figure 9.1 lists a few of the `Describe` object properties. The box (labeled "describe data objects") represents the `Describe` object and lists a few of its universal properties. The darker boxes inside this one represent some of the specialized property groups. For example, `FeatureClass` type objects have six properties, plus they have access to `Dataset` and `Table` properties and `RasterDataset` type objects have five properties, plus they have access to Dataset and Raster Band properties. Only a few of the properties are listed in the boxes as a reminder of the many available properties. The ArcGIS Resources documentation contains the complete property lists for each data type.

9.3.3 Using Specialized Properties

Care must be taken when using the specialized properties. If you try to use the wrong type of property, an error will occur. The following code attempts to use the format property but it fails because 'park.shp' is a ShapeFile data type so it doesn't have this specialized property:

```
>>> fcFile = 'C:/gispy/data/ch09/park.shp'
>>> desc2 = arcpy.Describe(fcFile)
>>> desc2.dataType
```

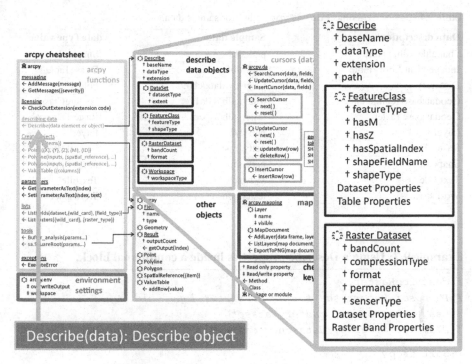

Figure 9.1 The `arcpy` `Describe` function returns a `Describe` object. These boxes show a subset of the `Describe` object properties.

```
u'ShapeFile'
>>> desc2.format
Traceback (most recent call last):
File "<interactive input>", line 1, in <module>
AttributeError: DescribeData: Method format does not
exist
```

For this reason, the specialized properties are usually used inside conditional constructs which check the `dataType` of the object before using these properties. The `dataType` property returns a string value, an alias that `arcpy` uses for the given type of data. Table 9.4 lists the `dataType` values for ten sample datasets. Others can be found in the ArcGIS Resources pages for the data types, linked to the 'Describe (arcpy)' page. Example 9.7 checks the value of the `dataType` property instead of just using the format property on any `Describe` object. If the second argument is 'getty.tif', the script prints 'Image format: TIFF'. If the input is 'park.shp', the script prints nothing.

Table 9.4 `Describe` object `dataType` values for sample data.

Data description	Sample data	dataType value
Shapefile '.dbf' table	park.dbf	'ShapeFile'
Independent '.dbf' table	site1.dbf	'DbaseTable'
Directory	C:/gispy/data/ch09	'Folder'
Geodatabase feature class	tester.gdb/data1	'FeatureClass'
Geodatabase raster	tester.gdb/aspect	'RasterDataset'
Layer file	park.lyr	'Layer'
Raster Dataset	getty_rast	'RasterDataset'
Shapefile	park.shp	'ShapeFile'
Text File	xy1.txt	'TextFile'
Workspace	tester.gdb	'Workspace'

Example 9.7: Using a Describe properties inside a conditional block.

```
# describeRaster.py
# Purpose: Report the format of raster input file.
# Usage: workspace, raster_dataset
# Example: C:/gispy/data/ch09 getty.tif

import arcpy
arcpy.env.workspace = arcpy.GetParameterAsText(0)
data = arcpy.GetParameterAsText(1)
desc = arcpy.Describe(data)
if desc.dataType == 'RasterDataset':
    print 'Image format: {0}'.format(desc.format)
```

9.3.4 Compound vs. Nested Conditions

Sometimes you may want to check more than one condition before calling a geo-processing tool. For example, to use the Smooth Line (Cartography) tool, you can check the `dataType` and the `shapeType`. You can do this with a compound statement, but any specialized conditions (such as `shapeType`) must be placed after the more general conditions (such as `dataType`) because the conditions in a compound statement are evaluated left to right. This code causes an error:

```
>>> desc = arcpy.Describe('C:/gispy/data/ch09/getty.tif')
>>> desc.dataType
u'RasterDataset'
>>> if desc.shapeType == 'Polyline' and \
dsc.dataType in ['FeatureClass', 'Shapefile']:
```

```
...        print 'Smooth line'
...
Traceback (most recent call last):
File "<interactive input>", line 1, in <module>
AttributeError: DescribeData: Method shapeType does not exist
```

How could you reorder the compound condition to correct this problem? In Example 9.8, 'smoothLineCompound.py' correctly uses a compound statement to only call the Smooth Line tool for 'FeatureClass' or 'ShapeFile' types with a shapeType of 'Polyline'.

Example 9.8: Using a Describe properties inside a conditional block.

```
# smoothLineCompound.py
# Usage: workspace, features_with_line_geometry
# Example 1: C:/gispy/data/ch09 trails.shp
# Example 2: C:/gispy/data/ch09 park.shp

import arcpy

arcpy.env.overwriteOutput = True arcpy.env.workspace =
arcpy.GetParameterAsText(0) data =
arcpy.GetParameterAsText(1)
outFile = 'C:/gispy/scratch/smoothOut'
desc = arcpy.Describe(data)
if desc.dataType in ['FeatureClass' ,'ShapeFile'] and \
                    desc.shapeType == 'Polyline':
    result = arcpy.SmoothLine_cartography(data, outFile,
                                    'BEZIER_INTERPOLATION')
    print 'Smooth line created {0}.'.format(result.getOutput(0))
```

Compound conditions work well as long as exactly the same behavior is desired for both conditions. However, suppose we want to tailor warning messages depending on which condition fails. Then we need to use nested conditional constructs. The pseudocode for this would look like this:

```
# GET input
IF input data feature class or shapefile THEN
    IF the shape type is line THEN
        CALL the smooth line tool.
    ELSE
        Warn the user about shape type requirements.
    ENDIF
ELSE
    Warn the user about the data type requirements.
ENDIF
```

The nested conditions allow the script to use one `ELSE` block for each condition. Example 9.9 shows the corresponding script.

Example 9.9: Using Describe properties inside a nested conditional block.

```
# smoothLineNested.py
# Usage: workspace, features_with_line_geometry
# Example 1: C:/gispy/data/ch09 trails.shp
# Example 2: C:/gispy/data/ch09 park.shp
# Example 3: C:/gispy/data/ch09 getty.tif

import arcpy arcpy.env.overwriteOutput = True

arcpy.env.workspace =arcpy.GetParameterAsText(0)

data = arcpy.GetParameterAsText(1)
outFile = 'C:/gispy/scratch/output'
desc = arcpy.Describe(data)

if desc.dataType in ['FeatureClass' ,'ShapeFile']:
    if desc.shapeType == 'Polyline':
        result = arcpy.SmoothLine_cartography(data, outFile,
                                    'BEZIER_INTERPOLATION')
        print 'Smooth line created {0}.'.format(result.getOutput(0))
    else:
        print 'Warning: shape type is {0}. Smooth Line only works
            on Polyline shape types. '.format( desc.shapeType)
else:
    print "Warning: Input data type must be 'FeatureClass' or 'ShapeFile'."
    print 'Input dataType:', desc.dataType
```

9.3.5   Testing Conditions

Scripts that have conditional constructs, should be tested with input that fulfill each of the conditions they are built to handle. For example, 'smoothLineNested.py', in Example 9.9, considers three outcomes. We can test for each of these by varying the input.

1. Input data type is FeatureClass or Shapefile and shape type is Polyline.

   Input: C:/gispy/data/ch09 trails.shp

   Output:
   ```
   Smooth line created C:\gispy\scratch\output.shp.
   ```

2. Input data type is FeatureClass or Shapefile but shape type is not Polyline.

   Input: C:/gispy/data/ch09 park.shp

   Output:
   ```
   Warning: shape type is Polygon. SmoothLine only works on Polyline
   shape types.
   ```

3. Input data type is neither FeatureClass nor Shapefile.

   Input: C:/gispy/data/ch09 getty.tif

   Output:
   ```
   Warning: Input data type must be 'FeatureClass' or 'ShapeFile'.
   Input dataType: RasterDataset
   ```

9.4   Required and Optional Script Input

Conditional constructs can also be used to handle optional script arguments. The script must somehow check whether or not optional arguments have been given and set a default value if necessary. For example, the workflow represented by the following pseudocode optionally takes one argument, a directory path, and lists the files in the given directory or the files in the script directory:

```
IF argument found THEN
    SET working directory to argument.
ELSE
    SET working directory to script directory.
ENDIF
List files in working directory.
```

This can be implemented using the sys module argv list, as in Example 9.10. Since sys.argv is a Python list, if a script uses an index that is not available, an IndexError will be reported:

```
>>> import sys
>>> sys.argv[1]
Traceback (most recent call last):
    File "<interactive input>", line 1, in <module>
IndexError: list index out of range
```

To avoid this, you can check the length of the list before trying an index for an optional argument. Recall that the first item in this list (`sys.argv[0]`) is the script file name:

```
>>> sys.argv[0]
'C:\\gispy\\sample_scripts\\ch09\\scriptPathOptionalv1.py'
```

This means that the length of the list is the number of user arguments plus one. Therefore, if the user passes in one argument, the list length is 2.

Example 9.10

```
# scriptPathOptionalv1.py
# Purpose: List the files in the given directory or
#          the current directory.
# Usage: {directory_path}
import sys, os

if len(sys.argv) == 2:
    workingDir = sys.argv[1]
else:
    # Get the script location
    scriptPath = os.path.abspath(__file__)
    workingDir = os.path.dirname(scriptPath)

print '{0} contains the following files:'.format(workingDir)
print os.listdir(workingDir)
```

Alternatively, this can be implemented with the `GetParameterAsText` function, as in Example 9.11. The default value returned by this function is an empty string (`' '`). Since an empty string is considered false, and all other strings are considered true, we can use the value of `arcpy.GetParameterAsText(0)` as the conditional expression. Both versions of this script work with or without an argument.

Example 9.11

```
# scriptPathOptionalv2.py
# Purpose: List the files in the given directory or
#          the current directory.
# Usage: {directory_path}
import arcpy, os

if arcpy.GetParameterAsText(0):
    workingDir = arcpy.GetParameterAsText(0)
else:
    # Get the script location
    scriptPath = os.path.abspath(__file__)
```

```
        workingDir = os.path.dirname(scriptPath)
print '{0} contains the following files:'.format(workingDir)
print os.listdir(workingDir)
```

Conditional constructs can also be used to enforce required arguments. The km/mile converter in Example 9.12 requires two arguments, a numerical distance and a distance unit. All arguments are passed into `sys.argv` as string types. The first argument is a numerical value, so we need to use the built-in `float` function to get the numerical value (cast the value). The unit is converted to all lower case when it's tested (using `unit.lower()`), so that the input is not case sensitive.

Example 9.12: Simple distance converter.

```
# distanceConvertv1.py
# Purpose: Converts km to miles and vice versa.
# Usage: numerical_distance, unit_of_measure
# Example: 5 km

import sys

dist = float(sys.argv[1]) # Cast string to float
unit = sys.argv[2].

mileList = ['mi','mi.','mile','miles']

if unit.lower() in mileList:
    output = dist*1.6
    print '{0} {1} is equivalent to {2} kilometers(s).'.format(
                                            dist, unit, output)
else:
    output = dist*.62
    print '{0} {1} is equivalent to {2} mile(s).'.format(
                                        dist, unit, output)
```

If either argument is missing, the script will throw an `IndexError` exception. Since the user might not understand this, it could be helpful to instead print a message about how the script is designed to be run. Example 9.13 adds this information and also makes the second argument optional. If no user arguments are given, the script warns the user and exits. A `sys` module function named `exit` is used to exit the script. If one user argument is given, the script sets a default distance unit, otherwise the unit is set to the second argument. The remainder of the script is the same as version 1.

Example 9.13: Distance converter with input checking.

```
# distanceConvertv2.py
# Purpose: Converts km to miles and vice versa.
# Usage: numerical_distance, {unit_of_measure}
# Example: 5 km
```

```
import sys

numArgs = len(sys.argv)

# If no user arguments are given, exit the script and warn the user.
if numArgs == 1:
    print 'Usage: numeric_distance {distance_unit (mi or km)}'
    print 'Example: 5 km'
    sys.exit(0) # exit the script

# If only one user argument is given, set the unit to miles.
if numArgs < 3:
    unit = 'miles'
    print '''Warning: No distance unit provided.
    Assuming input is in miles.'''
else:
    # Get the unit provided by the user
    unit = sys.argv[2]

# Get the numeric distance (cast string to float).
dist = float(sys.argv[1])

# Perform conversion.
mileList = ['mi','mi.','mile','miles']

if unit.lower() in mileList:
    output=dist*1.6
    print '{0} {1} is equivalent to {2} kilometers(s).'.format(
                                        dist, unit, output)
else:
    output = dist*.62
    print '{0} {1} is equivalent to {2} mile(s).'.format(
                                    dist, unit, output)
```

When handling several user input cases, it's a good idea to test an example of each input. Table 9.5 shows test cases for 'distanceConvert2.py'. The last test case (2.2 bananas) gives an example of how the script could be made more robust. This improvement is left as an exercise.

Table 9.5 Five test cases for Example 9.13.

Input	Output
(No arguments)	Usage: numeric_distance {distance_unit (mi or km)} Example: 5 km
2.3	Warning: No distance unit provided. Assuming input is in miles 2.3 miles is equivalent to 3.68 kilometers
5 km	5.0 km is equivalent to 3.1 miles
10 MI	10.0 MI is equivalent to 16.0 kilometers
2.2 bananas	2.2 bananas is equivalent to 1.364 mile(s).

9.5   Creating Output Directories

Sometimes we want to store output files in a different workspace than the input file workspace. For example, to create a backup copy of files using the same name in another directory or to group a batch of output files in a directory on their own, we need to send the output to another directory. In some cases, the output directory may exist; in other cases, the script might create it for us. If this isn't handled carefully, the script could try to write output in a directory that doesn't exist or it could try to create a directory that already exists. Either of these would cause the script to fail. `os` module functions can be used to avoid these problems. The `os.path.exists` function checks if a file or directory exists and the `os.mkdir` creates a new directory as shown in the following code:

```
>>> myDir = 'C:/gispy/data/ch09/happyGoat'
>>> os.path.exists(myDir)
False
>>> os.mkdir(myDir)
>>> os.path.exists(myDir)
True
>>> os.mkdir(myDir)
Traceback (most recent call last):
File "<interactive input>", line 1, in <module>
WindowsError: [Error 183] Cannot create a file when that file already
exists: 'C:/gispy/data/ch09/happyGoat'
```

Calling `os.mkdir` on an existing directory causes an error. To avoid this error, use these two functions together in a conditional construct and the `not` logical operator as follows:

```
if not os.path.exists(myDir):
    os.mkdir(myDir)
```

If the directory exists, it won't call `os.mkdir`. Place these lines of code in the script before the output directory is used, as shown in Example 9.14. This example uses `arcpy.env.workspace` to set the location of the file to be copied. But it uses a full path file name for the output file. The call to `os.path.join` creates this full path name by joining the output directory path with the name of the file being copied. In this way, a duplicate of the original with the same name is created in the backup directory.

The `os` module commands work for directories, but not necessarily for some specialized Esri structures. `arcpy` has equivalent commands for these structures which can be used in the same way. The following code uses `arcpy` commands to create a file geodatabase if it doesn't already exist:

```
>>> outPath = 'C:/gispy/data/ch09/'
>>> gdbName = 'happyHorse.gdb'
```

```
>>> if not arcpy.Exists(outPath + gdbName):
...        arcpy.CreateFileGDB_management(outPath, gdbName)
...
<Result 'C:/gispy/data/ch09\\happyHorse.gdb'>
```

The Create File GDB (Data Management) tool requires the geodatabase path and geodatabase name as two separate parameters; Otherwise, the approach is the same.

Example 9.14

```
# copyFilev2.py
# Purpose: Copy a file.
# Usage: source_directory destination_directory file_to_backup
# Example: C:/gispy/data/ch09/ C:/gispy/scratch/backup park.shp
#          The example works even if the C:/gispy/scratch/backup
#          directory doesn't exist yet.

import arcpy, os

arcpy.env.workspace = arcpy.GetParameterAsText(0)
outputDir = arcpy.GetParameterAsText(1)
fileToCopy = arcpy.GetParameterAsText(2)

if not os.path.exists( outputDir ):
    os.mkdir( outputDir )

outputFile = os.path.join(outputDir, fileToCopy)
arcpy.Copy_management( fileToCopy, outputFile )

print 'source =', os.path.join(arcpy.env.workspace, fileToCopy)
print 'destination =', outputFile
```

Printed output:

```
source = C:/gispy/data/ch09\park.shp
destination = C:/gispy/scratch/backup\park.shp
```

9.6   Key Terms

if, elif, else, Python keywords   Compound conditional expressions
Conditional constructs   The os.path.exists function
Boolean expressions   The os.mkdir function
Conditional expressions   The CreateFileGDB tool

9.7 Exercises

General Instructions: The `Describe` method, along with other geoprocessing tools can be useful for performing batch processing on a geodatabase or a directory. The next discussion covers batch processing. For now, we can perform some fundamental actions that can be extended with batch processing.

1. **conditionalSound.py** Inspect the sample script 'conditionalSound.py' shown below. Then predict what sounds will play with each of these input conditions:

   (a) No arguments.
   (b) C:/gispy/data/ch09/park.shp
   (c) C:/gispy/data/ch09/tree.gif
   (d) C:/gispy/data/ch09/jack.jpg
   (e) C:/gispy/data/ch09/xy1.txt

   Write down your answers and then verify them by running the sample script 'conditionalSound.py'. Did you get them all correct? Explain any mistakes you made.

```
conditionalSounds.py
5    import winsound, os, sys, time, arcpy
6
7    scriptPath = os.path.abspath(__file__)
8    mydir = os.path.dirname(scriptPath) + '/'
9
10  -def playSound(soundfile):
11        # Function to play the input sound file using the winsound module.
12        winsound.PlaySound(soundfile, winsound.SND_FILENAME|winsound.SND_ASYNC)
13        # Wait 1.5 seconds
14        time.sleep( 1.5 )
15
16    # If sys.argv list length is one, no argument was passed.
17  -if len(sys.argv) < 2:
18        playSound( mydir + 'wah_wah_wah.wav' )
19
20  -else:
21        fileName = sys.argv[1]
22
23        # Get the describe object
24        desc = arcpy.Describe(fileName)
25        dataType = desc.dataType
26
27  -    if dataType == 'RasterDataset':
28            playSound( mydir + 'haha.wav' )
29
30  -        if desc.Format == 'GIF':
31                playSound( mydir + 'pukpukpuk.wav' )
32
33  -        else:
34                playSound( mydir + 'doh.wav' )
35
36  -    elif dataType == 'ShapeFile':
37            playSound( mydir + 'oh.wav' )
38
39
40    playSound( mydir + 'yeehaa.wav' )
```

2. For each of the sample scripts named in bold, modify the script as described in each part while maintaining the same functionality as the original version of the script.

   (a) **cond1.py** Use if, elif, and else keywords, to replace the sequential if statements.

   (b) **cond2.py** Remove the unnecessary conditional branch.

   (c) **cond3.py** Rewrite the nested conditional as a compound conditional.

3. **box.py** Write a script that determines if a point is inside of the box bounded by the points $(0,0)$ and $(1,1)$. The script should take two required arguments, an x and a y coordinate. Hint: Remember that script arguments are strings.

Input	Output
0.2 0.4	(0.2,0.4) is in the box.
–0.1 0.5	(–0.1,0.5) is outside the box.
0 0	(0.0,0.0) is in the box.

4. **ski.py** Season pass fees for a North Carolina ski resort are as listed in the table below. Write a script that takes the age of the skier as input (one required argument) and prints the season pass fee as output. Use conditional if, elif, and else keywords in this simple script. Hint: Remember that script arguments are strings.

Age range	Season pass fee ($)
Ages 6 and under	30
Ages 7–18	319
Ages 19–29	429
Adult (ages 30 plus)	549

5. **describeData.py** Write a script to take one required argument, the full path file name of a dataset as input. Using conditional constructs along with the Describe object, the script should determine if the dataType is ShapeFile or RasterDataset. In the first case, it should print the shapeType; in the second, it should print the format. If it is neither of these, it should print the dataType. Example input and resulting output:

Input	Output
C:/gispy/data/ch09/park.shp	Polygon
C:/gispy/data/ch09/tree.gif	GIF
C:/gispy/data/ch09/xy1.txt	TextFile

6. **gridToPoly.py** Given a full path file name of a GRID raster, sample script 'gridToPoly.py' converts GRID format raster datasets to polygon feature classes. It also handles three other user input scenarios. Give four input condition that will test each branch of the code. Use sample data in the 'C:/gispy/data/ch09' directory and fill in the input-output table to list the input and the corresponding printed output:

Input	Output

7. **tableSelect.py** Practice using an ArcGIS tool that makes a selection. The park cover types in 'park.shp' include 'woods', 'orch', and 'others'. Fire risk experts would like a table including only those polygons with a cover type of 'woods'. Write a script to perform a Table Select that will generate this table and name the output table 'wooded.dbf'.

8. **currency.py** Write a script that converts between US dollars and Euros. (Use a rate of 1 US=0.7 E for this exercise.) The script should take one numerical required argument and one optional argument (the currency, E or US). If the user gives two arguments, perform the conversion. If the user gives only one argument, a number, assume the number is given in US dollars, warn the user, and perform the conversion. If the user gives no arguments, print a statement explaining how to run it. Examples of test cases and resulting output are shown below:

Input	Output
45.32 E	45.32 Euros is equivalent to 64.74 US Dollars
55	Since you didn't specify a currency, I'm assuming US to Euros. 55 US Dollars is equivalent to 38.5 Euros
100 US	100.00 US Dollars is equivalent to 70.00 Euros
(No arguments)	Usage: number {currency (US or E)} Example: 100 US

9. **temperatureConvert.py** Write a temperature conversion script to convert between Celsius and Fahrenheit, using the following equations for conversion:
$$F = 1.8 * C + 32$$
$$C = (0.56) * (F - 32)$$
The script should take one numerical required argument and one optional argument (the scale, F or C). If the user gives two arguments, perform the conversion. If the user gives only one argument, a number, assume the number is given in Fahrenheit, warn the user, and perform the conversion. If the user provides no arguments, print a statement explaining how to run it. Several test cases and resulting output are shown below:

Input	Output
32 F	32 Fahrenheit is equivalent to 0.0 Celsius
100 C	100 Celsius is equivalent to 212.0 Fahrenheit
55	Since you didn't specify a scale, I'm assuming F to C. 55 Fahrenheit is equivalent to 12.88 Celsius
(No arguments)	Usage: number {unit (C or F)} Example: 32 F

10. **distanceConvertv3.py** Sample script, 'distanceConvertv2.py', only tests if the input lower-cased unit is one of the four strings in the miles list (['mi', 'mi.', 'mile', 'miles']). Any other unit is assumed to stand for kilometers. Input of '2.2 bananas' gives output of '2.2 bananas is equivalent to 1.364 mile(s)'. Modify the script so that if the lowercased unit is not in the miles list, nor the kilometers list (['km', 'km.', 'kilometer', 'kilometers']), the script prints a warning and does not perform a conversion. Example input and resulting output:

Input	Output
(No arguments)	Usage: numeric_distance {distance_unit (mi or km)}
	Example: 5 km
2.3	Warning: No distance unit provided. Assuming input is in miles
	2.3 miles is equivalent to 3.68 kilometers
5 km	5.0 km is equivalent to 3.1 miles
10 MI	10.0 MI is equivalent to 16.0 kilometers
2.2 bananas	Warning: unit must be either miles or kilometers

11. **compactBranch.py** Write a script that takes one argument, a full path dataset file name, as input. If the `dataType` is 'Workspace' and the workspaceType is 'LocalDatabase', it compacts the file, and reports the file size before and after compacting. Otherwise, it warns the user that it could not perform the compact operation. Example input and resulting output:

Input	Output
C:/gispy/data/ch09/cities.mdb	File size BEFORE compact: 1552384
	File size AFTER compact: 397312
C:/gispy/data/ch09/xy1.txt	Input data must be a personal or file geodatabase.
	Could not compact: xy1.txt

12. **bufferBranch.py** Write a script to take a required full path file name argument and an optional buffer distance argument. Perform Buffer (Analysis) on the file only if it is a Shapefile type (If the filename ends with '.shp', assume it is a valid shapefile). If it's not a shapefile, warn the user that it will not buffer the file. Use the input buffer distance, if it is given; otherwise, use a default buffer distance of 1 mile. Place the output in the same directory as the input, and report the output file name. Example input and resulting output:

Input	Output
C:/gispy/data/ch09/park.shp	No buffer distance given. Used default buffer
	distance of 1 mile
	Buffer output: C:/gispy/scratch/parkBuffer.shp
C:/gispy/data/ch09/xy1.txt "8000 meters"	Input data format must be shapefile. Could not
	buffer input file C:/gispy/scratch/xy1.txt
C:/gispy/data/ch09/park.shp "200 Meters"	Buffer distance: 200 Meters
	Buffer output: C:/gispy/scratch/parkBuff.shp

Chapter 10
Repetition: Looping for Geoprocessing

Abstract Advances in technology are enabling the collection and storage of massive amounts of GIS data. To learn from this data we need to conduct analysis efficiently. Batch processing is a powerful scripting capability, saving time by automating repetitive tasks. Chapter 8 discussed the three basic workflow structures (sequences, decision-making, and repetition) in terms of pseudocode and Chapter 9 presented the Python syntax for decision-making and describing GIS data. This chapter focuses on Python repetition structures and looping syntax, Python FOR-loops and WHILE-loops, looping with the range function for geoprocessing, nested looping, and listing directory contents. Then the chapter concludes with a tip about debugging whitespace glitches.

Chapter Objectives
After reading this chapter, you'll be able to do the following:

- Implement WHILE-loops and FOR-loops in Python.
- Identify the three key iterating variable components in a WHILE-loop.
- Explain how Python FOR-loops work.
- Repair infinite loops.
- Call a geoprocessing tool in a WHILE-loop to vary a numerical parameter.
- Automatically generate numerical lists.
- Loop with the range function.
- Branch and loop within loops.
- List the files in a directory.
- Geoprocess each file in a list.
- Correct indentation errors.

10.1 Looping Syntax

Python has two structures for performing repetition, WHILE-loops and FOR-loops, which use the Python keywords while and for. They provide two slightly different approaches. FOR-loops are used more often in geoprocessing scripts; however, both looping techniques are used, so it is helpful to be familiar with both. We'll start

L. Tateosian, *Python For ArcGIS*, DOI 10.1007/978-3-319-18398-5_10

with WHILE-loops which provide an easier introduction to looping because Python WHILE-loops have the looping mechanisms exposed; whereas, the workings of Python FOR-loops are less explicit. Many programming languages share a similar WHILE-loop structure with Python; whereas, the structure of Python FOR-loops is less common; both will seem easy, once you gain some practice.

10.1.1 WHILE-Loops

The code in Example 10.1 uses a Python WHILE-loop and line numbers are shown on the left. Can you predict what this script will print?

Example 10.1

```
1    # simpleWhileLoop.py
2    x = 1
3    while x <= 5:
4        print x
5        x = x + 1
6    print 'I'm done!'
```

A WHILE-loop checks a Boolean expression placed on the same line as the while keyword. In Example 10.1, it checks if x<=5. The 'condition', like the ones used in decision-making statements, evaluates to either be true or false and the statements in the indented code block that follows (such as, lines 4 and 5) are only performed so long as the condition is true. After it runs the last indented step, it evaluates the value of the condition again (true or false?). If the condition is true, it repeats the indented steps from the top again (line 4, then line 5). If the condition is false, it moves to the first dedented line of code after the WHILE-loop (line 6). We call the indented lines of code the 'WHILE-loop code block'. The WHILE-loop repeatedly runs this code block as long as the test condition is still true. If the condition is never true in the first place, the indented code block will never be executed. Run 'simpleWhileLoop.py' and the results should look like this:

```
>>> 1
2
3
4
5
I'm done!
```

x is printed five times and "I'm done!" is only printed once, since this print statement is outside the WHILE-loop code block.

WHILE-loops rely on a test condition. If that condition never becomes false, this code can run in an infinite loop. The Python interpreter has to be interrupted in a special way if this happens. Comment out line 5 in Example 10.1 by placing a hash sign (#) at the beginning of the line. Predict what will be printed before you run it. Then follow the 'How to interrupt running code' instructions.

How to Interrupt Running Code

1. For PythonWin,

   (a) Right-click on the PythonWin icon that represents your running process, in the lower right corner of the screen.
   (b) Click "Break into running code". You may need to select this multiple times before the code actually stops running.

   (c) Once you successfully stop the running code, a 'KeyboardInterrupt' traceback is printed in the Interactive Window.

2. In PyScripter,

   (a) Click CTRL+F2
   (b) Click 'Yes' on the box that says 'The interpreter is busy. Are you sure you want to terminate?'

Removing line 5 from Example 10.1 creates an 'infinite loop' because the test condition is checking the value of x. When the value of x is not modified, the test condition never becomes false. If x is always 1, $x<=5$ is always true! WHILE-loop test conditions usually involve a variable that needs to be initialized outside the loop and updated inside the loop. Usually the update should occur after all the other WHILE-loop block statements. The WHILE-loop generally has this format:

```
initialize test condition variable     <-- component #1
while condition:                        <-- component #2
    code statement(s)
    update test condition variable.     <-- component #3
```

If any of the three labeled components—initializing, checking, or updating the test condition variable—are missing, the loop will not work as expected.

Notice the syntax elements, the colon and indentation. All Python code blocks— structures containing multiple lines of related code—require the colon and indentation. This includes conditional code blocks, WHILE and FOR-loop code blocks and other structures such as functions and classes, which are discussed in upcoming chapters. The colon always indicates that a block of related code will follow. A missing colon will cause a syntax error. A syntax error will also occur if the code block is empty or if the first line after the colon is not indented. As long as you type the colon, indentation is easy, because IDEs automatically indent the next line when you press the 'Enter' key to move to the next line.

Example 10.2 shows a WHILE-loop geoprocessing example with the row numbers on the left. This script generates three rasters containing randomized cell values. The looping variable, n, is initialized, checked, and updated on lines 10, 11, and 15. n is used to create a unique name for each output raster. The output data name is set at the beginning of the loop code block on line 12. The iterating variable, n, starts with the value 1, then becomes 2, and finally 3. Line 12 creates the names out1, out2, and out3. Notice that the general setup steps, importing modules, setting the environment properties, and checking out the Spatial Analyst Extension, are outside of the loop, before the repetition starts. These statements are only executed once; there is no need to repeat them. The repeated geoprocessing steps— naming, creating, and saving the output raster—are placed inside the loop where they belong.

Example 10.2: Use a WHILE-loop to create 3 rasters containing random values with a normal distribution.

```
 1   # normalRastsLoop.py
 2   # Purpose: Create 3 raster containing random values.
 3
 4   import arcpy
 5   arcpy.env.workspace = 'C:/gispy/data/ch10'
 7   arcpy.env.overwriteOutput = True
 8   arcpy.CheckOutExtension('Spatial')
 9
10   n = 1
11   while n < 4:
12       outputName = 'out{0}'.format(n)
13       tempRast = arcpy.sa.CreateNormalRaster()
14       tempRast.save(outputName)
15       n = n + 1
16   del tempRast
17   print 'Normal raster creation complete.'
```

10.1.2 FOR-Loops

FOR-loops are essential for processing data in Python. Suppose you have a list of files. You can perform batch processing by using a list to control the loop. In Example 10.3, 'simpleForLoop.py' prints the names of each file in a list. Upcoming examples will substitute geoprocessing for printing and the `arcpy` module has functions for getting lists of ArcGIS data (discussed in Chapter 11), so that the data list won't need to be hard-coded.

Example 10.3: Basic FOR-loop.

```
1  # simpleForLoop.py
2  dFiles = ['data1.shp', 'data2.shp','data3.shp','data4.shp']
3  for currentFile in dFiles:
4      print currentFile
5  print 'I'm done!'
```

Printed output:

```
data1.shp
data2.shp
data3.shp
data4.shp
I'm done!
```

When you first see the Python FOR-loop, it may appear mysterious because the mechanisms that make the looping occur are not explicit. The variable placed between the `for` and `in` keywords automatically *iterates* over the items in the list that follows the `in` keyword. In other words, it assumes the value of each item successively. To understand what's happening in Example 10.3, it may help to picture an arrow labeled `currentFile` moving through the list, as shown in Figure 10.1. `currentFile` iterates over the items in the `dFiles` list. It assumes the value of each item successively. Each time it assumes a new value, the indented code block is executed (the indented code block only consists of line 4 in Example 10.3). When there are no remaining unused values in the list, the looping is complete and the next dedented line of code is executed (line 5 in Example 10.3).

To summarize the discussion of Example 10.3, the Python FOR-loop structure repeats the indented code block as many times as there are items in the list. Example 10.3 executed `print currentFile` four times. The Python keywords `for` and `in` are paired together. Whenever you use the `for` keyword, you must use the `in` keyword along with it. The syntax for Python FOR-loops is as follows:

```
for iteratingVar in sequence:
    code statement(s)
```

Figure 10.1 Python automatically updates the variable, currentFile in Example 10.3, to each item in the list.

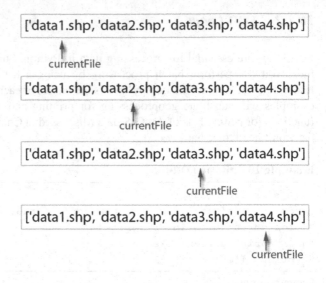

As with the WHILE-loop, the colon and indented code block are required syntax. The *iterating variable*, iteratingVar, is assigned the successive values in the sequence. Python data types that consist of a collection of items are *sequence data types*. Examples include Python strings, tuples, and lists. Strings are made of characters; Tuples and lists are made of comma separated items. The sequence data object we use most frequently is a Python list, though other types such as Python strings and tuples can also be placed after the in keyword.

Example 10.3 used a variable named currentFile as the iterating variable, but there's nothing special about this name. Though you can use any variable name, most often you don't want to use one that is already being used for some other purpose because this will change that variable's value. Any variable name you place between the for and in keywords automatically gets assigned each value in the list as the loop iterates. Suppose you're fond of elephants. The following code shows that you can name your looping variable elephant (though we usually select a term that is more indicative of the items in the list):

```
>>> dFiles = ['data1.shp', 'data2.shp','data3.shp','data4.shp']
>>> for elephant in dFiles:
...     print elephant
...
data1.shp
data2.shp
data3.shp
data4.shp
```

Notice the triple dots in the PythonWin Interactive Window code sample above. These dots are used in the Interactive Window to reinforce that the current line of code is part of a code block. They appear automatically when a code block (following a colon) is entered in the Interactive Window. In fact, when you enter the last line of this code block, you must click the 'Enter' key again to escape from the loop code block. Only then will the code be executed.

Example 10.4 shows a FOR-loop geoprocessing example with the row numbers on the left. This script connects the points in each input file to form polyline output files. When a script creates output inside a loop, the output name needs to be unique at each iteration; otherwise, a file with the same name will be written over each time and the output in the end will be a single file. In order to build a unique output name during each iteration of the loop, you need to use something that is being updated during each iteration. In Example 10.4 the currentFile variable is being updated at each iteration, so we need to use this in the output file name.

> **Note** In general, when geoprocessing a set of files within a loop, the iterating variable is used to build the output file name, since this variable is changing each time.

In Example 10.4, the Points To Line (Data Management) tool is called on line 14. The other lines inside the code block are concerned with creating the unique output file names. To do so, they use the name of the input file as part of the name for the output file. Line 11 gets the base name of the input (data1, data2, ...). Line 13 adds 'Line.shp' to create names like 'data1Line.shp', 'data2Line.shp', and so forth. This is a basic example of geoprocessing within a loop. Environment variables only need to be set once, so they are set prior to the loop (just like in Example 10.2). In Example 10.4, a list has to be initialized prior to the loop, because this is used to create the loop. Next comes the loop statement followed by the indented code block. This code sets up the output variable name and then performs geoprocessing.

Example 10.4: Use a FOR-loop to perform a point to line conversion on a hard-coded file list.

```
1   # point2Line.py
2   # Purpose: Create a set of line features from a
3   #          set of point features in a list.
4   import arcpy
5   arcpy.env.workspace = 'C:/gispy/data/ch10'
6   arcpy.env.overwriteOutput = True
7
8   theFiles = ['data1.shp', 'data2.shp','data3.shp','data4.shp']
9   for currentFile in theFiles:
10      # Remove file extension from the current name.
```

```
11 |     baseName = currentFile[:-4]
12 |     # Create unique output name. E.g., 'data1Line.shp'.
13 |     outName = baseName + 'Line.shp'
14 |     arcpy.PointsToLine_management(currentFile,outName)
15 |     print '{0} created.'.format(outName)
```

Examples 10.1–10.4, show number-based iteration and list-based iteration. There are other uses of these structures, but these are the most common. The examples used WHILE-loops for number-based iteration and FOR-loops for list-based iteration. In fact, a FOR-loop can also perform number-based iteration by using a numeric list. Rather than hard-coding a list of numbers, the built-in range function, which returns lists of numbers, can be used to automatically generate numeric lists. For example, the following code returns a Python list containing the integers from 1 to 5:

```
>>> range(1,6)
[1, 2, 3, 4, 5]
```

You can either set a variable to the return value or embed the range function directly in the FOR-loop statement itself, as follows.

```
>>> for i in range(1,6):
...     print i
...
1
2
3
4
5
```

Example 10.5 employs the range function in a FOR-loop which buffers data. The numbers one through five are used to vary the buffer distance from one to five miles. The numbers are also used to create output names. The iterating variable forms the unique portion of the output names.

Example 10.5: FOR-loop using the range function to update the linear unit.

```
# bufferLoopRange.py
# Purpose: Buffer a park varying buffer distances from 1 to 5 miles.

import arcpy
arcpy.env.workspace = 'C:/gispy/data/ch10'
arcpy.env.overwriteOutput = True
inName = 'park.shp'
for num in range(1, 6):
    # Set buffer distance based on num ('1 miles', '2 miles', ...)
    distance = '{0} miles'.format(num)
    # Set output name ('buffer1.shp', 'buffer2.shp', ...)
```

```
outName - 'buffer{0}.shp'.format(num)
arcpy.Buffer_analysis(inName, outName, distance)
print '{0} created.'.format(outName)
```

10.2 Nested Code Blocks

Scripts often need to use nested codes structures—such as conditional blocks inside FOR-loops (Example 10.6) or FOR-loops inside FOR-loops (Example 10.7). Notice how the indentation steps inward as the nesting occurs. In Example 10.6 the if, elif, and else keywords are aligned at one level and the print keywords contained within these blocks are aligned at the next level. The 'emotaLoop.py' script gets its name from the emoticons it prints. This script loops through each file name in the list and determines which face to print, based on the file extension. The string endswith method, which returns True or False, is used to check the file extension.

Example 10.6

```
# emotaLoop.py
# Purpose: Nest conditions inside a loop to print an emoticon
#          for each file name.
theFiles = ['crops.csv', 'data1.shp', 'rast','xy1.txt']

for f in theFiles:
    if f.endswith('.shp'):
        print '     ;]    ' + f
    elif f.endswith('.txt'):
        print '     :(    ' + f
    else:
        print '     :o    ' + f
```

Printed output:

```
    :o      crops.csv
    ;]      data1.shp
    :o      rast
    :(      xy1.txt
```

Example 10.7, 'scatting.py', uses nested loops to print a rhythmic pattern. Can you predict the output by reading the code? Spaces are embedded in the print statements to reflect the level of nesting. Those outside the FOR-loops have no spaces. Those inside the first FOR-loop are preceded by four spaces. Those inside the nested FOR-loop are proceeded by eight spaces. With this spacing inserted in the print statements, the output indentation mimics indentation of the lines of code that generated it.

Example 10.7

```
# scatting.py
# Purpose: Use nested loops to scat.
print '\nskeep-de'
for i in range(2):
    print '    beep'
    for j in range(3):
        print '        bop'
print 'ba-doop!'
```

Printed output:

```
skeep-de
    beep
        bop
        bop
        bop
    beep
        bop
        bop
        bop
ba-doop!
```

10.3   Directory Inventory

FOR-loops can be used with Python built-in os module functions listdir and walk to navigate data and directories. We demonstrate the simple listdir function here and the more complex os.walk function in Chapter 12. The listdir function lists the files in an input directory, such as sample data directory 'pics'. The following code prints a list of the files in 'C:/gispy/data/ch10/pics':

```
>>> import os
>>> theDir = 'C:/gispy/data/ch10/pics'
>>> # os.listdir returns a list of the files
>>> theFiles = os.listdir(theDir)
>>> theFiles
['istanbul.jpg', 'istanbul2.jpg', 'italy', 'jerusalem', 'marbleRoad.
jpg', 'spice_market.jpg', 'stage.jpg']
```

The output is a Python list, so a FOR-loop can be used to iterate through the files as follows:

```
>>> for fileName in theFiles:
...       print fileName
...
istanbul.jpg
istanbul2.jpg
italy
jerusalem
marbleRoad.jpg
spice_market.jpg
stage.jpg
```

Example 10.8 lists the files in 'C:/gispy/data/ch10' and creates a copy of any text files in the directory. The geoprocessing workspace is set and then this value is passed into the listdir function. Only arcpy module functionality is affected when the arcpy.env.workspace property is set. The os module does not consider this environment setting, so this value must still be passed to the listdir function. 'copyLoop.py' uses the string endswith method to find the files with a 'txt' extension, but this trick won't always work. For example, suppose you want to copy any files that are rasters. It could be difficult to check all of the extensions. Some may not even have extensions, such as the Esri GRID raster files. In fact, the need to process a specific type of file is so common in the GIS workflow that the arcpy package has functions for getting lists of specific data types within a directory. These functions are introduced in the next chapter.

Example 10.8: List files with the os module and geoprocess the files.

```
# copyLoop.py
# Purpose: Make a copy of each ASCII .txt extension file.

import arcpy, os
arcpy.env.workspace = 'C:/gispy/data/ch10'
outDir = 'C:/gispy/scratch/'
theFiles = os.listdir(arcpy.env.workspace)
for fileName in theFiles:
    if fileName.endswith('.txt'):
        outName = outDir + fileName[:-4] + 'V2.txt'
        arcpy.Copy_management(fileName, outName)
        print '{0} created.'.format(outName)
```

Notice that the listdir function returns a list of base names, not full path file names. But os.path methods that return file information, such as size or modification date, need to know where the file is located. If only a base name is given, this method will look for the file in the os module current working directory. If it is not found there, it won't be able to find the file. For example, the os.path.exists function returns True if it determines that the file passed as an argument exists. The following code lists the files in a directory, then checks if they exist:

```
>>> import os
>>> theDir = 'C:/gispy/data/ch10/pics'
>>> # os.listdir returns a list of the files
>>> theFiles = os.listdir(theDir)
>>> for fileName in theFiles:
...         print os.path.exists(fileName)
...
False
False
False
...(and so forth)
```

The files were not deleted between listing them and checking for existence. Rather, the files were not found by the os.path.exists method, because the full path was not specified. To specify the full path, join the directory and file name:

```
>>> fullName = os.path.join(theDir, fileName)
>>> os.path.exists(fullName)
True
```

Example 10.9 uses the os.path.getmtime method to print a time stamp for the last time a file was modified. The os.path.join method is used inside the loop to create full path file names for the files. Example 10.8 set arcpy.env.workspace so that the arcpy.Copy method could locate the files. But the arcpy workspace has no effect on the os module current working directory. This is why you need to use the full path file names in Example 10.9, but not in Example 10.8. It is also possible to change the os module current working directory, using the os.chdir command. But it's important to realize that, this has no influence on the arcpy.env.workspace setting.

Example 10.9: Use os.path.join inside a loop to create full path file names.

```
# printModTime.py
# Purpose: For each file, print the time of most
#          recent modification.
# Input:   No arguments required.

import os, datetime

theDir = 'C:/gispy/data/ch10/pics'
theFiles = os.listdir(theDir)
```

```
for f in theFiles:
    fullName = os.path.join(theDir, f)
    # Get the modification time.
    print os.path.getmtime(fullName)
```

10.4   Indentation and the TabNanny

Chapter 4 introduced the PythonWin syntax check button ![icon] which checks the syntax and runs the TabNanny as well. Now that you're writing indented blocks of code, let's revisit the TabNanny, which checks for consistent spacing. The TabNanny inserts underline marks before indented code if inconsistencies in the indentation are found. By now you know that indentation within a block needs to be aligned. The TabNanny marks indicate a more subtle problem which sometimes occurs when code is copied from sources such as PDF files or Microsoft® software documents into PythonWin. For example, the following code was copied from a PowerPoint® presentation to PythonWin:

```
1  # tabNannyExample.py
2  decList = []
3 -for num in range(5):
4 -     decNum = num + 0.5
5  ------decList.append(decNum)
6  print decList
```

To understand the cause of the error, make 'Whitespace' visible in PythonWin (View > Whitespace). *Whitespace* consists of the invisible characters (such as single spaces and tabs) used to indent and separate elements in a line of code. PythonWin displays spaces as gray dots and tabs as arrows. This is how our example looks with Whitespace turned on:

```
1  # ·tabNannyExample.py
2  decList·=·[]
3 -for·num·in·range(5):
4 -····decNum·=·num·+·0.5
5  ·····→decList.append(decNum)
6  print·decList
```

The problem occurred because the Whitespace usage is inconsistent; Line 4 uses four spaces and line 5 uses one space and one tab. Replace the tab on line 5 with three spaces to repair the problem.

10.5   Key Terms

The while keyword   Sequence data objects
The for keyword   Nested looping
The for and in keyword pairing   The os.listdir function
Iterating variable

10.6   Exercises

1. **whileLoopino.py** Rewrite the following script, replacing the FOR-loop with a
   WHILE-loop.

   ```
   for n in range(-50,150,5):
       print n,
   ```

   Note: The comma at the end of the print statement in NOT a typo. It's a way to
   make consecutive print statements appear on the same line separated by a space.

2. **forLoopino.py** Rewrite the following script, replacing the WHILE-loop with a
   FOR-loop and use the built-in range function.

   ```
   index = 9
   while index <= 99:
       print index
       index = index + 10
   ```

3. **outline.py** Write a script that uses nested FOR-loops to print the output shown
   below. Use the built-in range function for the numerical (outer) loop and use a
   hard-coded list ['a', 'b', 'c'] for the inner loop. Prefix the inner loop print statements with a tab to indent them as shown.

   ```
   1) Hehe
       a) Hoho
       b) Hoho
       c) Hoho
   2) Hehe
       a) Hoho
       b) Hoho
       c) Hoho
   ```

3) Hehe
   a) Hoho
   b) Hoho
   c) Hoho
4) Hehe
   a) Hoho
   b) Hoho
   c) Hoho

4. **loopDistance.py** Write a script to find a set of distance tables from fire stations to schools in the feature classes in 'C:/gispy/data/ch10/county.gdb'. Use the Point Distance (Analysis) tool to determine the distance from the fire stations (input point features) to schools (near features) and use a WHILE-loop to run the tool with ten different search radius values: 500 meters, 1000 meters, ..., 5000 meters. Use 'C:/gispy/scratch' as the output directory for the ten output tables, instead of creating them inside the file geodatabase. Name the output files using the numerical distance as 'dist500.dbf', 'dist1000.dbf', 'dist1500.dbf', and so forth. Hard-code the input features, near features, and output directory in this script, so that no script arguments are needed.

5. **loopAggregate.py** To compare the results of using the Aggregate Polygons (Cartography) tool with various values for the aggregation distance parameter, write a script that aggregates polygons in 'park.shp' for ten different aggregation distances: 150 yards, 250 yards, ... , 1050 yards. Use the built-in range function and a FOR-loop. Vary the output names based on the loop iterator ('park150_agg.shp', 'park250_agg.shp', and so forth). Assume 'park.shp' resides in 'C:/gispy/data/ch10/' and place the output in 'C:/gispy/scratch'. No script arguments are needed.

6. **loopSimplify.py** To investigate the results of using the Simplify Line (Cartography) tool with various tolerance values, write a script that simplifies lines in 'parkLines.shp' with four different tolerance values. Use the built-in range function and a FOR-loop to perform line simplification for tolerance values: 30 feet, 60 feet, 90 feet, and 120 feet. Use the 'POINT_REMOVE' algorithm. Name output as 'simp30.shp', 'simp60.shp', 'simp90.shp', and 'simp120.shp', and place the output in 'C:/gispy/scratch'. No script arguments are needed.

7. **loopGetCount.py** Use the os.listdir method to get a list of the shapefiles in 'C:/gispy/data/ch10/'. Then for any shapefiles whose name contains the word 'out', case-insensitive, use the Get Count (Data Management) tool to determine the number of records in the attribute table. Use one script argument, a directory path. Report the results as shown in the example.

Example input: C:/gispy/data/ch10/archive

Example output:
```
linesOUT.shp has 530 entries.
outData.shp has 100 entries.
parkOutput.shp has 426 entries.
```

8. **loopFileSize.py** Use the `os.listdir` method to get a list of the files in a directory. Then use the `os.path.getsize` function to print the names of small files and their sizes. Allow the user to specify a maximum size in bytes. Use two script arguments, the directory path and the size threshold in bytes. Report the results as shown in the example.

Example input: C:/gispy/data/ch10/64

Example output:
```
data.txt is 42 bytes.
xy1.txt is 42 bytes.
xy_current.txt is 44 bytes.
```

Chapter 11
Batch Geoprocessing

Abstract Python enables you to perform geoprocessing tasks on batches of Esri format data files, fields, and workspaces. This chapter focuses on how to get lists of Esri data and use these lists together with FOR-loops for batch geoprocessing.

Chapter Objectives

After reading this chapter, you'll be able to do the following:

- List the file rasters, feature classes, workspaces, datasets, workspaces, and tables within a workspace.
- List files with a common type and/or substring in the name.
- List the fields in an attribute table.
- Get the name, type, and length of an attribute table field.
- List the printer names, tool, toolboxes, and environment settings for an ArcGIS install.
- Batch geoprocess lists of GIS data.
- Step through each line and watch variables to detect bugs.

11.1 List GIS Data

Python enables you to perform geoprocessing tasks on batches of data files, fields, and workspaces. The `os.listdir` function lists all the files in a directory, but we often need to perform batch processing on a specific GIS file type. The `arcpy` module has a set of listing methods to get lists of these items. The `arcpy` overview diagram shows two of these methods in the left column under 'listing'. The large box in Figure 11.1 shows a more extensive list of these methods. Categories I and II methods are used for listing data. As the names imply, these methods return lists of datasets, feature classes, files, fields, and so forth. This chapter discusses how to use these methods.

Using the Category I methods involves setting the workspace and then calling the method with the usual `object.method` format. In this case, the object is

© Springer International Publishing Switzerland 2015   187
L. Tateosian, *Python For ArcGIS*, DOI 10.1007/978-3-319-18398-5_11

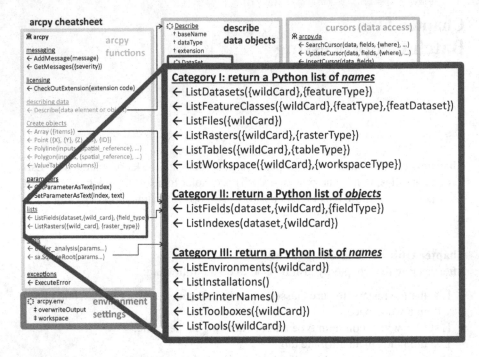

Figure 11.1 The arcpy cheat sheet shows two popular `arcpy` listing methods (`ListFeatureClasses` and `ListFields`) in the left column. The enlarged box on the right has a more complete list of listing methods.

arcpy. In the following code, the `ListFeatureClasses` method returns a list of the feature classes in the workspace:

```
>>> import arcpy
>>> arcpy.env.workspace = 'C:/gispy/data/ch11'
>>> # The ListFeatureClasses method returns a Python list of strings.
>>> fcs = arcpy.ListFeatureClasses()
```

The `arcpy.env.workspace` property must be set before `ListFeature Classes` is called to determine where it looks for feature classes. Category I data listing methods return a list of names in the current workspace. In the example above, the `ListFeatureClasses` method returns a Python list containing the names of feature classes in 'C:/gispy/data/ch11', as shown here:

```
>>> fcs
[u'data1.shp', u'park.shp', u'USA.shp']
```

Each item in the list is a Python unicode string data type. The u that precedes the string indicates that it is a unicode string. You can ignore the u and use unicode string data types as you use 'str' string data types. The following code prints the first item in the list and its data type:

```
>>> fcs[0]
u'park.shp'
>>> type(fcs[0])
<type 'unicode'>
```

Since fcs is a Python list, we can use it in a FOR-loop as follows:

```
>>> for fc in fcs:
...      print fc
...
park.shp
CT_historic_landmarks.shp
CT_national_trails.shp
NEROfires.shp
workzones.shp
```

The other Category I data listing methods, ListDatasets, ListFiles, ListRasters, ListTables, and ListWorkspaces are similar to the ListFeatureClasses method. These methods each list their data for the current arcpy workspace. This means you need to set the workspace before calling them and they will return a Python list of the names of items in that workspace.

Note The arcpy.env.workspace must be set *before* calling Category I data listing methods—ListDatasets, ListFeatureClasses, ListFiles, ListRasters, ListTables, or ListWorkspaces.

'Datasets', 'files', 'tables', 'rasters', and 'workspaces' are fairly broad terms. If you're not sure which types of files these methods will list, check the ArcGIS Resources site. Or, if you have specific data in mind, it's easiest to simply try it on sample data—set the workspace, call the data listing method, and print the returned list. Example 11.1 lists the Esri workspaces found in 'C:/gispy/data/ch11'. According to this function, a file geodatabase (e.g., 'C:/gispy/data/ch11\tester.gdb'), as well as a folder (e.g., 'C:/gispy/data/ch11\pics') is a workspace. The second loop lists the tables found in 'C:/gispy/data/ch11'. In this example, it returns some 'csv' files, 'txt' files, and 'dbf' files. Others items not listed here might qualify as well. All 'txt' files are listed, regardless of content ('loveLetter.txt' has no field headers or records). Not all 'dbf' (dBASE) tables are listed. Only independent dBASE tables in the workspace are listed. Dependent tables are ones associated with shapefiles. Even though 'park.dbf' is in 'C:/gispy/data/ch11', this table is not listed because it is part of a set of files that make up a shapefile.

Example 11.1: Call an `arcpy` listing method and loop through the results.

```
# listStuff.py
# Purpose: Use arcpy to list workspaces and tables.
import arcpy

arcpy.env.workspace = 'C:/gispy/data/ch11'

print '---Workspaces---'
workspaces = arcpy.ListWorkspaces()
for wspace in workspaces:
    print wspace

print '\n---Tables---'

tables = arcpy.ListTables()
for table in tables:
    print table
```

Printed output:

```
>>> ---Workspaces---
C:/gispy/data/ch11\pics
C:/gispy/data/ch11\rastTester.gdb
C:/gispy/data/ch11\tester.gdb

---Tables---
coords.csv
loveLetter.txt
xy1.txt
xy_current.txt
summary.dbf
```

11.2   Specify Data Name and Type

Most of the `arcpy` listing methods have optional parameters. The purple box in Figure 11.1 shows the function signatures, which list the arguments in parentheses behind each method name with optional arguments in curly braces. Most list methods have two optional arguments for filtering names and types. For example, the `ListRasters` signature includes no required arguments, but it has two optional arguments, `wild_card` and `raster_type`, as shown here:

```
ListRasters({wild_card},{raster_type})
```

If no arguments are used, all raster datasets in the current `arcpy` workspace are listed. The optional arguments allow you to restrict the search to a subset of the rasters based on names and/or types. The `wild_card` argument is a string name which can use asterisks as wild cards. A *wild card* stands for a string (or substring) that can have any value. Any string of characters can be substituted in the position where the asterisk appears in the string. For example, '`*am`' could stand for strings such as 'spam', 'ham', and 'am', but it could not stand for 'hamper' and the wild card string '`*elev*`' could stand for strings like 'elevation', 'relevant', 'peak_elev', and 'elev'.

The `wild_card` argument allows you to specify a data name substring using the asterisk as a placeholder for any string. Then if data has a semantic naming schema, e.g., any rasters related to elevation, might contain the substring 'elev', you can use this to select a subset of files for processing. The asterisk can be used anywhere in the string. Example 11.2 demonstrates wild card usage. Can you predict the output from each of these?

Example 11.2a lists all the rasters in the workspace using '`*`' as the wild card. This has the same affect as passing no arguments. Example 11.2b uses '`elev*`' as the wild card parameter; it specifies a prefix, by placing an asterisk at the end of the string. To specify a suffix, use an asterisk at the beginning. If no asterisk is used in a string, it looks for that exact string. Though it can be done, you would usually not want to use the wild card without asterisks. Example 11.2c uses 'elev' as the wild card. This returns only the raster named 'elev'. If this is the intended outcome, the `arcpy.Exists` function would be a better choice for checking one specific data element.

Example 11.2a–c: List rasters in the 'C:/gispy/data/ch11/rastTester.gdb' workspace and use wild_card arguments.

```python
# wildcards.py
# Purpose: Use a wild_card to selectively list files.

import arcpy
arcpy.env.workspace = 'C:/gispy/data/ch11/rastTester.gdb'

# a. Use '*' or empty parentheses to list ALL
#    the rasters in the workspace.
rasts = arcpy.ListRasters('*')
print 'a. All the rasters:'
print rasts
print

# b. List the rasters whose names START with 'elev'.
rasts = arcpy.ListRasters('elev*')
print 'b. elev* rasters:'
print rasts
print
```

```
# c. List a raster whose name is exactly 'elev'.
rasts = arcpy.ListRasters('elev')
print 'c. elev raster:'
print rasts
```

Printed output:

```
>>> a. All the rasters:
[u'elev', u'landcov', u'soilsid', u'getty_cover', u'landc197',
u'landuse', u'aspect', u'soils_kf', u'plus_ras', u'CoverMinus',
u'elev_srt', u'elev_sh', u'elev_ned', u'Int_rand1', u'Int_rand2',
u'landc196', u'TimesCOVER']

b. 'elev*' rasters:
[u'elev', u'elev_srt', u'elev_sh', u'elev_ned']

c. 'elev' raster:
[u'elev']
```

Many of the arcpy listing methods also have an optional argument for specifying a type, such as feature_type, raster_type, or table_type. The values for the type variable, are listed in the ArcGIS Resources site page for each method. For example, search the help for 'ListRasters (arcpy)' and view the 'Explanation' for the raster_type parameter in the parameter table. Valid raster types include 'BMP', 'GIF', 'IMG', and so forth. The raster types often match the file extension ('GRID' files are an exception; they have no extensions, since they are composite files made up of directories and supporting files).

To use the type parameter, put the type in quotation marks. The only tricky aspect to using the type argument is that, it is not the first optional argument. To use arguments that are not first in the code signature, you always have to specify the arguments that precede it; sometimes this means inserting a placeholder. For example, to list all the GIF type raster files in a directory, you don't want to restrict the file names, but you still have to use a placeholder for the wild_card argument. A string containing only an asterisk ('*') can be used as a placeholder in this case. Example 11.3 demonstrates various arguments for the ListRasters method. Can you predict the output for each part?

11.3a lists all the rasters in the workspace. 11.3b finds all of the 'GIF' type raster files in the workspace. Example 11.3c omits the wild_card parameter to demonstrate that if the placeholder is omitted, what was intended to be the raster_type argument would be interpreted as a wild_card. There are no rasters named 'GIF', so the list is empty. Example 11.3d lists all the raster files whose names start with 'tree'. Example 11.3e uses raster_type together with the wild_card argument to narrow this list to only the 'GIF' type 'tree*' rasters.

Examples 11.2 and 11.3 shows several ways to use wild_card and raster_type parameters with the ListRasters method; the other Category I methods follow the same patterns.

Example 11.3a–e: List rasters in the 'C:/gispy/data/ch11/' workspace using `wild_card` and `raster_type` arguments.

```python
# rasterTypes.py
# Purpose: Use a wildcard to selectively list files.
import arcpy
arcpy.env.workspace = 'C:/gispy/data/ch11/'

# a. Use empty parenthesis to list ALL the rasters
#    in the current workspace.
rasts = arcpy.ListRasters()
print 'a. All the rasters:'
print rasts
print

# b. List ALL the GIF type rasters.
rasts = arcpy.ListRasters('*', 'GIF')
print 'b. GIF rasters:'
print rasts
print

# c. List the raster whose name is GIF
rasts = arcpy.ListRasters('GIF')
print 'c. raster named GIF:'
print rasts
print

# d. List the rasters whose names start with 'tree'.
rasts = arcpy.ListRasters('tree*')
print 'd. tree* rasters:'
print rasts
print

# e. List the rasters whose names start with 'tree' which are GIF
#    type files.
rasts = arcpy.ListRasters('tree*', 'GIF')
print 'e. tree* GIF type rasters:'
print rasts
print
```

Printed output:

```
>>> a. All the rasters:
[u'jack.jpg', u'minus_ras', u'tree.gif', u'tree.jpg', u'tree.png',
u'tree.tif', u'tree2.gif', u'tree2.jpg', u'window.jpg']

b. GIF rasters:
[u'tree.gif', u'tree2.gif']
```

c. raster named GIF:
[]

d. tree* rasters:
[u'tree.gif', u'tree.jpg', u'tree.png', u'tree.tif', u'tree2.gif',
u'tree2.jpg']

e. tree* GIF type rasters:
[u'tree.gif', u'tree2.gif']

11.3 List Fields

Unlike Category I methods, Category II methods require one argument, an input
dataset—the items they are listing (such as fields) pertain to a dataset, not an entire
workspace. These methods are slightly more complicated than Category I because
they return lists of objects instead of string names. The ListFields method
returns a list of Field objects, not string names of fields. For example, the following code gets the list of Field objects for input file 'park.shp':

```
>>> # List the Field objects for this dataset.
>>> fields = arcpy.ListFields('C:/gispy/data/ch11/park.shp')
>>> fields
[<Field object at 0x12efa5f0[0x1241ba88]>, <Field object at
0x12efab70[0x1241bf68]>, <Field object at 0x12efa8b0[0x29888f0]>,
<Field object at 0x12efa770[0x2988fe0]>]
```

Each item in this list is a Field object (the numbers indicate location in memory). A Field object can be thought of as a compound structure that contains
multiple pieces of information about the field, that is, the field properties such as
name, type, length, and so forth (Figure 11.2 lists more of these properties). Usually,
we don't want to use the Field object, but rather the field name or some other
property. To access these properties, you need to use object.property format.
The following code prints the field names as seen in the attribute table:

```
>>> for fieldObject in fields:
...        print fieldObject.name
...
FID
Shape
COVER
RECNO
```

Other Field object properties can be accessed in the same manner, as shown in
Example 11.4.

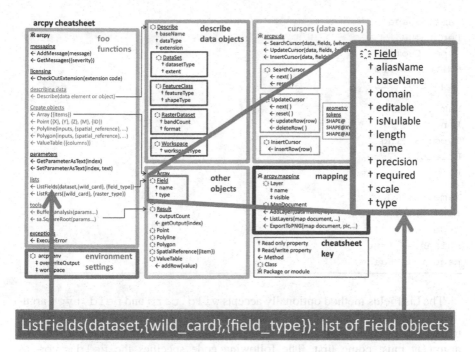

Figure 11.2 The `ListFields` method returns a list of `Field` objects. A `Field` object has the list of properties shown in the enlarged box on the right. These properties are read-only when they are returned by the `ListFields` method.

Example 11.4: List the `Field` object properties.

```
# listFields.py
# Purpose: List attribute table field properties.
import arcpy
arcpy.env.workspace = 'C:/gispy/data/ch11/'

fields = arcpy.ListFields('park.shp')
for fieldObject in fields:
    print 'Name: {0}'.format(fieldObject.name)
    print 'Length: {0}'.format(fieldObject.length)
    print 'Type: {0}'.format(fieldObject.type)
    print 'Editable: {0}'.format(fieldObject.editable)
    print 'Required: {0}\n'.format(fieldObject.required)
```

Printed output:

```
>>> Name: FID
Length: 4
Type: OID
Editable: False
Required: True
```

```
Name: Shape
Length: 0
Type: Geometry
Editable: True
Required: True

Name: COVER
Length: 5
Type: String
Editable: True
Required: False

Name: RECNO
Length: 11
Type: Double
Editable: True
Required: False
```

The ListFields method optionally accepts `wild_card` and `field_type` arguments in addition to the required dataset argument. Using these looks very similar to using the optional arguments for Category I methods, except that the one required argument must come first. The following code specifies the `field_type` as `'Double'` (and uses a placeholder for the `wild_card` parameter, so that the name is unrestricted):

```
>>> parkData = 'C:/gispy/data/ch11/park.shp'
>>> fields2 = arcpy.ListFields(parkData,'*', 'Double')
>>> # The list length shows how many Field objects were returned.
>>> len(fields2)
1
>>> fields2[0].name
'RECNO'
```

> **Note** The `ListFields` method returns a list of `Field` objects, not string names. Use the `object.property` format to access the name property.

11.4 Administrative Lists

GIS scripts use Category I and II listing methods most frequently. Category III methods, like Category I methods return lists of names. Category III methods provide access to ArcGIS software and computer system information for administrative purposes. For example, `ListEnvironments` returns a list of the environment settings, `ListInstallations` returns a list of the type of ArcGIS software installed on the system (server, desktop, engine), `ListPrinterNames` returns a list of printer names available to the system, `ListTools` returns a list of the ArcGIS tools, and `ListToolboxes` returns a list of the built-in ArcGIS toolbox names and aliases with the format 'toolbox_name(toolbox_alias)'. The `arcpy` package has additional listing methods for web mapping administration.

11.5   Batch Geoprocess Lists of Data

The `arcpy` listing methods provide lists of existing data files to batch process. With these methods, we can automatically perform any kind of geospatial processing on a batches of files by iterating over the list with a FOR-loop and tucking the geoprocessing inside the loop. The following code that deletes the 'GRID' datasets with the prefix `'out'`:

```
>>> # List all coverage, geodatabase, TIN, Raster, and CAD datasets.
>>> datasets = arcpy.ListDatasets('out*', 'GRID')
>>> for data in datasets:
...        arcpy.Delete_management(data)
```

Calling the delete tool only requires one argument—the file to be deleted. For tools that create output data, the code needs to update the output file name inside the loop; Otherwise, the looping won't generate multiple output files. It will just overwrite the same output file multiple times. The value of the iterating variable (the variable between the `for` and `in` keywords in the FOR-loop) changes during each iteration. Incorporating this value into the output variable name creates unique names.

Example 11.5: Batch buffer the feature class files in 'C:/gispy/data/ch11/'.

```
# batchBuffer.py
# Purpose: Buffer each feature class in the workspace.

import arcpy, os
arcpy.env.overwriteOutput = True
arcpy.env.workspace = 'C:/gispy/data/ch11/'

# GET a list of feature classes
fcs = arcpy.ListFeatureClasses()
for fc in fcs:
    # SET the output variable.
    fcBuffer = os.path.splitext(fc)[0] + 'Buffer.shp'
    # Call Buffer (Analysis) tool
    arcpy.Buffer_analysis(fc, fcBuffer, '1 mile')
    print '{0} created.'.format(fcBuffer)
```

Table 11.1 Batch processing pseudocode format.

```
GET a list of data elements (fields, files, workspaces, etc.)
FOR each item in the list
    SET output file name
    CALL geoprocessing
ENDFOR
```

Example 11.5 buffers each feature class in the workspace. Notice that the output file name is set *inside* the loop and *before* the geoprocessing. The pseudocode template for batch geoprocessing is shown in Table 11.1.

Example 11.5 places the output in the same workspace as the input. To use a different output workspace, the file needs to do two things differently. First, the script needs to ensure that the output workspace exists by creating it, if necessary. Second, the full path file name of the output needs to be used within geoprocessing calls; otherwise, the path of `arcpy.env.workspace` will be assumed. Example 11.6 differs from Example 11.5 by using 'C:/gispy/data/ch11' for the input directory and 'C:/gispy/scratch/buffers' for the buffer output directory. These paths are hard-coded in the example, but if they are passed in by the user, we can't be sure if the output directory ends in a slash. The `os.path.join` method handles either case by inserting a slash only if needed.

Example 11.6: Batch buffer files in 'C:/gispy/data/ch11/' and place them in the 'C:/gispy/scratch/buffers' directory.

```python
# batchBufferv2.py
# Purpose: Buffer each feature class in the workspace and
#     place output in a different workspace.
import arcpy, os
arcpy.env.overwriteOutput = True
# SET workspaces
arcpy.env.workspace = 'C:/gispy/data/ch11/'
outDir = 'C:/gispy/scratch/buffers'
if not os.path.exists(outDir):
    os.mkdir(outDir)
#GET a list of feature classes
fcs = arcpy.ListFeatureClasses()
for fc in fcs:
    # SET the output variable
    outName = os.path.splitext(fc)[0] + '_buffer.shp'
    fcBuffer = os.path.join(outDir, outName)
    # Call buffer tool
    arcpy.Buffer_analysis(fc, fcBuffer, '1 mile')
    print '{0} created in {1}.'.format(fcBuffer, outDir)
```

So far, we have seen loops which call a single geoprocessing tool and we have seen how to insert an existing block of code into a batch processing loop. Adding multiple geoprocessing steps inside a loop is very similar to what we have already done. The only additional concern may be determining which variables should be set inside the loop and which should be set prior to the loop. Example 11.7 lists data, performs multiple steps within a loop, and then performs an operation on the combined output.

The script lists `'stations*'` dBASE files, which are tables of (x,y) point locations and converts each one to a point feature layer. Then it connects the vertices of the feature layer with lines (This works like a 'connect the dots' game). When the looping is complete, the intersection of all the new line feature classes is computed. The result is a point file 'hubs.shp'. The polyline feature class names (`lineFile`) are set inside the loop because they need to be unique. The temporary point layer is not stored after each iteration, so that name can be reused. The `tempPoints` variable is set just once (before the loop). Make XY Event Layer and Points To Line (Data Management) tools are called once for each table, so they are inside the loop. The Intersect (Analysis) tool is called once outside the loop since that is a collective action on the entire assembled loop output.

Example 11.7

```
# tableLister.py
# Purpose: Create shapefiles from 'stations*' xy tables,
#     connect the points, and then find then
#     intersection of the lines.
# Usage:     workspace_containing_tables
# Example: C:/gispy/data/ch11/trains
import arcpy, os, sys
arcpy.env.workspace = sys.argv[1]
arcpy.env.overwriteOutput = True
tables = arcpy.ListTables('stations*', 'dBASE')

tempPoints = 'temporaryPointLayer'

for table in tables:
    # SET the output variable.
    lineFile = os.path.splitext(table)[0] + 'Line'
    # CALL geoprocessing tools.
    arcpy.MakeXYEventLayer_management(table, 'POINT_X',
                                      'POINT_Y', tempPoints)
    arcpy.PointsToLine_management(tempPoints, lineFile)
    print '\t{0}/{1} created.'.format(arcpy.env.workspace, lineFile)

# GET the list of lines and intersect the lines.
lineList = arcpy.ListFeatureClasses('stations*Line*')
arcpy.Intersect_analysis(lineList,'hubs','#', '0.5 mile', 'POINT')
print '{0}/hubs created.'.format(arcpy.env.workspace)
```

11.6   Debug: Step Through Code

Now that you have learned about block code structures for decision-making and looping and you can write more complicated scripts, you'll be able to create your own bugs! *Bugs* are errors or flaws in the code. *Debugging* is the process of locating and removing errors. Now is a good time to adopt some basic debugging techniques. We'll go over two debugging tips now—stepping through code and watching expressions.

First, you can use the debugger to *step through* the code, meaning you can run each line, advancing the cursor one statement at a time. This will help you understand the flow of the code. That is, this will reveal which statements are executed (in the case of branching) and the number of times each statement is executed (in the case of looping). There are three 'Step' buttons. Most of the time, we only use the one in the middle—the 'Step Over' button. Instructions for PythonWin are given, but any good Python IDE (such as PyScripter) will have equivalent buttons for stepping through code. Work through the following example to learn how to use the 'Step' buttons.

1. In PythonWin, make the debugging toolbar visible (View>Toolbars> Debugging). It looks like this:

2. Mouse slowly over the three 'Step' buttons to see their labels:

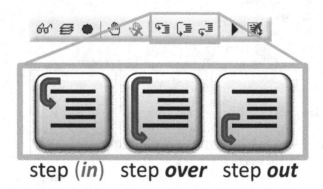

step (*in*)   step ***over***   step ***out***

3. Open 'batchBufferv2.py'.
4. To start stepping through the code, click the 'Step Over' button once.
5. Observe the position of the arrow in the margin. This tracks your location in the script.

   Specifically, the arrow points to the next line of code that has not yet been executed. The next time you click the 'Step Over' button, this line of code will be executed. When we started the debugging session, the arrow jumped immediately to line 4, because comment lines are not executed.

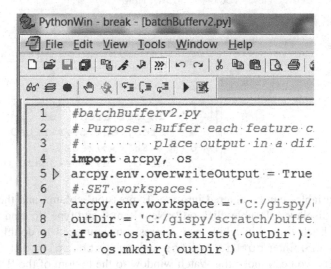

6. Click the 'Step Over' button a second time. If you haven't yet imported `arcpy` in this PythonWin session, this step could take a few moments, after which the arrow will point at line 5. This means line 4 has been executed.

7. Now this time, click the 'Step (in)' button instead of the 'Step over' button. This steps inside the `arcpy.env.workspace` statement. A script named '_base.py' opens and the cursor is inside an arcpy function because you stepped into the middle of this `arcpy` script. You should never alter any of these files or any of the built-in Python modules, so if you step into one of these you should step back out to your own script.

8. Click the 'Step out' button until you return to the 'batchBufferv2.py'. Sometimes you may need to click the 'Step out' button several times to return to your main script.

9. Now step over again until you reach line 12. The arrow jumps from line 9 to line 12 if the output directory has already been created.

10. Stop the debugging session using the 'Close' button on the right end of the debugging toolbar. This doesn't close the script. It only stops the debugging session. The arrow disappears and the feedback bar reports that the script returned exit code 0, meaning a normal exit.

The second tip we'll go over now is to use the 'Watch' window to watch the variable values update as you step through the script.

1. Start stepping through the code by clicking the 'Step over' button.

2. Open the Watch window, by clicking the 'Watch' button on the left end of the debugging toolbar. The window has two columns, 'Expression' and 'Value'. Initially, the Expression column says 'New Item' as shown here:

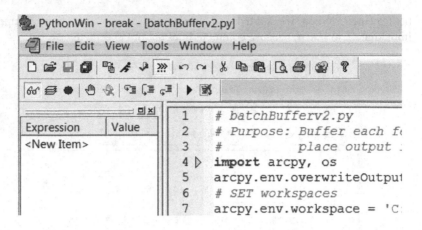

3. Script variables can be entered into the 'Expression' column and their current value is displayed. Double-click on '<New Item>' and type 'fc' and then click 'Enter'. Each time you add a new expression, you need to double-click on <New Item>. Enter `outDir` and `fcs` and `os.path.splitext(fc)[0]`. Optionally, you can move the Watch window to the bottom of the IDE (Shift + left-click-and-drag on the docking bar).

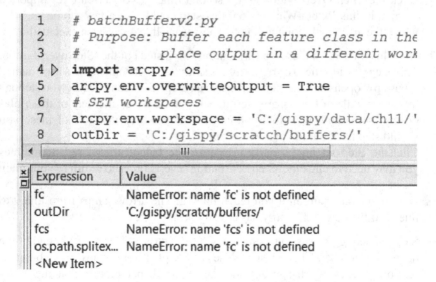

4. `fc` and `fcs` are not yet defined. These values will update as you step through the script. Keep an eye on the Watch window and step over each line until you reach line 12. The arrow jumps from line 9 to line 12 since the previous run created the 'buffer' directory.
5. Step again watching the value of `fcs` become defined as a list of feature classes after line 12 executes.
6. Step again and see the value of `fc` become defined as a feature class name after line 13 executes.

```
12    fcs = arcpy.ListFeatureClasses()
13   -for fc in fcs:
14          # SET the output variable
15 ▷        outName = os.path.splitext(fc)[0] + '_buf
◄                      ||                              
```

Expression	Value
fc	u'data1.shp'
outDir	'C:/gispy/scratch/buffers/'
fcs	[u'data1.shp', u'park.shp', u'USA.shp']
os.path.splitex...	u'data1'

7. Step over lines 14–18 and watch how the arrow moves back up to line 13 and then the value of fc is updated. You can also watch your output being created one by one in the 'buffer' directory.
8. Step through the loop until the script completes or stop the debugging session using the 'Close' button 🖾

With these examples, you have practiced two debugging techniques:

- **Stepping**: Step for line-by-line inspection of the code. 'Step Over' to start execution of a script.
- **Watching**: The watch window allows you to view the values of variables as you step through the code. Double click <New Item> and replace it with a variable name, such as x. Before the variable is defined, the value will appear as "Name Error: name 'x' is not defined". This value is updated whenever a new value is assigned to the variable.

Stepping through code and watching expressions helps programmers become familiar with how Python runs code and how the debugger works. Even advanced programmers often need to slow down to closely inspect the code. We have only introduced a few of the buttons on the debugging toolbar. The remaining debugging buttons will be discussed in Chapter 13.

11.7   Key Terms

The ListFeatureClasses, List Raster, ListTables, and List Fields methods
Asterisks as wild cards

The arcpy Field object
Step though code
Step, step over, step out buttons
Watch window

11.8   Exercises

1. **rasterNames.py** Write a script that uses an `arcpy` listing method to get a list of all the JPG type rasters in the input workspace. Then the script should join the list into a semicolon delimited string and print that string. Use the Python string method named `join` to create the semicolon delimitated string. Print the string. The script should take one argument, the workspace.

   Example input: C:/gispy/data/ch11/pics

   Example output:
   ```
   istanbul.jpg;istanbul2.jpg;marbleRoad.jpg;spice_market.jpg;stage.jpg
   ```

2. **batchCount.py** Write a script that uses an `arcpy` listing method twice to get a list of the Point and Polygon type feature classes in a workspace whose name contains an input substring. Then use a FOR-loop to print the record count result derived from calling the Get Count (Data Management) tool. The script should take two arguments, the input workspace and the desired substring.

   Example input1: C:/gispy/data/ch11/tester.gdb oo

   Example output1:
   ```
   schools has 8 entries.
   smallWoods2 has 55 entries.
   ```

   Example input2: C:/gispy/data/ch11/tester.gdb ne

   Example output2:
   ```
   workzones has 7 entries.
   ```

3. **batchPostOffice.py** For each point feature class in the input workspace whose name contains the string 'data', generate the postal service districts around each point by calling the Create Thiessen Polygons (Analysis) tool. Place the output in a user specified directory and append `'Postal'` to the output names. The script should take two arguments, the input and output workspaces.

   Example input: C:/gispy/data/ch11/tester.gdb C:/gispy/scratch

   Example output:
   ```
   C:/gispy/scratch/data1Postal.shp created.
   C:/gispy/scratch/data2Postal.shp created.
   C:/gispy/scratch/data3Postal.shp created.
   C:/gispy/scratch/ptdata4Postal.shp created.
   ```

4. **batchSimplifyPoly.py** Write a script that uses the Simplify Polygon (Cartography) tool to simplify all the polygon feature classes in the input work-

space. Use the POINT_REMOVE algorithm for polygon simplification and use a degree of simplification tolerance of 50. Place the output in a user specified directory and append 'Simp' to the output names. (An accompanying '*_Pnt. shp is generated for each input file as well). The script should take two arguments, the input and output workspaces.

Example input: C:/gispy/data/ch11/tester.gdb C:/gispy/scratch

Example output:

```
C:/gispy/scratch/simp/parkSimp.shp created.
C:/gispy/scratch/regions2Simp.shp created.
C:/gispy/scratch/simp/workzonesSimp.shp created.
C:/gispy/scratch/regions1Simp.shp created.
C:/gispy/scratch/smallWoods2Simp.shp created.
```

5. **batchPoly2Line.py** Batch convert the polygon feature classes in the input workspace to line files using the Polygon to Line (Data Management) tool. Have the script place the output in a user specified output directory. The script should create this directory if it doesn't already exist. Append 'Lines' to the output file name. The script should take two arguments, the input and output workspaces.

Example input: C:/gispy/data/ch11/tester.gdb C:/gispy/scratch/lineOut

Example output:

```
C:/gispy/scratch/lineOut/parkLine.shp created.
C:/gispy/scratch/lineOut/regions2Line.shp created.
C:/gispy/scratch/lineOut/workzonesLine.shp created.
C:/gispy/scratch/lineOut/regions1Line.shp created.
C:/gispy/scratch/lineOut/smallWoods2Line.shp created.
```

6. **quickExport.py** Write a script that uses an `arcpy` listing method to get a list of all the feature classes in the input workspace and then calls the Quick Export (Data Interoperability) tool to export these files to comma separated value files in output directory, 'C:/gispy/scratch'. Modify the example, `"GIF,c:/data/"` for the output parameter. This example exports the input to GIF image format and places the output in 'C:/data' directory. Look at the ArcGIS Resources site for the Quick Export tool and observe how the parameter table indicates that the tool can take a Python list as the first argument. Call the function only once (without using a loop). The script should take one argument, the input workspace. When the tool call is complete, the script should again call an `arcpy` listing method to get a list the tables in the output directory and print the list.

Example input: C:/gispy/data/ch11/

Example output:
```
C:/gispy/scratch contains:
[u'data1.csv', u'park.csv']
```

7. **tableToDBASE.py** Write a script that uses an `arcpy` listing method to get a list of the tables in the input workspace and then calls the Table to dBASE (Conversion) tool to convert the tables into dBASE format and place them in the given output directory. (Files that are already dBASE format will simply be copied). Look at the ArcGIS Resources site for the Table to dBASE tool and observe how the parameter table indicates that the tool can take a Python list as the first argument. Call the function only once (without using a loop). The script itself should take two arguments, the input and output workspaces. When the tool call is complete, the script should use an `arcpy` listing method to get a list the tables in the output directory and print the list.

Example input: C:/gispy/data/ch11/ C:/gispy/scratch

Example output:
```
C:/gispy/scratch contains:
[u'coords.dbf', u'loveLetter.dbf', u'summary.dbf',
u'xy1.dbf', u'xy_current.dbf']
```

8. **batchFieldNames.py** Write a script that lists the rasters in a workspace whose name contain a specified substring. Also list the field names for each raster. The script should take two arguments, the workspace and the desired substring. Indent the field names using a tab (`'\t'`) to match the example output.

Example input: 'C:/gispy/data/ch11/rastTester.gdb' 'COVER'

Example output:
```
>>> getty_cover
    OBJECTID
    VALUE
    COUNT

CoverMinus
    OBJECTID
    VALUE
    COUNT
    Location

TimesCOVER
    OBJECTID
    Value
    Count blades
```

9. **freqFieldLoop.py** The Frequency (Analysis) tool creates a dBASE table with frequency results for a given field. Practice using nested looping; Write a script that calls a Frequency (Analysis) tool on all the 'string' type fields in the shapefiles in a given directory. Allow the user to input the workspace name. Use the field name and 'freq.dbf' as part of the frequency table output name, as shown in the examples below.

Example input: C:/gispy/data/ch11 C:/gispy/scratch

Example output:
```
>>> C:/gispy/scratch/park_COVERfreq.dbf created.
C:/gispy/scratch/USA_STTE_NAMEfreq.dbf created.
C:/gispy/scratch/USA_SUB_REGIONfreq.dbf created.
C:/gispy/scratch/USA_STATE_ABBRfreq.dbf created.
```

Frequency analysis yields a table with the count for the occurrence of each unique value. The file called 'C:/gispy/scratch/park_COVERfreq.dbf' looks like this:

OID	Frequency	Cover
0	151	orch
1	62	other
2	213	woods

Though the help lists the first parameter, 'in_table' as a 'Table View or Raster layer', this parameter can be specified as the name of a shapefile, as in the following example:

```
arcpy.Frequency_analysis('park.shp', 'out.dbf', 'COVER')
```

10. **batchRatioField.py** Write a script to perform sequential geoprocessing steps inside a loop as follows. Copy each polygon type feature class in an input directory to an output directory, using the Copy (Data Management) tool. Then add a new field named 'ratio', using the Add Field (Data Management) tool, and calculate the field as follows. Use the Calculate Field (Data Management) tool with a Python expression that calculates the ratio of the shape area to the 'rating' field value. E.g., for area=60000 and rating=10, ratio=6000. Assume the 'rating' field exists in each feature class.

Example input: C:/gispy/data/ch11/trains C:/gispy/scratch

Example output:
```
regions1.shp copied to C:/gispy/scratch/regions1.shp
New field ratio added to regions1.shp and calculated as
!shape.area!/!rating!
```

```
regions2.shp copied to C:/gispy/scratch/regions2.shp
New field ratio added to regions2.shp and calculated as
!shape.area!/!rating!
```

11. **batchPoint2Segment.py** Write a script that performs sequential geoprocessing steps inside a loop to create a line type feature class in an output directory for each point type feature class in an input directory. First, call the Points to Line (Data Management) tool to create a line feature class with one line feature that connects all the points. Use names such that the output from this tool creates 's1Line.shp' for a point file named 's1.shp'. Then call the Split Line (Data Management) tool to split the line at the vertices. Use names such that the output from this tool creates 's1Segment.shp' for a point file named 's1.shp'. Finally, delete the intermediate line output files using the Delete (Data Management) tool, such as 's1Line.shp' so that only the line segment files remain.

Example input: C:/gispy/data/ch11/trains C:/gispy/scratch

Example output:
```
C:/gispy/scratch/s1Line.shp created.
C:/gispy/scratch/s1Segment.shp created.
C:/gispy/scratch/s1Line.shp deleted.
C:/gispy/scratch/s2Line.shp created.
C:/gispy/scratch/s2Segment.shp created.
C:/gispy/scratch/s2Line.shp deleted.
```

Chapter 12
Additional Looping Functions

Abstract Most GIS scripting projects are motivated by the desire to automate repetitive tasks. The Python programming language has a number of special functions and structures designed to facilitate various common demands in repetitive workflows. In this chapter, we'll highlight some additional useful techniques for GIS scripting, related to Python FOR-loops, including list comprehension, the built-in enumerate and zip functions, and the os.walk function.

Chapter Objectives
After reading this chapter, you'll be able to do the following:

* Modify lists efficiently.
* Get the index and the value of a list within a FOR-loop.
* Loop through two or more lists simultaneously.
* Traverse subdirectories.

12.1 List Comprehension

Often you want to change a list by making the same type of change to every item in the list. Suppose, for example, you want to change all the strings in a list to lower-case. Do you know how to do this? Examples in Chapter 11 consumed the values in a list, but did not change them. A FOR-loop, gives access to the items in a sequence, but it does not supply a way to modify the list itself. To change all the strings in a list to lower case, you might try applying the lower method within a loop like this:

```
>>> # Mixed-case list of field names.
>>> listA = ['FID', 'Shape', 'COVER', 'RECNO']
>>> for field in listA:
...     field = field.lower()
...     print 'field =', field
...
field = fid
```

```
field = shape
field = cover
field = recno
>>> listA
['FID', 'Shape', 'COVER', 'RECNO']
```

However, the code above changes only the field variable value, but listA remains unchanged. Instead, you can save the changes in another list. The following code creates an empty list, listB, loops through the original list, listA, uses the string lower method, and appends the results to listB:

```
>>> listA = ['FID', 'Shape', 'COVER', 'RECNO']
>>> listB = []
>>> for field in listA:
...        lowName = field.lower()
...        listB.append(lowName)
...
>>> listB
['fid', 'shape', 'cover', 'recno']
```

This code works for the intended purpose. In the end, listB holds a lower case version of listA. However, the task of making the same change to every item in a list is so common that Python has a special syntax called *list comprehension* for handling this. List comprehension efficiently creates a list based on the values of an input list. The following code uses a list comprehension to more efficiently accomplish the same result as the loop above:

```
>>> listA = ['FID', 'Shape', 'COVER', 'RECNO']
>>> listB = [field.lower() for field in listA] # List comprehension
>>> listB
['fid', 'shape', 'cover', 'recno']
```

The syntax for list comprehension looks like this:

```
[expression for iteratingVar in sequence]
```

The list comprehension has a similar format to a FOR-loop statement, except for the expression and the square braces on either end, both of which are required. The iterating variable (iteratingVar) and the sequence follow the same guidelines as in the FOR-loop statement. The expression is a statement that usually involves the iterating variable (iteratingVar). This is where the action happens. The expression dictates what change is being made to the list. The choice of expression depends on the data in the list and the desired outcome. In the example

above, the expression is field.lower(). This works because each item in the list
is a string. In the following example, a numerical expression (i*2) is used to double
the list values:

```
>>> listA = range(1,6)
>>> listA
[1, 2, 3, 4, 5]
>>> listB = [i*2 for i in listA]
>>> listB
[2, 4, 6, 8, 10]
```

The examples above start with listA and store the results in listB. List comprehension can also be used to modify the original list. This example uses the
numerical expression i**2 to square each of the numbers in listA and uses an
assignment statement so that the results are stored in listA:

```
>>> listA
[1, 2, 3, 4, 5]
>>> listA = [i**2 for i in listA]
>>> listA
[1, 4, 9, 16, 25]
```

The following example strips the leading zeros from ID numbers which are
stored as strings:

```
>>> IDlist = ['000004', '000345', '003000', '000860']
>>> IDlist = [ID.lstrip('0') for ID in IDlist]
>>> IDlist
['4', '345', '3000', '860']
```

List comprehension can also be nested and use conditional constructs; however,
list comprehension may not be the best choice in some cases where the logic is
complex, as the list comprehension may become difficult for the reader to
interpret.

12.2   The Built-in enumerate Function

The design of FOR-loops, gives easy access to the items in a sequence, but it does
not supply an index for the current item. The Python built-in enumerate function
provides a way to get both the index and the item concurrently in a FOR-loop.

The enumerate function takes one argument, a sequence, such as a list and can
be used within a FOR-loop to get both the index and the value of each item as you

loop through the list. To use the `enumerate` function in a FOR-loop, you need to place *two* comma-separated variables between the `for` and `in` keywords as shown in the following code, which prints the iterating variables:

```
>>> listA
['FID', 'Shape', 'COVER', 'RECNO']
>>> for i, v in enumerate(listA):
...     print i, v
...
0 FID
1 Shape
2 COVER
3 RECNO
```

The iterating variables `i` (for index) and `v` (for value) are automatically assigned values as the FOR-loop iterates. The first time through the loop, the index is zero and the value is 'FID'. The second time through the loop, the index is one and the value is 'Shape', and so forth.

The syntax for enumerating in a FOR-loop is similar to that of a standard FOR-loop, however, two iterating variables are required and the portion following the `in` keyword is the `enumerate` function with the sequence as its argument:

```
for iterVar1, iterVar2 in enumerate(sequence):
    code statement(s)
```

One useful application of the `enumerate` function is to check the input arguments, as shown in Example 12.1. When you are building GUIs (Chapters 22 and 23) it will be very useful to print the arguments, to double-check their order and correctness. Since `sys.argv` is a Python list, it is a sequence which we can pass to the `enumerate` function. Example 12.1 uses iterating variable names `index` and `arg`, since these are meaningful in this context. Place this code at the beginning of any script to check that the input arguments are being passed correctly.

Example 12.1: Use the built-in `enumerate` function to print system arguments.

```
# argPrint.py
# Purpose: Print args with built-in enumerate function.
# Usage: Any arguments.
# Example input: 500 miles
import sys
for index, arg in enumerate(sys.argv):
    print 'Argument {0}: {1}'.format(index, arg)
```

Printed output:

```
Argument 0: C:\gispy\sample_scripts\ch12\argPrint.py
Argument 1: 500
Argument 2: miles
```

Another use of the `enumerate` function is to use the index for naming geoprocessing output. Example 10.4 (in Chapter 10) performed point to line operations on `theFiles` and used the input file names to create unique output names. Example 12.2 also calls the Points to Line tool on these files; However, the index, garnered from calling the `enumerate` function in the FOR-loop statement is used to create unique output file names, Line0.shp, Line1.shp, Line2.shp, and Line3.shp.

Example 12.2: Use enumerate to create unique output names.

```
# point2LineV2.py (a modification of 'points2Line.py' from Ch. 10)
# Purpose: Use the enumerate function in the loop to create
#          unique output names based on a number.
# Usage: No arguments needed.
import arcpy
arcpy.env.workspace = 'C:/gispy/data/ch12'
arcpy.env.overwriteOutput = True
outDir = 'C:/gispy/scratch/'

dFiles = ['data1.shp', 'data2.shp','data3.shp','data4.shp']
for index, currentFile in enumerate(dFiles):
    # Create a unique output names, 'Line1.shp,', 'Line2.shp'...
    outputName = 'Line{0}.shp'.format(index)
    arcpy.PointsToLine_management(currentFile,
                                  outDir + outputName)
    print outputName
```

12.3   The Built-in `zip` Function

The `enumerate` function allows you to gather both the index and the value in a FOR-loop through a single list. In some applications, you may need to loop through more than one list simultaneously. The built-in `zip` function supports this process by returning a list of tuples:

```
>>> listA = ['FID', 'Shape', 'COVER', 'RECNO']
>>> listB = ['OID', 'Geometry', 'String']
>>> zip(listA,listB)
[('FID', 'OID'), ('Shape', 'Geometry'), ('COVER', 'String')]
```

The length of the list returned by zip is the length of the shortest input list (listB in this example). The zip function can be used to loop through multiple lists simultaneously, by using one iterating variable for each list. In the following code, the variable named a takes on the values in listA and b takes on the values in listB:

```
>>> for a, b in zip(listA, listB):
...        print 'a: ', a ,' b: ', b
...
a: FID b: OID
a: Shape b: Geometry
a: COVER b: String
```

12.4 Walking Through Subdirectories

FOR-loops can be used with Python built-in os module functions listdir and walk to navigate data and directories. As discussed in Chapter 10, the listdir function lists the files in an input directory, but does not list files in subdirectories. The os function called walk provides a handy way to process files distributed in multiple levels and subdirectories. To use the walk function in a FOR-loop, you need to place *three* comma-separated variables between the for and in keywords as shown in Example 12.3 which prints the values of the iterating variables root, dirs, and files. We use the name root to stand for *root directory*, which is the top-most directory in a hierarchy. dirs stands for subdirectories in the root directory and files stand for files in the root directory. The iterating variables are automatically assigned values as the FOR-loop iterates.

The listdir function allows a script to list the files immediately inside of a given directory; whereas, the walk allows a script to walk recursively to every file under a given directory. It does this by using an internal list of subdirectories. This list is not exposed to the user, but in order to explain how the walk works, we'll name it toDoList and keep track of its contents. The walk adds each subdirectory that it encounters to the to-do list and it visits each one in the list. It removes a directory once it has been visited. When the to-do list is emptied, the walk ends.

In Example 12.3, 'walkPics.py' is walking through 'pics', a directory of travel pictures from Turkey, Italy, and Israel. Figure 12.1 depicts the 'pics' directory tree. The 'pics' directory contains Turkish photos and has two subdirectories, 'Italy' and 'Jerusalem', which, themselves contain more items. Compare the directory tree in Figure 12.1 with the output in Example 12.3. The first time through the loop, the print statements yield this:

```
>>> root = 'C:/gispy/data/ch12/pics'
dirs = ['italy', 'jerusalem']
files = ['istanbul.jpg', 'istanbul2.jpg', 'marbleRoad.jpg',
spice_market.jpg', 'stage.jpg']
```

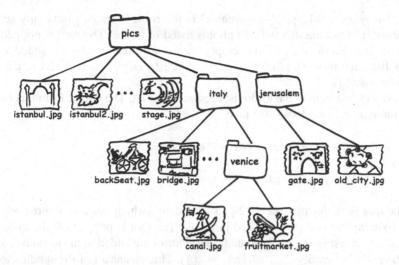

Figure 12.1 The 'C:/gispy/data/ch12/pics' directory tree.

Initially, `root` is the input directory, 'C:/gispy/data/ch12/pics'. The directory list is ['italy', 'jerusalem'] because 'pics' contains two subdirectories 'italy' and 'jerusalem'. The list of files is ['istanbul.jpg', 'istanbul2.jpg', 'marbleRoad.jpg', 'spice_market.jpg', 'stage.jpg'], because these photos are the files that reside directly under 'C:/gispy/data/ch12/pics'. The full paths of the subdirectories in the `dirs` list are added as the first entries in the to-do list (`toDoList = ['C:/gispy/data/ch12/pics/italy', 'C:/gispy/data/ch12/pics/jerusalem']`).

After the first loop, the `walk` function starts working through the to-do List; The first directory in the to-do list, becomes the new root (and it is popped off the to-do list (`toDoList = ['C:/gispy/data/ch12/pics/jerusalem']`) and the processes repeats. The output from the second time through the loop looks like this:

```
root = C:/gispy/data/ch12/pics\italy
dirs = ['venice']
files = ['backSeat.jpg', 'bridge.jpg', 'ct.jpg', 'florence.jpg']
```

The new root is 'C:/gispy/data/ch12/pics/italy' since this was the first item in the to-do list. Now the `dirs` list only has one entry, 'venice', because 'italy' contains only one subdirectory. The photographs directly inside 'C:/gispy/data/ch12/pics/italy' are listed in `files`. Next, the directory in the `dirs` list is added to the front of the to-do list (`toDoList = ['C:/gispy/data/ch12/pics/italy/venice', 'C:/gispy/data/ch12/pics/jerusalem']`). This first item in the to-do list becomes the root. The output from the third time through the loop looks like this:

```
root = C:/gispy/data/ch12/pics\italy\venice
dirs = []
files = ['canal.jpg', 'fruitMarket.jpg']
```

'C:/gispy/data/ch12/pics\italy\venice' is the root, it doesn't have any subdirectories; it only contains the two photos listed in `files`. The root is popped off the to-do list and the `dirs` list is empty, so no additional entries are added to the to-do list this time (`toDoList = ['C:/gispy/data/ch12/pics/jerusalem']`).

Next, the first item in the to-do list becomes the root. The output from this fourth time through the loop looks like this:

```
root = C:/gispy/data/ch12/pics\jerusalem
dirs = []
files = ['gate.jpg', 'old_city.jpg']
```

The root is 'C:/gispy/data/ch12/pics\jerusalem', which has no subdirectories; it only contains the two photos listed in `files`. The root is popped off the to-do list and the `dirs` list is empty, so no additional entries are added to the to-do list, leaving the to-do list empty (`toDoList = []`). This means all of the subdirectories have been visited, so the walk is complete.

Example 12.3: Walk through subdirectories.

```
# walkPics.py
# Purpose: Demonstrate the os.walk function.
import os
myDir = 'C:/gispy/data/ch12/pics'
for root, dirs, files in os.walk(myDir):
    print 'root = {0}'.format(root)
    print 'dirs = {0}'.format(dirs)
    print 'files = {0}\n'.format(files)
```

Printed output:

```
>>> root = C:/gispy/data/ch12/pics
dirs = ['italy', 'jerusalem']
files = ['istanbul.jpg', 'istanbul2.jpg', 'marbleRoad.jpg', 'spice_
market.jpg', 'stage.jpg']
root = C:/gispy/data/ch12/pics\italy
dirs = ['venice']
files = ['backSeat.jpg', 'bridge.jpg', 'ct.jpg', 'florence.jpg']
root = C:/gispy/data/ch12/pics\italy\venice
dirs = []
files = ['canal.jpg', 'fruitMarket.jpg']
root = C:/gispy/data/ch12/pics\jerusalem
dirs = []
files = ['gate.jpg', 'old_city.jpg']
```

The format for using the `walk` function in a FOR-loop is similar to that of the `enumerate` function FOR-loop; However, three iterating variables are required and the portion following the `in` keyword is the `os.walk` function with a directory path as its argument:

```
for iterVar1, iterVar2, iterVar3 in os.walk(dirPath):
    code statement(s)
```

Though the `os.walk` function iterates through three pieces of information (the root directory, a list of subdirectories, and a list of files), we're mostly interested in the lists of files, so that we can do geoprocessing. Notice that the file names are given as Python lists. The following code simply prints the `files` variable:

```
>>> import arcpy, os
>>> myDir = 'C:/gispy/data/ch12/pics'
>>> for root, dirs, files in os.walk(myDir):
...     print files:
...
['istanbul.jpg', 'istanbul2.jpg', 'marbleRoad.jpg', 'spice_market.
jpg', 'stage.jpg']
['backSeat.jpg', 'bridge.jpg', 'ct.jpg', 'florence.jpg'] ['canal.
jpg', 'fruitMarket.jpg']
['gate.jpg', 'old_city.jpg']
```

To access the individual files one at a time, we need to use a nested loop through each one of these lists as shown in the following code:

```
>>> for root, dirs, files in os.walk(myDir):
...     for f in files:
...         print f
...
istanbul.jpg
istanbul2.jpg
marbleRoad.jpg
spice_market.jpg
stage.jpg
backSeat.jpg
bridge.jpg
ct.jpg
florence.jpg
canal.jpg
fruitMarket.jpg
gate.jpg
old_city.jpg
```

Notice that the list of file names gives only the base name, not the full path file name, so `arcpy` won't know where to find the files:

```
>>> for root, dirs, files in os.walk(myDir):
...        for f in files:
...             print arcpy.Exists(f)
...
False
False
False
...(and so forth)
```

The `arcpy Exists` function returns `False` for these files, though we know they exist! When using `arcpy` on files within a `walk` loop, we need to set the workspace to the file path, the path which the `root` variable stores. The following code updates the workspace so that `arcpy` can locate each file:

```
>>> for root, dirs, files in os.walk(myDir):
...        arcpy.env.workspace = root
...        for f in files:
...             print arcpy.Exists(f)
...
True
True
True
...(and so forth)
```

Depending on what type of geoprocessing is being performed, you may need to filter the file types using another level of nesting. Example 12.4 nests a conditional construct inside the inner file loop so that it prints only files ending in 'txt'.

Example 12.4: Walk and print the full path names of the files with 'txt' extensions.

```
# walkTXT.py
# Purpose: Walk and print the full path file names of
#        'txt' extension files in the input directory.
# Usage: input_directory
# Example: C:/gispy/data/ch12/wTest
import arcpy, os, sys
mydir = sys.argv[1]
for root, dirs, files in os.walk(mydir):
    arcpy.env.workspace = root
    for f in files:
        if f.endswith('.shp'):
            # Print the full path file name of f
            print '{0}/{1}'.format(root,f)
```

Example 12.4 uses a string method to filter the data, but we can also use the `arcpy` listing methods for more precise selection. Example 12.5 buffers only the point feature classes in the directory tree by calling the `ListFeatureClasses` method each time it reaches a new workspace. Then it loops through the `fcs` list instead of the `files` list.

Example 12.5: Walk through the directories and buffer each shapefile.

```
# osWalkBuffer.py
# Purpose: Walk and buffer the point shapefiles.
# Usage: input_directory output_directory
# Example: C:/gispy/data/ch12/wTest C:/gispy/scratch
import arcpy, os, sys
rootDir = sys.argv[1]
outDir = sys.argv[2]
arcpy.env.overwriteOutput = True
for root, dirs, files in os.walk(rootDir):
    arcpy.env.workspace = root
    fcs = arcpy.ListFeatureClasses('*', 'POINT')
    for f in fcs:
        # Set output name and perform geoprocessing on f
        outfile = outDir + '/' + os.path.splitext(f)[0] + \
                  'buffer.shp'
        arcpy.Buffer_analysis(f, outfile, '1 mile')
        print '{0}/{1}  buffer ouput: {2}'.format(root, f, outfile)
```

The `os.walk` function does not automatically crawl specialized compound Esri nested data structures, such as file geodatabases. Versions of ArcGIS 10.1 and higher provide a walk function in the data access module (da) built upon the `os.walk` function designed specifically to handle Esri data structures. Sample script 'arcpy-WalkBuffer.py' shows how to call this method. For additional examples, search online in ArcGIS Resources for 'walk (arcpy.da)'.

12.5   Key Terms

List comprehension
The string `lstrip` method
Built-in `enumerate` function
Built-in `zip` function
The `os.walk` function
The `arcpy.Exists` function

12.6   Exercises

1. **fieldTypes.py** Write a script that gets a list of the `Field` objects for an input file with an attribute table, uses Python list comprehension to create a list of field names, and prints that list. It should take one argument, the input file.

   Example input: C:/gispy/data/ch12/park.shp

   Example output:
   ```
   >>> [u'OID', u'Geometry', u'String', u'Double']
   ```

2. **listComprehension.py** Modify the code in the Chapter 12 sample script 'listComprehension.py' on the lines that say `### modify this`. The code should use list comprehension to create a new list based on each given list. Output should always appear as shown below, except for the last two, which are dynamic, changing based on the content of 'C:/gispy/data/ch12/comp', which currently contains the files 'Jack.jpg', 'site1.dbf', and 'xy1.txt'. The first one is completed as an example.

   ```
   >>> 1. All cap field names: ['FID', 'SHAPE', 'COVER', 'RECNO']
   2. Rounded float values: [3.0, 1.0, 4.0, 5.0, 5.0]
   3. The reciprocal values: [0.125, 0.25, 0.25, 1.0, 0.2, 0.25, 0.25,
   0.5]
   4. No slashes: ['Fire_TypeProtection-Type', 'Time_Date',
   'Condition_Status_Role']
   5. Output files: ['Jackout.jpg', 'site1out.dbf', 'xy1out.txt']
   6. File extensions: ['.jpg', '.dbf', '.txt']
   ```

3. **enumList.py** Rewrite the code in sample script 'enumList.py'. Replace the `WHILE`-loop with a `FOR`-loop and use the built-in `enumerate` function to get both the indices and the values in the list. The printed output should be identical to what the original code prints.

4. **fieldProperties.py** Write a script that gets a list of the `Field` objects for an input file with an attribute table and prints the name, data type, and length of each field. Use the built-in `enumerate` function to print the index number for each field (Field 0, Field 1,...). It should take one argument, the input file. The printed output should follow the format of the following example:

   Example input: C:/gispy/data/ch12/park.shp

   Example output:
   ```
   >>> Field 0: Name=FID, Type=OID, Length=4
   Field 1: Name=Shape, Type=Geometry, Length=0
   Field 2: Name=COVER, Type=String, Length=5
   Field 3: Name=RECNO, Type=Double, Length=11
   ```

5. **MPHzip.py** Add code to sample script, 'MPHzip.py', so that it uses the built-in `zip` function to find and print the average speed in miles/hr for five errands

based on distance and time data collected by a GPS unit. The distance for each trip is given in miles in `distanceList` and the corresponding times are given in minutes and seconds in the same order in `timeList`. The time format separates minutes from seconds with a colon. Use string methods and built-in functions inside the zip loop to extract the numeric temporal values.

```
distanceList = [0.04, 0.05, 0.91, 0.16, 18]
timeList = ['7m:13s', '11m:29s', '16m:48s', '3m:26s', '120m:0s']
```

The output should look like this:

```
>>> Distance: 0.04       Time: 7m:13s        Speed: 0.33 miles/hr
Distance: 0.05       Time: 11m:29s       Speed: 0.26 miles/hr
Distance: 0.91       Time: 16m:48s       Speed: 3.25 miles/hr
Distance: 0.16       Time: 3m:26s        Speed: 2.80 miles/hr
Distance: 18       Time: 120m:0s       Speed: 9.00 miles/hr
```

6. **nameWalk.py** Use `os.walk` to search the files in the input directory and all of its subdirectories and print those file names with base names that contain two or more occurrences of the letter 'a'. The script should take one argument, the input directory path.

   Example input: C:/gispy/data/ch12/pics

   Example output:
```
>>> C:/gispy/data/ch12/pics/marbleRoad.jpg
C:/gispy/data/ch12/pics\italy/backSeat.jpg
C:/gispy/data/ch12/pics\italy\venice/canal.jpg
```

7. **countWalk.py** Use `os.walk` to perform a Get Count (Data Management) tool call on all the feature classes in the input directory and all of its subdirectories. The script should take one argument, the input directory path. Print the count results as shown in the example.

   Example input: C:/gispy/data/ch12/wTest

   Example output:
```
>>> C:/gispy/data/ch12/wTest/data1.shp has 100 entries.
C:/gispy/data/ch12/wTest\trains/regions1.shp has 426 entries.
C:/gispy/data/ch12/wTest\trains/s2.shp has 8 entries.
C:/gispy/data/ch12/wTest\tSmall.gdb/c1 has 100 entries.
C:/gispy/data/ch12/wTest\tSmall.gdb/trail has 1 entries.
C:/gispy/data/ch12/wTest\tSmall.gdb/regions1 has 1 entries.
```

Chapter 13
Debugging

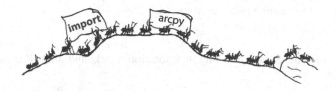

Abstract Even experienced programmers create *bugs*—coding errors that take time to correct. Syntax has to be precisely correct and programming is a complicated task. Computers do exactly what the program says; they are unforgiving. Fortunately, there are tools for locating, understanding, and handling coding errors. Integrated Developments Environments (IDEs), such as PythonWin, have syntax checking and debugging tools. The `arcpy` package has functions for gathering information about errors and the Python language itself has functions and keywords for handling errors gracefully. This chapter covers error debugging and related topics. The three types of programming errors are: *syntax errors*, *exceptions*, and *logic errors*.

Chapter Objectives
After reading this chapter, you'll be able to do the following:

- Check for syntax errors.
- Recognize exceptions.
- Run code in debug mode.
- Watch variables as code is running.
- Step through code one line at a time.
- Set stopping points (breakpoints) at crucial points within the code.
- Run the code to a breakpoint.
- Set conditions for stopping at a breakpoint.
- Design input to test the code.

© Springer International Publishing Switzerland 2015   223
L. Tateosian, *Python For ArcGIS*, DOI 10.1007/978-3-319-18398-5_13

13.1   Syntax Errors

Syntax errors are usually caused by mistakes in spelling, punctuation, or indentation. Spotting syntax errors becomes easier with practice. See how quickly you can locate the two syntax errors in the following code:

```
# buggyCode1.py
import os, sys
outputDir = os.path.dirname(sys.argv[1]) + '\outputFiles/
if not os.path.exists(outputDir)
    os.mkdir(outputDir)
```

The end of line 3 is a missing a quotation mark and line 4 is missing a parenthesis.

Chapter 4 introduced the PythonWin syntax check button 🕮, which is a bit like the spelling/grammar checker in a word processing editor. Word processors can automatically identify errors such as misspellings and run-on sentences; Proof readers are needed to catch other writing errors such as misplaced modifiers or split infinitives. Similarly, certain types of programming errors are more difficult to automatically detect than others.

Syntax errors are errors that can be detected just based on parsing the text. *Parsing* is breaking the text down into a set of components that the system can recognize. When the 'Run' button is pushed, the IDE first attempts to parse the code. Syntax errors confound the parsing process. If a syntax error is detected, the parser stops and the script will not run. Remember—the parser jumps to the line or near the line where the syntax error occurred to provide a clue.

13.2   Exceptions

The parser can only detect violations of syntax rules. If the syntax rules are adhered to, Python will run (or attempt to run) the script. If it encounters an error when attempting to execute a line of code, it throws an exception and stops the run at that point. Chapter 2 discussed Python built-in exceptions such as NameErrors and the traceback errors they print. Can you see why the following script throws a NameError exception?

```
# buggyCode2.py
import arcpy, sys, os
arcpy.env.workspace = sys.argv[1]
fc = arcpy.ListFeatureClasses()
for fc in fcs:
```

```
# Append Buffer to the output name.
fcBuffer = os.path.splitext(fc)[0] + 'Buffer.shp'
# Call buffer tool with required input,
#    output, and distance arguments.
arcpy.Buffer_analysis(fc, fcBuffer, '1 elephant')
```

Don't be distracted by the elephant—we'll get to that in a minute. Look earlier in the script. Python executes each line of code in order. Before it even gets close to the elephant, it reaches an undefined variable, throws an exception, stops the run, and prints a traceback with the following exception:

```
NameError: name 'fcs' is not defined
```

The variable 'fcs' is used on line 5, but it's not defined. If you have already run a script that assigned a value to 'fcs' during this PythonWin session, you won't see this error. Remember to test scripts with a fresh PythonWin session before sharing code. Modify line 4 to define 'fcs' as follows:

```
fcs = arcpy.ListFeatureClasses()
```

Now we're ready to talk about the elephant. Save and run the script again. This time it reaches the Buffer tool call, tries to create a buffer, can't understand the buffer distance argument, and prints a traceback with the following exception:

```
ExecuteError: Failed to execute. Parameters are not valid.
ERROR 000728: Field 1 elephant does not exist within table
Failed to execute (Buffer).
```

The required buffer distance parameter of the Buffer tool can be specified as a linear unit or a field name (for a field whose values specify a buffer distance for each feature). Since 'elephant' is not a recognized unit of measure, it searches for a field named '1 elephant'. No such field exists and so it fails to execute and throws an `ExecuteError`. The `ExecuteError` is a special error thrown by `arcpy` whenever a geoprocessing tool encounters an error.

Both of the exceptions thrown by 'buggyCode2.py' can be corrected by changing the code (by defining `fcs` and specifying a valid linear unit). Other factors that can cause exceptions to be thrown may be beyond the programmer's control. For example, the script may encounter a corrupted file in the list which it can't buffer. Python structures for handling these situations will be presented in Chapter 14.

13.3   Logic Errors

Errors do not always generate exceptions. A program containing logic errors can run smoothly to completion and have inaccurate results. Because logic errors are not detectable by the computer, we have to understand the problems we're solving and inspect the code closely when we perceive unexpected results. The following code for normalizing the time-series dates contains a logic error:

```
# buggyCode3.py
# normalize data time steps
timeSteps = [2011, 2009, 2008, 2005, 2004, 2003, 2001, 1999]
# normalize to values between 0 and 1
maxv = max(timeSteps)
minv = min(timeSteps)
r = maxv - minv
# list comprehension with buggy expression
normalizedList = [v - minv/r for v in timeSteps]
print normalizedList
```

A set of numbers are normalized by subtracting the minimum value and dividing by the range. The output should be numbers between 0 and 1, but the results are the following list:

```
>>> [1845, 1843, 1842, 1839, 1838, 1837, 1835, 1833]
```

'buggyCode3.py' runs without errors because the code doesn't violate any syntax rules and it doesn't throw any exceptions; it also gives the wrong result. Can you identify the two mistakes in the list comprehension expression? First, the missing parentheses in the numerator cause the arithmetic operation to be performed in an unintended order. The division is performed first. In this case, our minimum value is 1999 and the range is 12, so 1999 is divided by 12, giving 166. This value (166) is then subtracted from each time step (For example, for the first time step, the result is 2011 - 166 = 1845). The subtraction must be wrapped in parentheses to force it to be calculated first. Replace the list comprehension with the following line of code:

```
normalizedList = [(v - minv)/r for v in timeSteps]
```

Running the script with the modified list comprehension prints the following:

```
>>> [1, 0, 0, 0, 0, 0, 0, 0]
```

The results are mostly zeros when we would expect decimal values ranging between 0 and 1. This brings us to the second mistake; Integer division causes the division remainders to be discarded. Cast the numerator or denominator to solve this problem. Replace the list comprehension with the following line of code:

```
normalizedList = [(v - minv)/float(r) for v in timeSteps]
```

and the truly normalized list is printed as follows:

```
>>> [1.0, 0.8333333333333334, 0.75, 0.5, 0.4166666666666667,
0.3333333333333333, 0.16666666666666666, 0.0]
```

> **Tip** Use a small tractable sample dataset for preliminary testing to catch logic errors early.

The small set of numbers in the time step list made it easy to evaluate the results. When you're trying new code, it's good practice to test the functionality incrementally on a very small dataset so that you can predict the desired outcome and compare it to the results produced by the code. Had this been run on a large input file of dates, unearthing the bugs would take longer. In fact, you may not notice the problem until you perform other calculations on the output. Even then you may not notice the mistake. Further calculations might mask the error. Unlike syntax errors or exceptions, your only means of detecting logic errors is inspecting the results. By testing each piece as you build it, you can be more confident that the overall result is correct.

> **Tip** Test code incrementally as you build each new piece of functionality.

Sometimes errors are not revealed by one test. It's good practice to test multiple scenarios. The following script is supposed to remove names from a list, but the code is poorly designed:

```
# buggyLoop.py
# Remove feature classes whose names do not contain the given tag.
tag = 'zones'
fcs = [u'data1',u'data2',u'data3',u'fireStations',
       u'park_data',u'PTdata4',u'schoolzones',
       u'regions1',u'regions2',u'workzones']
print 'Before loop: {0}'.format(fcs)
for fcName in fcs:
    if tag in fcName:
        fcs.remove(fcName)
print 'After loop: {0}'.format(fcs)
```

This code removes items containing the word 'zones'. The script prints the following:

```
>>> Before loop: [u'data1', u'data2', u'data3', u'fireStations',
u'park_data', u'PTdata4', u'schoolzones', u'regions1', u'regions2',
u'workzones']
After loop: [u'data1', u'data2', u'data3', u'fireStations', u'park_data',
u'PTdata4', u'regions1', u'regions2']
```

But when `tag = 'zones'` is replaced with `tag = 'data'`, the code no longer works as expected:

```
>>> Before loop: [u'data1', u'data2', u'data3', u'fireStations',
u'park_data', u'PTdata4', u'schoolzones', u'regions1', u'regions2',
u'workzones']
After loop: [u'data2', u'fireStations', u'PTdata4', u'schoolzones',
u'regions1', u'regions2', u'workzones']
```

In this case, the `remove` statement confounds the proper course of iteration. Each time an item is removed, `fcName` skips the next item when it gets updated. Step through this code with the debugger and watch `fcName` and `fcs` to view this peculiar behavior.

In fact, making alterations to a list while you're looping through it is never a good idea. The loop needs to use a second list to store the results, as in the following code:

```
print 'Before loop: {0}'.format(fcs)
fcsOut = []
for fcName in fcs:
    if tag not in fcName:
        fcsOut.append(fcName)
print 'After loop: {0}'.format(fcsOut)
```

Logic errors come in many forms, but they all arise from the same basic problem: your code is doing something other than what you desire and expect. Once you perceive a mistake in the results, it may take time to discover the source. Many beginning programmers avoid using IDE's built-in debugger and try to use the "poor man's debugger"—debugging by simply printing variable values throughout the code. This is effective some of the time, but in many cases, using the debugging functionality reveals the problem more quickly. The remainder of this chapter discusses debugging toolbar buttons and how to use breakpoints to improve debugging technique.

13.4   PythonWin Debugging Toolbar

When code raises exception errors or contains logic errors, you often need to look more closely at variable values. With an IDE such as PythonWin, code can be run in the standard way, using the 'Run' button or it can be run in debug mode. Chapter 11 showed how to use the debugger to step through the code one line at a time and watch variable values change as the code executes. Recall these points from the debugger discussion in Chapter 11:

- To make the toolbar visible: View>Toolbars>Debugging (Some buttons are grayed out, but become active when a debugging session commences)

- Click the 'Step over' button ⤵≣ to start debugging and step one line at a time through the code.

- Generally, you don't want to click the 'Step (in)' button ⤴≣, but if you do by mistake, click the 'Step out' button ⤶≣ until you return to your script. We'll revisit these two buttons in Chapter 15.

- The arrow ▷ points at the current line of code while you are stepping using the debugger. This points to the next line of code that has not yet been executed.

- Click the 'Watch' button 𝟞𝟞 to open the Watch panel and double-click on 'New Item' to add a variable to the watch list.

- Click the 'Close' button 🖹 to stop the debugging session at any time before the script completes.

So far, you're familiar with half the debugging toolbar buttons. Figure 13.1 depicts the toolbar buttons clustered by related functionality (watch, breakpoint, step, and go/close). Table 13.1 explains each cluster, listing each button's name and usage. You have already used the 'Watch' button, the three 'Step (in)' buttons, and the 'Close' button. Now we'll address the unfamiliar buttons in Table 13.1.

The 'Stack view' button ⊜ opens a panel that displays the values of all objects that are currently defined within the PythonWin session. Each object expands to

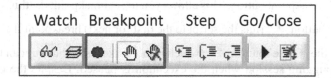

Figure 13.1 Debugging toolbar buttons clusters.

Table 13.1 PythonWin debugging toolbar buttons.

Icon	Key	Name	Usage
		Watch	Display the value of Python code expressions during a debugging session.
		Stack view	Displays the values of all the currently available Python objects.
		Breakpoint list	Lists each breakpoint and set breakpoint conditions.
	F9	Toggle breakpoint	Place the cursor on a line of code and click this button to toggle a breakpoint on this line.
		Clear breakpoints	Removes all breakpoints.
	F11	Step	Execute one line of code. Step inside any functions called in this next line of code.
	F10	Step over	Execute the next line of code without stepping inside any functions called in this line of code.
	Shift+F11	Step out	Step out of a function.
	F5	Go	Run in debug mode to the first (or next) breakpoint.
	Shift+F5	Close	Stop the debugging session without executing another line of code. This does not close PythonWin or any scripts.

show the contained objects. This complex tree structure may not be of interest to beginners. The next three buttons relate to using breakpoints, which are extremely useful for programmers of all skill levels.

13.4.1   Using Breakpoints

Breakpoints are a means to stop the script at a specified line in the program. Once you've reached that line, you can investigate the values of your variables and step through the code to see how it's working. The 'Toggle breakpoint' button 🖑 is the most crucial breakpoint button. This allows you to set breakpoints within your code.

The 'Go' button ▶ runs the script to the first (or next) breakpoint it encounters.

Conversely, the 'Close' button 📝 (which is more like a stop button) can be used to stop the debugging session when the execution is paused. Despite the name, this button does not close PythonWin nor any of the scripts. It only closes (or stops) the debugging session.

The 'Go' button can be used together with breakpoints to fast-forward the debugger to areas of code that need close inspection. The following code for performing frequency analysis on string fields has an error that is difficult to pinpoint without debugging:

```
# buggyFreq.py
# Purpose: Find frequency of each value in each string field.

import arcpy
arcpy.env.overwriteOutput = True
arcpy.env.workspace = 'C:/gispy/data/ch13/smallTest.gdb'
featureList = arcpy.ListFeatureClasses()

for inputFile in featureList:
    fields = arcpy.ListFields(inputFile, '*', 'String')
    for field in fields:
        fieldName = field.name
    outTable = inputFile + fieldName + 'freq'
    arcpy.Frequency_analysis(inputFile, outTable, fieldName)
    print 'Output table created: {0}'.format(outTable)
```

A frequency table tallies the number of times each value occurs in a field. Running this script creates two frequency tables, one for the 'cover' feature class ('coverCOVERfreq') and one for the 'fires' feature class ('firesAuthoriz_1 freq'):

- smallTest.gdb
  - cover
  - coverCOVERfreq
  - fires
  - firesAuthoriz_1freq
  - park

But then it crashes with the following error:

```
ERROR 000728: Field Authoriz_1 does not exist within table
Failed to execute (Frequency).
```

This code has an indentation error that is causing some lines of codes to be executed at the wrong time. It's often useful to put breakpoints inside loops or conditional constructs. To see the error in this script, place a breakpoint on line 11: Move the cursor to line 11 (any position on the line will do) and then click the 'Toggle breakpoint' button. A breakpoint circle appears in the margin:

```
 9  - for inputFile in featureList:
10          fields = arcpy.ListFields(inputFile, '*', 'String')
11 ○-      for field in fields:
12              fieldName = field.name
13          outTable = inputFile + fieldName + 'freq'
```

Use the 'Go' button ▶ to run the code to the breakpoint. The cursor stops on line 11 and won't execute this line of code until debugging is resumed:

```
 9  - for inputFile in featureList:
10          fields = arcpy.ListFields(inputFile, '*', 'String')
11 ◑-      for field in fields:
12              fieldName = field.name
13          outTable = inputFile + fieldName + 'freq'
```

> **Tip** Set watches on input variables.

Display the watch panel 🕶 and set watches for the inputFile, fieldName, and outputTable variables. You can also shift the Watch window from its default position on the left to stretch along the bottom of PythonWin. Now resume debugging by the stepping over line 11 ⬓. Continue stepping over, one line at a time, and watch the values change until the script stops. You'll first see the frequency table created for the 'COVER' field in the 'cover' feature class.

The next feature class, 'fires', has three string fields 'FireName', 'SizeClass', and 'Authoriz_1'. A frequency table is only created for the last one because of the indentation error. Lines 13–15 should be indented to be inside of the field loop. Instead, the only field name that reaches the Frequency (Analysis) tool is the last field name assigned before leaving the loop.

The next feature class is 'park', which has no string fields. Regardless of this, the Frequency (Analysis) tool is called using the most recent field name ('Authoriz_1'). Since 'park' has no such field, an error occurs. When the error is thrown, the debugging session curtails.

```
 9  - for inputFile in featureList:
10          fields = arcpy.ListFields(inputFile, '*', 'String')
11 ●-      for field in fields:
12              fieldName = field.name
13          outTable = inputFile + fieldName + 'freq'
14 ▷ |      arcpy.Frequency_analysis(inputFile, outTable, fieldName)
15          print 'Output table created: {0}'.format(outTable)
```

Expression	Value
inputFile	u'park'
fieldName	u'Authoriz_1'
outTable	u'parkAuthoriz_1freq'

The second button in the breakpoint cluster is called 'Breakpoint list' ●. This button toggles a panel with a list of breakpoints and allows you to specify breaking conditions. The Breakpoint List panel, like the Watch panel, is only available while a debugging session is in progress.

Tip Set conditional breakpoints on iterating variables and use values from traceback messages.

Using the breakpoint on line 11 of 'buggyfreq.py' improved efficiency by running the code directly to the section that was causing the error. This could be refined even further. Based on the error thrown by 'buggyfreq.py', the trouble only began when the field name became 'Authoriz_1'. To specify this condition, commence a new debugging session by clicking the 'Step over' button ⌐≡ and then click the 'Breakpoint list' button ●. The panel shows a list of breakpoint locations. (Though this example only uses one breakpoint, multiple breakpoints can be set.) Click in the 'Condition' column next to the breakpoint 'buggyfreq.py:11'. Set the condition to check when the field name is 'Authoriz_1'. Conditions are Python expressions, so two equals signs are used to check for equality.

Condition	Location
fieldName == 'Authoriz_1'	buggyfreq.py: 11

Now run the condition to this breakpoint, using the 'Go' button and note that it runs past the fields 'cover', 'FireName', and 'SizeClass' without stopping. The field name is now 'Authoriz_1'. Click 'Go' again and execution stops again, because the code reaches this point again with the field name of 'Authoriz_1'. Setting conditional breakpoints can be very useful when dealing with a high number of repetitions in a loop.

The third button in the breakpoint cluster removes all the breakpoints 🖐. Note that the Toggle breakpoint button 🖐 removes a single breakpoint on the line where the cursor resides. If you intend to only remove one break point, use the toggle breakpoint button. The symbols on these two buttons might lead one to believe that they are exact opposites, but they're not. Set several more breakpoints in the script and then experiment to see the difference.

Now that we're familiar with the buttons, well resolve the 'buggyfreq.py' example. To repair this script, indent lines 13–15. Indent multiple lines of code efficiently by selecting the lines and then pressing 'Tab'. Then rerun it to create four tables without throwing errors:

```
>>> Output table created: coverCOVERfreq
Output table created: firesFireNamefreq
Output table created: firesSizeClassfreq
Output table created: firesAuthoriz_1freq
```

> **Tip** Set breakpoints inside loops, conditional constructs, and other structures that control the flow.

Debugging is useful not only for finding bugs, but for understanding the flow of a working script. It allows you to methodically step through and watch variables to discover how code execution proceeds. In summary, this example introduced three additional debugging techniques:

- **Setting breakpoints**: Place your cursor on a line of code to set a breakpoint for that line. A breakpoint allows you to run to that point in the code without stepping through all the previous lines of code one by one. Before you run the code, a breakpoint appears as a circle in the margin next to the line. When you start running in debug mode, it turns pink.
- **Running to breakpoints:** Start the script running in debug mode with the 'Go' button to run to the first breakpoint. If there is more than one break point, it will run to the next one when selected again. If there are no breakpoints, it will run through the entire script non-stop.
- **Setting breakpoint conditions:** Specify stopping conditions for a breakpoint so that it runs directly to the iteration that is causing trouble.

13.5 Running in Debug Mode

So far we have started a debugging session in PythonWin by either selecting 'Step over' or 'Go'. You can also start a debugging session via the 'Run Scripts' dialog box. Your preference may vary depending on the bugs, but here's a list of the options:

Option 1: Make the debugging toolbar viewable. Click 'Step over' (or F10) and start stepping through the code.

Option 2: Make the debugging toolbar viewable. Set breakpoints and click 'Go' (or F5).

Option 3: Click the 'Run' button 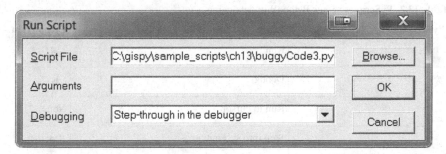 and choose 'Step-through in the debugger' on the 'Run Script' form.

The debugging toolbar will automatically appear. Otherwise, this achieves essentially the same effect as option #1. Whichever mode you choose in the combo box remains until you choose another mode or restart PythonWin. In other words, to stop running in debug mode you'll have to choose 'No debugging' in this combo box.

Option 4: Click the 'Run' button 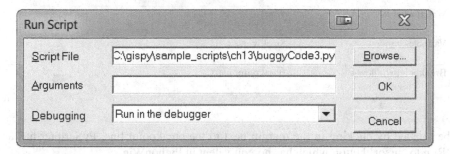 and choose 'Run in the debugger' on the 'Run Script' form. This achieves the same effect as option #2, but it's persistent.

There is only one button for stopping a debugging session, but there are several ways in which a session can end:

1. The 'Close' button (or Shift + F5) is selected.
2. The code throws an exception.
3. The code runs to completion.

13.6   PyScripter Debugging Toolbar

The toolbars in most IDEs have very similar buttons and functionality. As an example, Figure 13.2 shows the PyScripter debugging toolbar (visible by default). Table 13.2 lists the PyScripter buttons, their PyScripter names, and their equivalents in PythonWin. PyScripter has a 'Toggle breakpoint' button, but breakpoints can also

Figure 13.2 PyScripter toolbar.

Table 13.2 PyScripter versus PythonWin debugging.

PyScripter	Name	PythonWin
▷	Run	
	Debug	▶
▷I	Run to cursor	None
	Step into subroutine	
	Step over next function call	
	Step out of the current subroutine	
	Abort debugging	
	Toggle breakpoint	
	Clear all breakpoints	
Watch window	Watch	
Variables window	Stack view	
Breakpoint window	Breakpoint list	

be toggled by clicking in the margin next to the breakpoint line. PyScripter has a
'Run to cursor' button which has no equivalent in PythonWin.

PythonWin provides the Watch , Stack view , and Breakpoint list
buttons on the debugger toolbar to toggle the corresponding panels open/closed during debugging sessions.

PyScripter provides this functionality with windows that are open by default.
The 'Watches', 'Variables', and 'Breakpoints' window tabs appear along the bottom
of the application:

Watches	Type	Value
● fc	str	'park.shp'
fcs	n/a	n/a

🔍 Watches | 🔍 Variables | ● Breakpoints | Output | Python Interpreter

13.7 Debugging Tips

Debugging skills improve with experience. Here are a few tricks to keep in mind as you get started.

1. If the program is processing data files, use a very simple test file while you're debugging your code. If you don't have one already, create one. For example, if your data is a 1000 by 3000 table, create a 5 by 8 table by using the first few values in the larger table. Run your code on this until you think it's working. Then try it on the larger file.

2. Vary your test input. Code that is functioning correctly for one input, may not be working for another. It's impossible to test every scenario, but here are a few rules of thumb.

   (a) For complex code that takes data as input, test using multiple input files.
   (b) For code that hinges on conditional structures, use input that tests each of the branches in the conditional blocks.

3. Place breakpoints inside loops and conditional constructs. These structures are powerful, but at the same time, they can lead to confusion.

4. Watch variables where they are (supposed to be) changing, by inserting breakpoints or print statements in the code.

5. When working with ArcGIS data, beware of ArcCatalog locks that may throw exceptions and prevent a script from overwriting existing data.

6. Remember that you can 'Break into running code' in PythonWin (or Ctrl+F2 in PyScripter), if the script gets into an infinite loop as described in the Section 10.1 "How to Interrupt Running Code" instructions.

13.8 Key Terms

Bugs	Debugging toolbars
Syntax errors	Debug mode
Parsing	Watches
Exceptions	Breakpoints
Logic errors	Breakpoint conditions
Debugger	

13.9 Exercises

1. Use the sample script 'numGames.py' to practice using breakpoints.
   (a) Set breakpoints on lines 5, 7, 11, and 13 (just inside the loop and at each print statement).

(b) Set the argument to the number 6 and select 'Run in the debugger' in the 'Run script' dialog.

(c) Add x to the Watch window.

(d) Investigate the script's behavior by clicking 'Go' to run from each breakpoint to the next until the script exits (You'll see the script name 'returned exit code 0' in the feedback bar).

(e) Set the argument to the number −5 and run in the debugger.

(f) Remove the breakpoint on line 5 by clicking the 'Toggle breakpoint' button while the arrow is on line 5.

(g) Select 'Go' four more times. Take a screen shot of the IDE that shows the script, the watch window, the breakpoints, and the arrow where it is now.

(h) Remove all the breakpoints by clicking the 'Clear breakpoints' button.

(i) Start it running again with the 'Go' button.

(j) Stop the infinite loop by clicking on the PythonWin icon in the applications tray and selecting 'Break into running code'. (When this succeeds, you will see a `KeyboardInterrupt` exception in the Interactive Window)

(k) Set breakpoints on lines 5, 7, 11, and 13 again and experiment with various numerical input values. Run to the breakpoints to observe the behavior and determine solutions to the following:

   i. Find an example of an input integer that triggers exactly three print statements.

   ii. Find an example of an input integer that triggers exactly one print statement.

   iii. Find an example of an input integer that triggers exactly two print statements.

   iv. Define the set of numbers that will result in an infinite loop.

   v. Define the set of integers that will result in no print statements.

2. **buggyMerge.py** The sample script 'buggyMerge.py' should merge all the independent dBASE tables in a directory, but it contains a logic error. It works correctly the first time it's run, but if you run it again the merge fails. Set a breakpoint on line 10, use the watch window to watch the `tables` variable, set a breakpoint inside the conditional statement on line 15, and use the debugger to discover and repair the error (so that it can be rerun without breaking).

3. **buggyConditional.py** The sample script 'buggyConditional.py' contains three bugs. Read the following description of how it is supposed to work, then repair the errors, using the debugger.

   • The script takes five arguments:

     One feature class: `arcpy.GetParameterAsText(0)`
     Three numeric values: (Z, AF, T) `arcpy.GetParameterAsText(1), (2), & (3)`
     One output field name: `arcpy.GetParameterAsText(4)`

- The output field is added, if it doesn't already exist.
- The conditional constructs calculate the output field value as follows:
  If the first condition is met, the output value is the sum of T, Z, and AF.
  If the second condition is met, the output value is the sum of T and Z.
  If the third condition is met, the output value is the sum of T and AF.
  If none are met, the output value is set to T.
- Then the output field is set to the output value using the Calculate Field (Data Management) tool.

In the Watch window, set watches on the input variables and set a breakpoint inside each conditional branch (lines 17, 20, 22, 24, 26). The following sample input should create a field named 'veld' with the value 9 in each row:

C:/gispy/data/ch13/tester.gdb/c1 -2 3 6 veld

Also be sure to test with positive and negative value combinations for Z and AF.

Chapter 14
Error Handling

Abstract Data corruption and locking can cause geoprocessing scripts to crash and throw exceptions. Other influences, such as user input values, can also cause scripts to crash. The topic of this chapter is using error handling structures to control script behavior when exceptions occur. Error handling can suppress those alarming trace-back messages that exceptions throw and provide a smoother way to proceed.

Chapter Objectives

After reading this chapter, you'll be able to do the following:

- Handle potential errors from user input.
- Use error handling keywords, `try` and `except`.
- Anticipate named exceptions.
- Print geoprocessing messages.
- Avoid crashing in the midst of a loop.
- Identify when to use error handling instead of conditional blocks.

Chapter 13 discussed syntax errors, exceptions, and logic errors and how to debug logic. The focus of debugging is to remove errors in the code. The programmer needs to remove syntax errors and test the code to expose and remove logic errors. Exception errors, however, can come in two forms. Some exceptions are due to errors in the code; Others are due to outside factors, such as user input.

In the first case, the programmer needs to repair them. In the second case, the programmer should write code to anticipate and handle the error. This chapter presents Python structures for handling exceptions. To clarify the difference between exceptions that should be repaired and exceptions that should be handled, take the following map algebra script as an example:

```python
# multRast.py
import arcpy, sys

# Set multiplication factor
factor = float(sys.argv[1])

arcpy.env.overwriteOutput = True
arcpy.env.worspace = 'C:/gispy/data/ch14/rastTester.gdb'
arcpy.CheckOutExtension('Spatial')

# Get raster list & multiply each by a factor.
rasterList = arcpy.ListRasters()
for rasterImage in rasterList:
    rasterObj = arcpy.Raster(rasterImage)
    rastMult = rasterObj * factor
    rastMult.save(rasterImage + '_out')
del rastMult
```

'multRast.py' should apply a constant multiplier to all the rasters in the directory, but it throws the following exception:

```
for rasterImage in rasterList:
TypeError: 'NoneType' object is not iterable
```

This error occurs because of a sneaky typo that caused a series of cascading events leading to this exception. Have you spotted it? The problem occurs because 'workspace' is spelled wrong in the following line of code:

```
arcpy.env.worspace = 'C:/gispy/data/ch14/rastTester.gdb'
```

The script didn't give an error, when it reached this line. It just silently created a new (useless) variable `arcpy.env.worspace`. Since the value of `arcpy.env.workspace` was not set by this script, the default value—empty string—is used for the `arcpy` workspace setting.

When `ListRasters` is called without setting the workspace, it returns the Python built-in constant `None`, meaning the `rasterList` variable is set to `None`. Then the script tries to loop over the raster list, which is when the exception is finally thrown ('NoneType' object is not iterable).

This exception must be avoided by correcting the typo. However, other exceptions could occur due to user input. This script is designed to take a numerical argument. But suppose users are dealing with rasters representing elevation in meters. They might try to run the script with an argument such as "0.001 meters". Given this input, the script throws the following exception:

```
factor = float(sys.argv[1])
ValueError: invalid literal for float(): 0.001 meters
```

Python can't cast the input to a float, since it contains non-numeric components. This is the other flavor of exception—it is not occurring because of an error in the code. It is due to an outside input that is beyond the control of the programmer. In this example, the script stops running abruptly when the exception is thrown. However, the script can be modified to handle this exception more gracefully, that is, it can be *caught*. This chapter starts with some simple scripts to introduce Python error handling syntax and then demonstrates geoprocessing script applications.

14.1   try/except Structures

Exceptions can be caught by using Python `try` and `except` keywords to group lines of code into blocks. The code that could generate the exception is placed within the `try` block and contingency code (what to do in case of trouble) can be placed inside an `except` block. `try` and `except` blocks require colons and indentation, just like other Python code blocks such as conditional blocks and looping blocks. When the exceptions are caught, no traceback messages are printed in the Interactive Window (unless the code is run in debug mode); Instead the script can print a more intelligible message or it can perform some alternative action. For example, when a user passes "5 meters" as an argument into the following script, they receive a clear message and the code exits gracefully:

```
# doubleMyNumber.py
import sys

try:
    number = float(sys.argv[1])
    product = 2*number
    print 'The doubled number is {0}'.format(product)
except:
    print 'An error occurred.'
    print 'Please enter a numerical argument.'
print 'Good bye!'
```

Here's how it works:

- If an exception occurs in the `try` block, the rest of the `try` block is skipped and execution jumps to the `except` block. For example, if the input is "5 meters", the script prints the following:

```
>>> An error occurred.
Please enter a numerical argument.
Good bye!
```

- If no exceptions occur, the except block is skipped. For example, if the input is 5, the script prints the following:

```
The doubled number is 10.0
Good bye!
```

When any exception is thrown, the execution moves to the `except` block. This means the same action is taken regardless of whether the input is "5 meters" or the user does not supply an argument. The behavior in case of exceptions can be controlled more precisely by using named exceptions.

14.1.1   Using Named Exceptions

Traceback errors report the name of the exception, such as a `TypeError` or `ValueError`. Exceptions names can be used with `except` blocks to provide special handling for specific errors by placing the name of the exception behind the `except` keyword. For example, the following code catches a `ValueError`:

```
# doubleMyNumberV2.py
import sys

try:
    number = float(sys.argv[1])
    product = 2*number
    print 'The doubled number is {0}'.format(product)
except ValueError:
    print 'Input must be numerical.'
print 'Good bye!'
```

This code catches ValueError exceptions. If the user enters "5 meters", the script prints:

```
>>> Input must be numerical. Good bye!
```

Scripts use named exceptions to anticipate particular vulnerabilities. To do so, the programmer needs to determine which name to use. The 'Python Library Reference' can be used to look up built-in exceptions; But creating code samples that cause an exception is also a good approach. For example, the following code throws a `ZeroDivisionError` exception:

```
>>> 1/0
Traceback (most recent call last):
    File "<interactive input>", line 1, in <module>
ZeroDivisionError: integer division or modulo by zero
```

And the following code throws a `ValueError` exception:

```
>>> float('5 meters')
Traceback (most recent call last):
    File "<interactive input>", line 1, in <module>
ValueError: invalid literal for float(): 5 meters
```

The examples demonstrate that a `ZeroDivisionError` should be used as the named exception to catch errors where division by zero might occur and a `ValueError` should be used as the named exception to catch errors generated by calling the built-in `float` function on a non-numeric input.

The `except ValueError` statement in 'doubleMyNumbersV2.py' handles `ValueError` exceptions only. Exceptions with other names will not be caught in this `except` block. For example, if the user does not supply an argument, the script prints an `IndexError` when it tries to read the user argument:

```
number = float(sys.argv[1])
IndexError: list index out of range
```

To handle more than one type of named error, a script can use multiple `except` blocks.

14.1.2   Multiple except Blocks

When named errors are used, a `try` block can have multiple `except` blocks. Then, if an error occurs, the execution jumps to the corresponding named exception block. For example, the following script uses two named `except` blocks:

```
# slopeTry.py
import sys

rise = sys.argv[1]
run = sys.argv[2]
```

```
try:
    print 'Rise: {0} Run: {1}'.format(rise, run)
    slope = float(rise)/float(run)
    print 'Slope = rise/run'
except ZeroDivisionError:
    slope = 'Undefined (line is vertical)'
except ValueError:
    print 'Usage: <numeric rise> <numeric run>'
    slope = 'Not found'

print 'Slope:', slope
```

Here's how this one works:

- If no exceptions occur, the `try` block is completed and both the except blocks are skipped. For example, if the input is 1 3, the script prints the following:

```
>>> Rise: 1 Run: 3
Slope = rise/run
Slope: 0.333333333333
```

- If a `ZeroDivisionError` occurs in the `try` block, the rest of the `try` block is skipped and execution jumps to the `except ZeroDivisionError` block. For example, if the input is 5 0, the script prints the following:

```
>>> Rise: 5 Run: 0
Slope: Undefined (line is vertical)
```

- The script prints the statement before the division and then when the division fails, the `ZeroDivisionError` block sets the slope to 'Undefined (line is vertical)'.
- If a `ValueError` occurs in the `try` block, the rest of the `try` block is skipped and execution jumps to the `except ValueError` block. For example, if the input is one three, the script prints the following:

```
>>> Rise: one Run: three
Usage: <numeric rise> <numeric run>
Slope: Not found
```

- The script prints the statement before the division and then when the conversion to float fails, execution jumps to the `ValueError` block and prints script usage instructions and sets the slope to 'Not found'.

'slopeTry.py' handles three cases: successful slope computation, division by zero, and failure to cast to float. The last line of code prints the `slope` variable. Since the exceptions were caught, this line is executed in all three cases, so the script contains an assignment statement for `slope` in all three blocks.

'slopeTry.py' demonstrates using two named exception blocks. In fact, many other variations are possible as long as they comply with the following syntax rules:

- Any number of named `except` blocks can be used with one `try` block.
- At most one unnamed `except` block can be used with a single `try` block.
- An unnamed `except` block can be used along with named `except` blocks.
- When an unnamed `except` block is used with named `except` blocks, the unnamed except is placed last.
- A `try` block needs at least one `except` block (or a `finally` block, not discussed here, can be used as an alternative to an `except` block).

14.1.3   Error Handling Gotcha

Catching exceptions can shield users from encountering mysterious traceback messages. However, suppressing the Traceback messages can also potentially lead to pitfalls during code development. For example, the try/except structure in the following code hides an error that should be corrected by the programmer:

```
# cubeMyNumber.py
import sys
try:
    number = float(sys.argv[1])
    cube = number**3
    print 'The cubed number is {0}'.format()  # missing argument
except:
    print 'Input must be numerical.'
print 'Good bye!'
```

This script takes one numeric argument; however, even if the user provides a valid argument, an exception is raised. For example, if the input is 5, the script prints the following:

```
>>> Input must be numerical argument.
Good bye!
```

Something is wrong with the code, but in this case, suppressing the traceback message makes the problem more difficult to find. Can you find what is really wrong with the code? The message in the exception block anticipates that the user enters a non-numeric argument, but some other exception is causing the code to jump to the `except` block and print the 'numerical argument' message which doesn't help to uncover the problem. The Python traceback module can be used during code development to overcome this glitch.

The Python built-in `traceback` module can force traceback messages to print. The traceback `print_exc` method prints the most recent exception. To use the `print_exc` traceback method, import the `traceback` module and then call the `print_exc` method inside the except clause as in the following code:

```
# cubeMyNumberV2.py
import sys, traceback
try:
    number = float(sys.argv[1])
    cube = number**3
    print 'The cubed number is {0}'.format()
except:
    print 'Input must be numerical.'
    traceback.print_exc()
print 'Good bye!'
```

If this script is run with a valid argument, the raised exception is printed. For example, if the input is 5, the script prints the following:

```
Input must be numerical.
Traceback (most recent call last):
File 'C:\\example_scripts\cubeMyNumberV2.py', line 6, in <module>
print 'The doubled number is {0}'.format() #missing arg
IndexError: tuple index out of range
Good bye!
```

This traceback helps us to see that the string `format` is called incorrectly; it is missing a needed argument and throws an `IndexError` exception. It is true that an error occurred, but this is not the kind of error that should be handled with try/except blocks in the script. It should instead be repaired by the programmer. Line 6 should be corrected as follows:

```
print 'The cubed number is {0}'.format(cube)
```

This 'gotcha' is less likely to occur when using named exceptions, although, as 'cubeMyNumber.py' demonstrates, confusion can still occur.

> **Note** The amount of code placed inside an exception handling block should be minimal so that other exceptions are not masked.

14.2   Geoprocessing and Error Handling

Now that you know the basic approach for Python error handling, you can apply it to geoprocessing. This section describes how to get feedback from geoprocessing calls, how to use a named exception to catch geoprocessing errors, and how to use these components within a batch geoprocessing script.

14.2.1   *Getting Geoprocessing Messages*

When a tool is run from ArcToolbox, a Geoprocessing Window is opened. This window shows progress as the tool works and when it stops running, it displays a success or failure message, as in the following screen shot of a successful Get Count (Data Management) tool run:

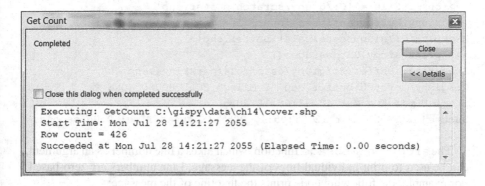

When geoprocessing tools are called by a Python script, a success or failure message is also generated. The `arcpy` package provides functions (`GetMessages`, `GetMessage`, and `GetMessageCount`) for accessing these messages. The `arcpy GetMessages` function returns the message from the most recent tool call. If no tools have been called, it returns an empty string.

```
>>> arcpy.GetMessages()
''
```

No arguments are required and it is usually used with a print statement. The following code calls the Get Count (Data Management) tool and the `GetMessages` function:

```
>>> inputFile = 'C:/gispy/data/ch14/cover.shp'
>>> count = arcpy.GetCount_management(inputFile)
>>> print arcpy.GetMessages()
Executing: GetCount C:/gispy/data/ch14/cover.shp
Start Time: Mon Jul 28 20:30:10 2055
Row Count = 426
Succeeded at Mon Jul 28 20:30:10 2055 (Elapsed Time: 0.00 seconds)
```

This message contains the same information printed in the Geoprocessing Window when the tool is called from ArcToolbox. It reports the tool name and arguments, when the tool was called, any values it returns (e.g., Row Count = 426), and successful completion.

The first line of the message contains a list of all arguments used to call the function (both those specified explicitly and the default values used during the run). In the following code, only the three required arguments are used to call the Buffer tool, but the message also shows the default values that were used for the optional arguments:

```
>>> inputFile = 'C:/gispy/data/ch14/parkLines.shp'
>>> outputFile = 'C:/gispy/scratch/buffer.shp'
>>> arcpy.Buffer_analysis(inputFile, outputFile, '1 mile')
<Result 'C:/gispy/data/ch14\\buffer.shp'>
>>> print arcpy.GetMessages()
Executing: Buffer C:/gispy/data/ch14/parkLines.shp
C:/gispy/scratch\buffer.shp "1 Miles" FULL ROUND NONE #
Start Time: Tue Mar 12 21:05:10 2055
Succeeded at Tue Mar 12 21:05:10 2055 (Elapsed Time: 0.00 seconds)
```

The arcpy GetMessage function, which takes a line number as an argument, can be used to print individual lines of the message. Line numbers are zero-based. For example, the following code prints the first line of the message:

```
>>> print arcpy.GetMessage(0)
Executing: Buffer C:/gispy/data/ch14/park.shp C:/gispy/scratch\
buffer.shp "1 Miles" FULL ROUND NONE #
```

The message line count varies depending on the geoprocessing tool. The GetMessageCount function returns the line count:

```
>>> arcpy.GetMessageCount()
3
```

The line count minus one is the last valid index. The following code prints the last line of the message:

```
>>> print arcpy.GetMessage (arcpy.GetMessageCount() - 1)
Succeeded at Tue Mar 12 21:05:10 2055 (Elapsed Time: 0.00 seconds)
```

If the most recent tool call failed, the message reports failure. For example, the following code calls the Get Count tool with no arguments, which throws an ExecuteError exception:

```
>>> count = arcpy.GetCount_management()
Traceback (most recent call last):
File "<interactive input>", line 1, in <module>
```

```
File "C:\Program Files (x86)\ArcGIS\Desktop10.1\arcpy\arcpy\
management.py",
line 13637, in GetCount
raise e
ExecuteError: Failed to execute. Parameters are not valid.
```

This time the GetMessages function returns a failure message:

```
>>> print arcpy.GetMessages()
Executing: GetCount #
Start Time: Tue Mar 12 20:43:25 2055
Failed to execute. Parameters are not valid.
ERROR 000735: Input Rows: Value is required
Failed to execute (GetCount).
Failed at Tue Mar 12 20:43:25 2055 (Elapsed Time: 0.00 seconds)
```

The failure message generated by GetMessages is usually easier to interpret than the traceback message thrown by failed geoprocessing tools. For example, in addition to stating that the parameters are not valid, the message from the failed Get Count tool run states that a value is required for the input rows. Messages like this can be used along with error handling to provide informative feedback in geoprocessing scripts.

14.2.2   The arcpy Named Exception

When arcpy geoprocessing tool calls fail, they throw an ExecuteError. This is not one of Python's built-in exceptions; Instead, this is a special exception thrown by arcpy. Since it's an arcpy property, it can be referred to with dot notation as arcpy.ExecuteError and it can be used in a named except block as in Example 14.1.

Example 14.1

```
# bufferTry.py
import arcpy, sys, os
arcpy.env.overwriteOutput = True
try:
    inputFile = sys.argv[1]
    buff = os.path.splitext(inputFile)[0] + 'Buff.shp'
    arcpy.Buffer_analysis(inputFile, buff, '1 mile')
    print '{0} created.'.format(buff)
except arcpy.ExecuteError:
    print arcpy.GetMessages()
except IndexError:
    print 'Usage: <full path shapefile name>
```

In case a geoprocessing tool fails, the tool failure information can be printed by calling GetMessages in the arcpy.ExecuteError exception block. For example, if "C:/gispy/scratch/bogus.shp" (a non-existent file) is used as input, the Buffer tool call throws an exception. The code execution jumps to the arcpy.ExecuteError block and the following message is printed:

```
Executing: Buffer C:/bogus.shp C:\bogusBuff.shp "1 Miles" FULL
ROUND NONE #
Start Time: Tue Mar 12 22:33:06 2055
Failed to execute. Parameters are not valid.
ERROR 000732: Input Features: Dataset C:/gispy/scratch/bogus.shp
does not exist or is not supported
Failed to execute (Buffer).
Failed at Tue Mar 12 22:33:06 2055 (Elapsed Time: 0.00 seconds)
```

14.3   Catching Exceptions in Loops

When a script performs geoprocessing on a batch of files, unhandled errors will halt the process. For example, if a Buffer tool call fails on a file in a batch of hundreds, the files that come after that one in the list won't be processed, even if they are valid. One bad apple can spoil the bunch. A main advantage of using error handling in scripts is so that batch processing can continue even if a tool call fails for one file.

try/except blocks should both go inside the loop with the try block wrapped around the geoprocessing calls as in Example 14.2. This script catches geoprocessing tool errors by using a named exception.

Example 14.2

```
# bufferLoopTry.py
# Purpose: Buffer the feature classes in a workspace.
# Usage: No arguments needed.
import arcpy, os
arcpy.env.overwriteOutput = True
arcpy.env.workspace = 'C:/gispy/data/ch14'
outDir = 'C:/gispy/scratch/'
fcs = arcpy.ListFeatureClasses()
distance = '500 meters'

for fc in fcs:
    outFile = outDir + os.path.splitext(fc)[0] + 'Buff.shp'
    try:
        arcpy.Buffer_analysis(fc, outFile, distance)
```

```
    print 'Created: {0}'.format(outFile)
except arcpy.ExecuteError:
    print arcpy.GetMessage(2)
```

A geoprocessing tool call can fail for any number of reasons. Data might be locked or corrupted (e.g., a required shapefile support file is missing). Figure 14.1 shows an invalid shapefile 'dummyFile.shp' as it appears in ArcCatalog.

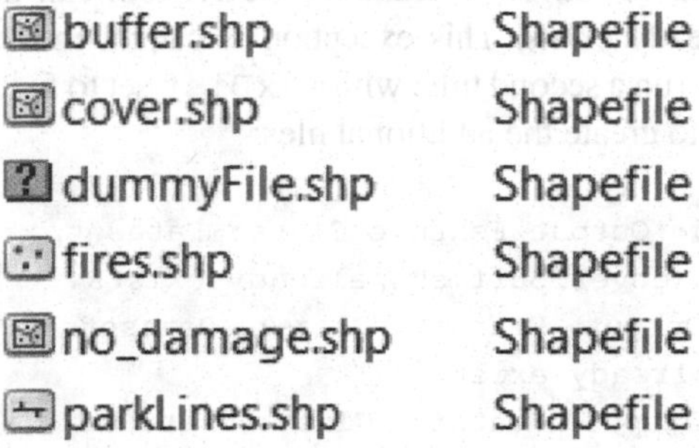

buffer.shp	Shapefile
cover.shp	Shapefile
dummyFile.shp	Shapefile
fires.shp	Shapefile
no_damage.shp	Shapefile
parkLines.shp	Shapefile

Figure 14.1   The ArcCatalog view of the shapefiles in the Chapter 14 data directory.

This file was created by renaming a text file to have an '.shp' extension. The following results are printed when Example 14.2 is run on these files:

```
>>> Created: buffer_buff.shp
Created: coverBuff.shp
ERROR 000229: Cannot open C:/gispy/data/ch14\dummyFile.shp.
Created: firesBuff.shp
Created: no_damageBuff.shp
Created: parkLinesBuff.shp
```

Every file is processed, except 'dummyFile.shp' and an error message is reported. Good programming style only places code inside a loop that needs to go inside the loop. In Example 14.2, the output file name must be updated inside the loop, but the distance remains constant. Along the same lines, the code that goes inside a try block should be minimized. In Example 14.2, the Buffer tool call must be inside of the try block, but the output file name can be set outside the try block.

Another aspect of arranging code comes into play when the batch processing occurs in a WHILE-loop. Since try/except blocks can cause the flow of the code to change, it's important to update the iterating variable at the end of the loop outside of both the try and except blocks. Take Example 14.3, which creates buffers at various distances, up to the maximum specified number of miles. An input of "C:/gispy/data/ch14/cover.shp" and "3" generates 1 mi., 2 mi., and 3 mi. buffer output files and prints the following output:

```
>>> Created: C:/gispy/scratch/cover1Buff.shp
Created: C:/gispy/scratch/cover2Buff.shp
Created: C:/gispy/scratch/cover3Buff.shp
```

The `overwriteOutput` property is set to False so that the script only creates files that don't already exist. This means the Buffer tool call throws an exception when an output already exists. This exception is caught in the named exception block. If the script is run a second time with `maxDist` set to 5, it prints three exceptions and moves on to create the additional files:

```
>>> ERROR 000725: Output Feature Class: Dataset
C:\gispy\scratch\cover1Buff.shp already exists.
ERROR 000725: Output Feature Class: Dataset C:\gispy\scratch\
cover2Buff.shp already exists.
ERROR 000725: Output Feature Class: Dataset C:\gispy\scratch\
cover3Buff.shp already exists.
Created: cover4Buff.shp
Created: cover5Buff.shp
```

The iterating variable is updated on the last line of code. Notice that this line is inside the `WHILE`-loop, but outside the `except` block. Placing it outside the `WHILE`-loop would cause infinite looping with the iterating variable stuck at 1; Placing it inside the `except` block would lead to infinite looping with the iterating variable stuck at the value of the first iteration where the Buffer tool call does not throw an error.

Example 14.3

```
# bufferLoopDistTry.py
# Purpose: Buffer the input file by the given distance.
# Usage: input_filename numeric_distance
# Example input: C:/gispy/data/ch14/cover.shp 3

import arcpy, sys, os
arcpy.env.workspace = os.path.dirname(sys.argv[1])
fc = os.path.basename(sys.argv[1])
outDir = 'C:/gispy/scratch/'
arcpy.env.overwriteOutput = False
maxDist = float(sys.argv[2])
i = 1
```

```
while i <= maxDist:
    try:
        outFile = outDir + os.path.splitext(fc)[0] + str(i) + \
                'Buff.shp'
        distance = str(i) + ' miles'
        arcpy.Buffer_analysis(fc, outFile, distance)
        print 'Created: ', outFile
    except arcpy.ExecuteError:
        print arcpy.GetMessage(3)
    i = i + 1
```

14.4 Discussion

Exceptions come in two flavors, ones which are caused by errors in the code and ones which are caused by some outside influence, such as input data. The first kind should always be resolved by the programmer; The second kind resides in a somewhat gray area. The need for error handling depends on the context. For example, if a script that requires numerical input is being run via a graphical user interface which limits the user input to numerical values, there's no need to handle a ValueError for that input. Some situations also require a decision between using conditional constructs and error handling.

Conditional blocks have certain similarities to try/except blocks. With both of these structures, the execution can be diverted depending on conditions. In some situations, conditional blocks could be used to achieve the same effect as try/except blocks. The try/except blocks should be used to handle exceptional cases—when something has gone wrong. When, on the other hand, both IF and ELSE alternatives are normal acceptable behavior, conditional blocks are more apt. For example, the following code checks if an argument has been provided and if not, it sets a default value:

```
import sys, arcpy
if len(sys.argv) > 1:
    arcpy.env.workspace = sys.argv[1]
else:
    arcpy.env.workspace = 'C:/gispy/data/ch14'
for rast in arcpy.ListRasters():
    print rast
```

The program is designed so that the user can provide an argument, but doesn't strictly need to provide an argument. The program continues with business as usual without arguments, taking the default value for the input workspace instead of an alternative provided by the user. The same result could be accomplished with a try/except block, as in the following code, which tries to access a user argument and catches an `IndexError`:

```
import sys, arcpy
try:
    arcpy.env.workspace = sys.argv[1]
except IndexError:
    arcpy.env.workspace='C:/gispy/data/ch14'
for rast in arcpy.ListRasters():
    print rast
```

In this case, though, conditional blocks are preferable. Nothing has gone wrong; the user has simply chosen to accept the default workspace.

In other scripts, an argument may be required. In this case, running the script without an argument would be considered an error. In fact, the script might need to exit without continuing. The following code uses try/except blocks more appropriately:

```
import sys, arcpy
try:
    arcpy.env.workspace = sys.argv[1]
except IndexError:
    print 'Usage: <input workspace>'
    sys.exit(0)
for rast in arcpy.ListRasters():
    print rast
```

The exception handling block prints instructions for using the script and then forces the script to exit with the `sys` module `exit` method (`sys.exit(0)`); This prevents the code from attempting to list the rasters on an unspecified workspace.

Note try/except blocks should be used judiciously based on desired functionality and expected usage.

Geoprocessing usually involves input data and since the quality of the data can't be guaranteed in most real-world projects, appropriate error handling is an important component of making scripts more reliable and robust.

14.5 Key Terms

try and except blocks   The arcpy.ExecuteError exception
Named exceptions   arcpy.GetMessage(index)
arcpy.GetMessages()   arcpy.GetMessageCount()
Catch exceptions

14.6 Exercises

1. **simplifyOops** Explore two common try/except mistakes as follows:

    (a) The purpose of sample script 'simplifyOops1.py' is to simplify polygons for all valid data within the input directory, using a FOR-loop with error handling. But it has some mistakes. Open the script and view the input directory to predict the output it will create.
    (b) Run 'simplifyOops1.py' with no debugging.
    (c) Run the script again using the 'step through in debugger' option.
    (d) Ideally, we want the script to create simplified polygons for all of the valid polygon files and print a warning if it fails for one of the files. try and except are in the wrong positions for this to happen. Repair the script.
    (e) Predict what will happen when you run the script. Then run it again with no debugging.
    (f) Run the script again stepping through in the debugger. Use the watch window to watch the value of fc. Three output shapefiles should be created.
    (g) Now we'll use a different sample script, 'simplifyOop2.py'. The purpose of this script is to simplify polygons for one data file using ten different minimum distance values within a WHILE-loop and with error handling. But this script has some mistakes. Open 'simplifyOops2.py' and predict the output it will produce.
    (h) Run 'simplifyOops2.py' and observe the output it creates.
    (i) Run the code again by stepping through in the debugger to see the flow.
    (j) Repair the mistake on the line that reads:
    minArea = '{0}foot'.format(x)
    (k) Now run the code by stepping through in the debugger again.
    (l) Watch the value of x in the watch window.
    (m) Repair the mistake on the line that reads: x = x + 1
    (n) Run the script again and confirm that there are ten output files.

2. **predictTry.py** The sample script, 'predictTry.py' contains multiple named exceptions, as well as conditional blocks. The script requires one argument, the name of an input file. Test your understanding of code flow with try/except blocks by predicting what will be printed by the script for each of the following input scenarios (then check your answers by running the script):

    (a) no arguments
    (b) C:/gispy/data/ch14/predict/bogus.shp (a file that doesn't exist)
    (c) C:/gispy/data/ch14/predict/tree.gif
    (d) C:/gispy/data/ch14/predict/coverPolygons.shp (a Polygon file)
    (e) C:/gispy/data/ch14/predict/firesPoints.shp (a Point file)
    (f) C:/gispy/data/ch14/predict/parkLines.shp (a Polyline file)

3. **dictionaryTry.py** The code in the try block of sample script, 'dictionaryTry. py' contains an error, but the except block simply prints the message 'An error occurred'. This message is too generic. Add code that uses the Python built-in

traceback module to force the script to print an exception in the exception handling block (Don't correct the error in the `try` block).

4. **fileOpenTry.py** The sample script, 'fileOpenTry.py' opens and reads the contents of a text file. Run the script with a valid text file argument (like example input1 below) to see it work correctly. Then run the script again using the name of a text file that doesn't exist (like example input2), and observe the resulting error. Add a named exception block to the script to handle this kind of exception. Next, print a message inside the exception handling block. Last, add code to force the script to exit by using the `sys` module `exit` function inside the exception handling block. The `###` comments in the script provide guidance.

Example input1: C:/gispy/data/ch14/cover.prj

Example output1: (only the first few characters are shown):
```
PROJCS["NAD_1927_StatePlane_Pennsylvania_Sou...
```

Example input2: C:/gispy/data/ch14/dummyFile.prj

Example output2:
```
Warning: could not open: C:/gispy/data/ch14/dummyFile.prj
```

5. **moduloTry.py** The modulo operator (represented by % in Python) finds the remainder of a division operation. For example, 10%6==4 because 4 is the remainder of 10/6. The sample script 'moduloTry.py' takes two input values and finds the modulo. Modify the script to implement error handling with named exception handling blocks to provide the following behavior:

Input	Printed output
25 4	a: 25 b: 4 c: 1.0
5 0	a: 5 b: 0 c: 5.0 mod 0 is undefined
woods 3	a: woods b: 3 Usage: <numeric value 1> <numeric value 2> c: Not found

6. **exportTry.py** Use the Quick Export (Data Interoperability) tool to export an ArcGIS format data file to a comma separated value file. The script should take two arguments, the full path file name of the data and an output directory. Use error handling in case the tool fails and use the `arcpy GetMessage` function indexes 4 and 3 to print selected portions of the geoprocessing message when the tool call fails. The script should have the following behavior:

Example input 1: C:/gispy/data/ch14/cover.shp C:/gispy/scratch

Example output 1:
```
Output created in: CSV, C:/gispy/scratch
```

Example input 2: C:/gispy/data/ch14/dummyFile.shp C:/gispy/scratch

Example output 2:
```
Failed to execute (QuickExport).
Feature class 'C:/gispy/data/ch14/dummyFile.shp' is invalid.
```

7. **boxTry.py** Use the Minimum Bounding Geometry (Data Management) tool to find the minimum bounding rectangles for the features in each shapefile in C:/gispy/data/ch14. Create output in 'C:/gispy/scratch'. The script needs no arguments. Use error handling in case the tool fails and use the `arcpy` `GetMessage` function with indexes 2 and 3 to print selected portions of the geoprocessing message when the tool call fails. If the user passes the Chapter 14 data directory, 'C:/gispy/data/ch14', (see Figure 14.1) into the script, it should print the following:

```
bufferBox.shp created.
coverBox.shp created.
ERROR 000229: Cannot open C:/gispy/data/ch14\dummyFile.shp
Failed to execute (MinimumBounding Geometry).
firesBox.shp created.
no_damageBox.shp created.
parkLinesBox.shp created.
```

Chapter 15
User-Defined Functions

Abstract Workflows often have sequences of common steps repeated within or across scripts. Functions allow programmers to group related steps of code, name them, and reuse them by calling them by name. This chapter discusses defining functions with required and optional parameters, returning values from functions, organizing code with functions, and managing variables inside and outside of functions.

Chapter Objectives

After reading this chapter, you'll be able to do the following:

- Define and call custom functions.
- Pass values into and out of functions.
- Write functions with optional arguments.
- Create functions which return multiple values.
- Describe when to employ a user-defined function.
- Explain why some functions have parameters and/or return values.

15.1 A Word About Function Words

By now, you are quite familiar with calling functions that are built-in (e.g., `float`, `len`, and `range`) or ones that are available when a module is imported (e.g., `os. path.dirname` or `arcpy.Describe`). As your own scripts become more complex, you will want to start writing your own functions. Functions provide a way to organize code so that it can easily be reused. As a quick preview, look at the following code, which defines a function named `printBird`:

```
def printBird():
    print """
     ,,,     ::.
    <*~)    ;;
    ( @}//
    _''
    """
```

The function definition starts with the keyword `def` (short for 'define'), followed by the function name, parentheses, and a colon. Function names follow the same rules as variable names (e.g., names can't be keywords, can't start with numbers, can't contain spaces, and so forth). Sometimes the parentheses are empty, sometimes not—more about that in a moment. The colon is required and signals the start of a block of code. The code block inside the definition is indented. These components are needed to define a function in Python. But this is just the definition. The function code block contains a set of related statements that are executed only when the function is called. Nothing is printed if you simply put the `printBird` function definition in a script and run the script. The function needs to be called to be executed. To call this function, use the name, followed by parentheses:

```
>>> printBird()
    ,,,      ::.
  <*~)      ;;
  (  @}//
    ''

>>>
```

The `printBird` example demonstrates how easy it is to define and call a simple custom function. Of course, user-defined functions can do much more than print punctuation art. This chapter provides more details about creating and using custom functions.

The term 'function' has several near synonyms in programming terminology—procedures, subroutines, and methods. The terms 'function', 'procedure', and 'subroutine' are used more or less interchangeably in Python. The term 'procedure' is used to indicate that the function does not explicitly return a value—more on returning values is coming up shortly. Other programming languages use the term 'subroutine' but this term is not used as frequently in Python. A 'method' is a special type of procedure associated with an object. The upcoming chapter on Python classes will reveal how custom methods are created and used. The term 'function' is used here since it is the most general term.

Your knowledge of using built-in functions is a good frame of reference for learning about custom functions. The vocabulary related to built-in functions also applies to custom functions:

- A code statement that invokes a custom function is referred to as a *function call*.
- Providing input for a custom function is referred to as *passing arguments* into the function.
- A term closely related to arguments, *parameters*, are the pieces of information that can be specified to a function.
- The *signature* of a custom function is a template for how to use the function and lists the required and optional parameters.
- Some custom functions come up with results from the actions they perform—they *return a value*.

Learning to write custom functions will deepen your understanding of these terms. Chapter 2 likened a function to a task assigned to a butler. You tell the butler your bidding and he goes off to make it happen. You don't have to concern yourself with the details of how he does it. You ask him to iron your shirt and he attends to the execution. You only need to know the task name (e.g., `ironApparel`). Functions organize code by grouping related statements together. Example 15.1 defines the `printArgs` function to loop through the script arguments and print them. The `printArgs` function relies on the script to import the `sys` module (outside the function). The import statement could also be placed inside the function (though Python style guidelines recommend importing modules at the start of a script instead of inside functions).

Example 15.1

```
# reportArgs.py
# Purpose: Print the script arguments.
import sys

def printArgs():
    '''Print user arguments.'''
    for index, arg in enumerate(sys.argv):
        print 'Argument {0}: {1}'.format(index, arg)

printArgs()
```

The last line of 'reportArgs.py' in Example 15.1 calls the function. The function must be defined before it is called; in other words, the call must come later in the script than the definition. Place a function call before the definition and a `NameError` exception is thrown.

15.1.1 How It Works

We'll use the three step buttons on the debugging toolbar (Figure 15.1) to explore how functions work. The "Step" button would be more aptly named 'Step in'. In this book, we'll refer to it as the "Step (In)" button. This button steps into a function that is being called. The 'Step over' button executes the function being called, but does not step inside. The 'Step out' button steps out of the current function. The black lines on the buttons represent lines of code with indentation and the arrows represent how the buttons control the debugging cursor.

Stepping through with the debugger shows that when Python encounters a function definition, it doesn't execute it; Rather, it stores the name of the function and only when it reaches a call to that function does it go inside the function code

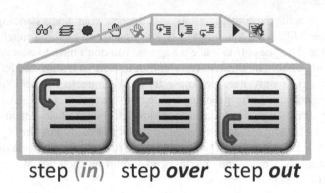

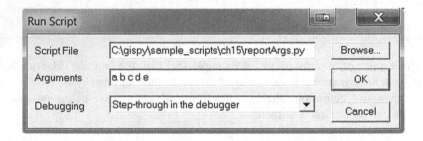

Figure 15.1 The three step buttons on the debug toolbar.

block and execute that code. To see this in action, run 'reportArgs.py' with the following steps:

1. Click the 'Run' button 🏃, enter arguments a, b, c, d, and e, and choose 'Step-through in the debugger' on the 'Run Script' form and click 'Okay' to start debugging.

Run Script		
Script File	C:\gispy\sample_scripts\ch15\reportArgs.py	Browse...
Arguments	a b c d e	OK
Debugging	Step-through in the debugger ▼	Cancel

2. Click the 'Step (in)' button ⬆≡ two times to reach the function call. Notice the arrow ▷ jumps from the function def statement to the function call without entering the function definition.

3. Click the 'Step (in)' button ⬆≡ one more time and the arrow jumps back to the function definition to start executing the function code.

4. This time, click the 'Step over' button ⬇≡ to step through the printArgs function. (If you continue to use the 'Step (in)' button, execution will go inside the 'winout.py' script when the script reaches the print statement and you will have to click the 'Step out' button several times ⬆≡ to return to 'reportArgs.py'.)

5. Click the 'Step over' button several times ⌇ until the first three arguments are printed.

6. Next click the 'Step out' button twice ⌐ to step out of the function and return to the function call. Notice that the remaining two arguments are printed in the 'Interactive Window' because the remainder of the function statements were executed before exiting the function.

7. Click the 'Step out' button one more time to exit the script.

The 'Step (in)' button steps into functions in the code; the 'Step over' button bypasses the details. Run 'reportArgs.py' again in debug mode only selecting the 'Step over' button. Notice that the debugger doesn't step through the execution of each line of code inside the function when the 'Step over' button is used. In contrast, the 'Step (in)' button demonstrates the flow more clearly—showing that the script does not run the function code block until it is called.

You may have noticed that the cursor jumped past the first line of code inside the function. This is not an executable line of code, but it serves a special purpose—it is a documentation string or docstring.

15.1.2 The Docstring

The string literal on the first line of code inside a function (such as, `'''Print user arguments.'''` in Example 15.1) is called a *docstring*. The docstring documents the function's purpose. A one line docstring suffices for simple functions, but more than one line can be used if needed. Triple quotes are necessary for multi-line docstrings and style guidelines encourage triple quotes for single line docstrings to maintain consistency.

The docstrings not only remind the authors themselves of the function's aim, but they facilitate sharing functions with others. Once the function definition is stored by Python (in other words, once the code has executed the `def` statement), the built-in `help` function can be used to print the docstring, as in the following code:

```
>>> help(printArgs)
Help on function printArgs in module__main__:
printArgs()
    Print user arguments.
```

The `printArgs` function prints arguments passed into the script by the user. Custom functions themselves can also be designed to take arguments, arguments that are passed to the function when it is called.

15.2   Custom Functions with Arguments

The butler can turn the lights off without further instructions, but if we want him to dim the lights, we should specify how low. Similarly, some functions require arguments while others don't. The arcpy GetMessages function, for example, automatically prints all the messages from the most recent tool call, but GetMessage requires a numerical index.

Example 15.2 consists of two similar functions, one with arguments (named delNamedFCS) and one without arguments (delBuffFCS). First, consider delBuffFCS which deletes each feature class in the current workspace whose name contains the word Buff. This is useful during script development. Suppose a geoprocessing script that buffers every shapefile listed in the workspace appends Buff to the input name. Each time the script runs, the output files from the previous run become input, doubling the number of input files and leading to output file names like 'firesBuffBuffBuff.shp'. Calling the delBuffFCS function restores the original workspace.

But suppose the script appends Out instead of Buff to each file name; Calling delBuffFCS won't help. The delNamedFCS function provides a more general solution, since it allows a string to be passed in to the function. This function deletes each feature class in the current workspace whose name contains the substring passed as an argument. The last line of Example 15.2 calls delNamedFCS with the argument Out. The delString variable in the definition list gets the value that is passed in to delNamedFCS when it is called. In this case the function deletes any file with the substring 'Out' in the name.

Example 15.2

```
# deleteFCS.py
# Purpose: Clear workspace of unwanted files.
# Usage: No arguments needed.
import arcpy

arcpy.env.workspace = 'C:/gispy/data/ch15/scratch'
def delBuffFCS():
    '''Delete feature classes with names containing "Buff".'''
    fcs = arcpy.ListFeatureClasses('*Buff*')
    for fc in fcs:
        arcpy.Delete_management(fc)
        print '{0} deleted.'.format(fc)

def delNamedFCS(delString):
    '''Delete feature classes with names containing delString.'''
    wildcard = '*{0}*'.format(delString)
    fcs = arcpy.ListFeatureClasses(wildcard)
    for fc in fcs:
        arcpy.Delete_management(fc)
```

```
    print '{0} deleted.'.format(fc)
delBuffFCS()
delNamedFCS('Out')
```

A custom function with arguments has the following general format:

```
def functionName (param1, param2, param3,...):
    '''Docstring describing the purpose.'''
    code statement(s)
```

The first line of the function definition (the one with the `def` statement) is the function signature. Each of the comma separated variable names inside the parentheses in the signature stands for a parameter. Parameters can be required or optional (more about that shortly). When the function is called, one value must be passed in for each required parameter in the signature and the values are separated by commas. The format for calling functions is as follows:

```
functionName(argument1, argument2, argument3,...)
```

Any number of arguments can be used in the signature. The following code defines a `batchBuffer` function which has four required parameters (workspace, featType, outSuffix, buffDistance):

```
# excerpt from batchBuff.py
def batchBuffer(workspace, featType, outSuffix, buffDistance):
    '''For a given workspace, buffer each
    feature class of a given feature type.'''
    arcpy.env.workspace = workspace
    fcs = arcpy.ListFeatureClasses('*', featType)
    for fc in fcs:
        fcParts = os.path.splitext(fc)
        outputName = fcParts[0] + outSuffix + fcParts[1]
        try:
            arcpy.Buffer_analysis(fc, outputName, buffDistance)
            print '{0} created.'.format(outputName)
        except:
            print 'Buffering {0} failed.'.format(fc)
```

To call the function, four arguments must be passed to the function. The `batch-Buffer` function requires string arguments, which can be given as string literals or variables. The following code calls the function, with four string literal arguments:

```
>>> batchBuffer('C:/gispy/data/ch15', 'Polygon', 'Buff', '1 mile')
```

The parameters in the signature take on the values in the order in which they appear. The workspace variable takes the value 'C:/gispy/data/ch15', the feat-Type variable takes the value 'Polygon', and so forth. Calling the function with less than the number of required parameters raises an exception:

```
>>> batchBuffer('C:/gispy/data/ch15', 'Polygon', 'Buff')
TypeError: batchBuffer() takes exactly 4 arguments (3 given)
```

The following code sets four string variables and then calls 'batchBuffer':

```
>>> wSpace = 'C:/gispy/data/ch15/tester.gdb'
>>> featureType = 'Point'
>>> outputSuffix = 'Ring'
>>> distance = '0.5 kilometers'
>>> batchBuffer(wSpace, featureType, outputSuffix, distance)
```

Notice that the argument variable names used in the function call are different from the parameter names in the signature (e.g., eworkspace and wSpace). This convention reduces confusion between function variables and non-function variables.

Note: Use names for variables inside function definitions that differ from those used outside the function definition.

The parameter data types are not specified in the signature, so it is important to use meaningful parameter names. Calling a function with the arguments of the wrong data type can lead to exception errors. For example, the 'batchBuffer' expects a string output suffix, not a numeric one. So the following code throws an exception:

```
batchBuffer('C:/gispy/data/ch15', 'Polygon', 5, '1 mile')
```

The TypeError exception occurs when the function tries to concatenate a string and an integer (fcParts[0] + outSuffix).

15.2.1   Script Arguments vs. Functions Arguments

The term 'arguments' is used both in reference to scripts and functions. This can lead to some confusion, so we'll take a closer look.

• We refer to arguments passed into the script when it is run as *script arguments*. When the PythonWin IDE is used to run a script, these are passed in to the script via the 'Arguments' text box in the 'Run Scripts' dialog window.

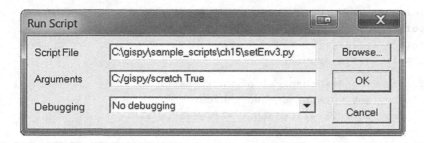

- The script arguments can then be retrieved inside the script using the `sys.argv` list or using the `GetParameterAsText` function in `arcpy` scripts.
- *Function arguments*, on the other hand, are those passed into a function when it is called in the script.

To clarify the difference, consider the three scripts in Example 15.3. 'setEnv1.py' takes no arguments whatsoever. It hard-codes the values for the environment settings. The `setEnviron1` function could be called repeatedly during a script to reset these properties after changes have been made. The second script, 'setEnv2. py' uses function arguments. Values are passed into the function to adjust these properties as needed. The third script, 'setEnv3.py', uses both script arguments and function arguments. The script arguments are used to set the `wSpace` and `over-write` variables. The last line of this script passes these variables in to `setEnviron3` as function arguments.

Example 15.3

```
# setEnv1.py
import arcpy

def setEnviron1():
    arcpy.env.workspace = 'C:/gispy/data/ch15'
    arcpy.env.overwriteOutput = True
setEnviron1()
------------------------
# setEnv2.py
import arcpy

def setEnviron2(workspace, overwriteVal):
    arcpy.env.workspace = workspace
    arcpy.env.overwriteOutput = overwriteVal

wSpace = 'C:/gispy/data/ch15/tester.gdb'
overwrite = False
setEnviron2(wSpace, overwrite)
------------------------
```

```
# setEnv3.py
import arcpy, sys
wSpace = sys.argv[1]
overwrite = sys.argv[2]

def setEnviron3(workspace, overwriteVal):
    arcpy.env.workspace = workspace
    arcpy.env.overwriteOutput = overwriteVal
setEnviron3(wSpace, overwrite)
```

15.2.2   Optional Arguments

Many functions have optional parameters as well as required parameters. For example, the `arcpy` Buffer function requires three arguments (input features, output name, and buffer distance), but it also has four optional arguments that can be used to further specify the buffer behavior. Custom functions can be designed to handle optional parameters. Optional parameters are given a default value in the signature with the following format:

```
def funcName(reqP1,reqP2,…,optP1=defaultV1,optP2=defaultV2,…):
    '''Docstring…'''
    code statement(s)
```

Parameters that are given a default value in the signature are optional—if the caller does not pass in an argument for that parameter, the default value is used. The following function has one required and one optional parameter:

```
def setEnviron4(workspace, overwriteVal = True):
    arcpy.env.workspace = workspace
    arcpy.env.overwriteOutput = overwriteVal
```

The function can be called using one argument or two. Either one of the following statements can be used to call the `setEnviron4` function:

```
>>> setEnviron4('C:/gispy/data/ch15', False)
>>> setEnviron4('C:/gispy/data/ch15')
```

Both calls have the same effect, except the first one sets the output overwriting property to `False` and the second one sets it to `True`. A function can take any number of required and optional arguments, but required arguments must come before optional ones in the signature, since arguments are mapped to parameter positions simply by the order in which they are passed.

The function examples so far in this chapter have printed information, performed geoprocessing, and modified properties, but they haven't returned values. Another common purpose for functions is to calculate or derive values and return them to the caller. The next section discusses returning values in custom functions.

15.3 Returning Values

We've used functions that simply print information or modify the environment (e.g., the built-in `help` function or the `arcpy` `Delete` function); And we've used others that return values (e.g., the built-in `round` function or the `arcpy` `ListRasters` function). Now we'll discuss how to return values in custom functions. In terms of our butler metaphor, he can perform tasks that modify our environment (e.g., dim the lights) and he can also perform tasks that generate a tangible result (e.g., bring tea).

To create a custom function that returns a value to the caller, use a `return` statement. The value that follows the `return` keyword is returned to the caller and can be stored in a variable with an assignment statement. A function that returns a value has the following general format:

```
def functionName (param1, param2, param3,...):
    '''Docstring ...'''
    code statement(s)
    return valueToBeReturned
```

The format for calling functions that return values is as follows:

```
variableName = functionName(arguments1, argument2, argument3,...)
```

The following code defines a function that returns the list of field names for a given dataset:

```
# excerpt from fieldHandler.py
import arcpy

def getFieldNames(data):
    '''Get a list of field names.'''
    fields = arcpy.ListFields(data)
        fnames = [f.name for f in fields]
        return fnames
```

When the function is called, the function code is executed and the `return` statement sends information back to the caller. The following code calls the `getFieldNames` function with the full path of an input dataset as an argument:

```
>>> names = getFieldNames('C:/gispy/data/ch15/park.shp')
```

The function gets a list of the `Field` objects and derives the names and returns the resulting list. An assignment statement stores the return value in the `names` variable. The following code prints the result:

```
>>> names
[u'FID', u'Shape', u'COVER', u'RECNO']
```

Return values can be any data type. The following code defines a function that returns a Python Boolean value (`True` or `False`):

```
def fieldExists(data, name):
    '''Check if a given field name already exists.'''
    fieldNames = getFieldNames(data)
    isThere = name in fieldNames
    return isThere

>>> result = fieldExists('C:/gispy/data/ch15/park.shp', 'COVER')
>>> result
True
```

Custom functions can call other custom functions that are defined within the same module. Both `getfieldNames` and `fieldExists` are defined in the 'field-Handler.py' script, so `fieldExists` can call `getFieldNames`. A custom function can also be called from another script, but the syntax is slightly different as discussed in the upcoming chapter on custom modules.

Whenever a `return` statement is reached within a function, the execution exits the function and goes back to where the function was called. This means that any lines of code that follow a `return` statement in the execution flow will not be executed. The 'countIntersection' function in Example 15.4 handles this incorrectly. The function calculates a temporary dataset, an intersection of features and then it calculates the number of features in the intersection file, and returns the value. Though the author may have intended to delete the temporary dataset before leaving the function, the code will never reach the Delete (Data Management) tool call because it occurs after the `return` statement.

Example 15.4

```
# oops.py
# Purpose: Count the number records in an intersections
#          between two datasets and delete the intersection file
```

```
#      (but the intersection output is not deleted since this line
#      of code is placed after the 'return' statement).
import arcpy
arcpy.env.workspace = 'C:/gispy/data/ch15/tester.gdb'
arcpy.env.overwriteOutput = True

def countIntersection(dataList):
    '''Calculate the number of features in the intersection.'''
    tempData = 'intersectOut'
    arcpy.Intersect_analysis(dataList, tempData)
    res = arcpy.GetCount_management(tempData)
    print '{0} created.'.format(tempData)
    return int(res.getOutput(0))
    #uh-oh! The deletion is not going to happen.
    arcpy.Delete_management(tempData)
    print '{0} deleted.'.format(tempData)

inputData = ['schools','workzones']
count = countIntersection(inputData)
print 'There are {0} intersections.'.format(count)
```

The examples given so far have used the `return` keyword only once. It can be used more than once in a function, though it's not recommended. Example 15.5 imports the Python built-in `datetime` module to deal with Gregorian calendar dates. The call to this function passes in the birth date April, 20, 2003. This is converted to a `datetime` object. It then finds this year's birthday and compares the birthday to today's date. If the birthday has already occurred this year, it returns the difference between the years; Else, it takes an additional year off and returns this value. The `calculateAge` function uses two `return` statements, but it could easily be rewritten to only use one. Execution leaves the function when it reaches a `return` statement. When multiple `return` statements are used, there is more than one way to exit the function. Style guidelines discourage multiple `return` statements because, ideally there should only be one way to exit a function—making the code easier to interpret.

Example 15.5

```
# age.py
# Purpose: Calculate age.
import datetime

def calculateAge(yr, mo, day):
    '''Calculate age based on the given birth date.'''
    # Get datetime objects for birth date and today.
    born = datetime.date(yr, mo, day)
```

```
today = datetime.date.today()
# Get this year's birthday and handle leap year exceptions.
try:
    birthday = born.replace(year=today.year)
except ValueError:
    birthday = born.replace(year=today.year, day=born.day-1)
# Return age.
if birthday < today:
    return today.year - born.year
else:
    return today.year - born.year - 1
print calculateAge(2012, 4, 20)
```

Usually, multiple `return` statements can be avoided by employing a variable within branches. For example, the last five lines of the `calculateAge` function can be replaced by the following lines of code which using only one `return` statement:

```
#Return age
if birthday < today:
    age = today.year - born.year
else:
    age = today.year - born.year - 1
return age
```

Both alternatives return the same value; However, the single return is preferred since the code can become difficult to follow in more complex functions when multiple exit locations are possible.

The examples given here store the return values in a variable and print the variable. Printing the return values helps with understanding what a custom function does, but it's important to note the difference between returning values and printing values within a function.

15.3.1 A Common Mistake: Where Did the None Come from?

In Python, all functions return something even if a `return` statement is not used. The `return` statement returns a value explicitly and should be used if the function is intended to return a value. However, if no `return` statement is used, the function returns None, which is a Python built-in constant that represents a null value. It's as if an implicit `return None` statement is added to the functions when no explicit one is used. The following script demonstrates a common mistake involving this phenomenon:

```
# except from returnVSprint.py
def positiveMinV1(numList):
    '''Find the minimum positive number in the list'''
    pos = []
    for val in numList:
        if val >= 0:
            pos.append(val)
    print min(pos)
theList = [8, 2.5, 0, 12, 5]
value = positiveMin(theList)
print value
```

This script prints the following output:

```
>>> 2.5
None
```

The function `positiveMinV1` correctly finds the minimum positive number in the list and prints it (2.5). But you may not have expected to see `None`. Can you spot the mistake? The function is called and the return value is being stored in the `value` variable. Since there is no explicit `return` statement, the function returns `None`, which is printed with the last line of code. Probably, the intention was not to print the value inside the function but rather to return the value as in the following code:

```
def positiveMinV2(numList):
    '''Find the minimum positive number in the list'''
    pos = []
    for val in numList:
        if val >= 0:
            pos.append(val)
    return min(pos)
```

> **Note** If a script prints a mysterious `None`, it often means that it's printing the return value of a function which does not contain an explicit `return` statement.

When authoring a function, determine if the purpose of the function is to return a value or to print information inside the function and design the function appropriately.

15.3.2   *Returning Multiple Values*

At times, you may want to return more than one value with a function. Suppose, for example, the function returns both the x and y coordinates of a point. Both x and y can be returned as separate values by using a comma to separate the values in the `return` statement. This returns a Python tuple. The following code defines a `midPoint` function which returns an x and a y value in the `return` statement:

```
# excerpt from returnMultVals.py

def midPoint(x1, y1, x2, y2):
    '''Calculate the midpoint of line segment (x1,y1), (x2,y2).'''
    xVal = (x1 + x2)/2.0
    yVal = (y1 + y2)/2.0
    return xVal, yVal
```

The calling function can use comma separated variable names to receive the return values, as in the following code which finds the midpoint of the line segment from (3,5) to (2,1):

```
>>> x, y = midPoint (3,5,2,1)
>>> x
3.0
>>> y
3.0
```

Alternatively, a single tuple variable can be used to store multiple returned values. Example 15.6 gets the current time before and after walking through the given subdirectory using the `arcpy.da.Walk` method. The count is found and printed for each file. Then the `diffTime` function is used to calculate the amount of time elapsed. The `diffTime` function returns a tuple of time components (weeks, days, hours, and so forth). The variable `t` is used to store the tuple when it is returned. Then a print statement indexes into the tuple to access the returned values.

Example 15.6

```
# walkCount.py
# Purpose: Walk and get the record count for
#     each file, where possible.
# Usage: inputdirectory
# Example input: C:/gispy/data/ch15
import arcpy, datetime, sys
mydir = sys.argv[1]
def diffTime(start, end):
    '''Calculate the difference between two datetime objects'''
    difference = end - start
```

```
        weeks, days = divmod(difference.days, 7)
        minutes, seconds = divmod(difference.seconds, 60)
        hours, minutes = divmod(minutes, 60)
        return weeks, days, hours, minutes, seconds
before = datetime.datetime.now()
for root, dirs, files in arcpy.da.Walk(mydir):
    for f in files:
        try:
            count = arcpy.GetCount_management(root + "/"+ f)
            print '{0}/{1}    Count = {2}'.format(root,f,count)
        except arcpy.ExecuteError:
            print arcpy.GetMessages()
after = datetime.datetime.now()

t = diffTime(before, after)

print 'Time elapsed: {0} weeks, {1} days,
{2}:{3}:{4}'.format(t[0],t[1],t[2],t[3],t[4])
```

When the number of return values is not known in advance, a list can be used to return the values, as shown in the getFieldNames function in sample script 'fieldHandler.py'.

15.4 When to Write Functions

To identify places where functions would be useful in your code, find related blocks of code that are repeated within scripts. For example, the following code uses several statements to print selected portions of the geoprocessing messages after two tools are called:

```
# scriptWithoutFunction.py
# Purpose: Call three tools (to find avg. nearest neighbor, intersection,
#          and get count) Print the results from avg. nearest neighbor
#          and get count without using a function.
import arcpy
arcpy.env.workspace = 'C:/gispy/data/ch15/tester.gdb'
res = arcpy.AverageNearestNeighbor_stats('schools')
resList = res.getMessages().split('\n')
for message in resList:
    if '...' not in message and 'Time:' not in message:
        print message
```

```
arcpy.Intersect_analysis(['schools','workzones'],'intersectOutput')

res = arcpy.GetCount_management('intersectOutput')
resList = res.getMessages().split('\n')
for message in resList:
    if '...' not in message and 'Time:' not in message:
        print message
```

Example 15.7 moves the repeated code in 'scriptWithoutFunction.py' into a function named `reportResults`. The script then calls the function twice, clearing the clutter so that it's easier to tell at a glance the main activity occurring in the script—calling three geoprocessing tools, `AverageNearsestNeighbor`, `Intersect`, and `GetCount`.

It may only make sense to call a function such as `printArgs` (in Examples 15.1) one time in a script—there is only one set of arguments coming into the script, so it's unlikely to be useful to print them more than once. However, the function still bundles those related code statements together, so that the details of the operation appear as one statement within the main flow of the code and the reader can chose to ignore these details to focus on understanding the overall purpose of the function. A well-named function and its docstring can serve as a sufficient shorthand to signify its purpose.

> Functions are a good way to organize code, grouping related statements together, so that they can be called one or more times from within the same script or from other scripts.

Also, although a function such as `printArgs` may only be useful once in each script, it is likely to be useful in many scripts. Grouping the code into a function makes it easier to grab the related block and insert it into another script. In fact, in an upcoming chapter, we'll discuss how to call a function that's in another script, so that you can write a function once and call it from many scripts.

Example 15.7

```
# scriptWithFunction.py
# Purpose: Call three tools (to find avg. nearest neighbor,
#          intersection, and get count) Print the results from avg.
#          nearest neighbor and get count using a function.

import arcpy
arcpy.env.workspace = 'C:/gispy/data/ch15/tester.gdb'

def reportResults(resultObj):
    '''Print selected result messages.'''
    resList = resultObj.getMessages().split('\n')
    for message in resList:
        if '...' not in message and 'Time:' not in message:
            print message
```

```
res = arcpy.AverageNearestNeighbor_stats('schools')
reportResults(res)

arcpy.Intersect_analysis(['schools','workzones'], 'intersectOutput')
res = arcpy.GetCount_management('intersectOutput')
reportResults(res)
```

15.4.1 Where to Define Functions

In addition to the constraint of defining functions before calling them, programmers follow a few other patterns so that it's easy to locate function definitions in the script. The standard pattern groups function definitions together near the beginning of the script. Generally, they should not be dispersed throughout the code. Figure 15.2 shows an example of poor organization (on the left) and improved organization on the right. The header comments should be first, followed by any imports. Next, all the functions should be defined (usually in alphabetical order by the function names). The main processing activities and any calls to the function should follow the function definitions. Generally, functions should not be defined inside a loop.

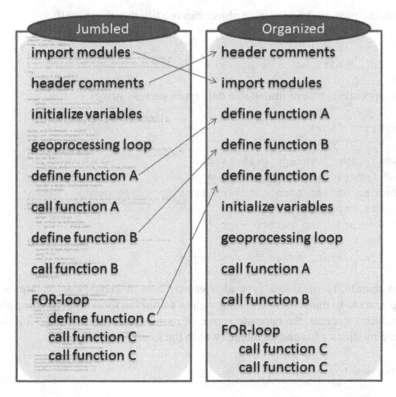

Figure 15.2 Example of poor code organization (on the left) and improved organization (on the right).

15.5   Variables Inside and Outside of Functions

When programming with functions, you need to be aware of several scenarios
where mutable variables work differently from immutable ones. This section gives
some examples and guidelines for avoiding confusion.

15.5.1   *Mutable Arguments Can Change*

Mutable and immutable variables behave somewhat differently when it come to
functions. We previously discussed the concept of mutability in the context of lists
and strings and how their methods work. Recall that indexing can be used to change
the values of items in a mutable sequence such as a list:

```
>>> myList = ['a','b','c']
>>> myList[0] = 'z'
>>> myList
['z', 'b', 'c']
```

Also, the methods of a mutable object can modify the object itself:

```
>>> myList.sort()
['b', 'c', 'z']
```

The opposite is true of immutable data types such as strings:

```
>>> myStr = 'abc'
>>> myStr[0] = 'z'
Traceback (most recent call last):
File "<interactive input>", line 1, in ?
TypeError: object does not support item assignment
>>> myStr.replace('a','z')
'zbc' #This is the return result.
>>> myStr
'abc' #The string value is unchanged.
```

The mutability of a data type also effects how it behaves as an argument.
Changes made to mutable function arguments within the function persist outside of
the function; whereas, the opposite is true of immutable arguments. The following
function modifies a list and an integer within the function:

```
def augment(myList, myInt):
        myList.append('some value')
        myInt = myInt + 1
```

The following code defines a list and a numeric variable and then calls the augment function:

```
>>> aList = ['first entry']
>>> num = 5
>>> print aList
['first entry']
>>> print num
5
>>> augment(aList, num)
```

After the function is called, the list is altered and the number is not:

```
>>> print aList
['first entry', 'some value']
>>> print num
5
```

The list is altered even if it is not returned to the caller. If this is an unintended result, you should use a deep copy which creates a new list object and then perform operations on the new list, as in the following example:

```
def augmentList(list1):
    list2 = list(list1)
    list2.append('some value')
    return list2
aList = ['first entry']
result = augmentList(aList)
```

In this case, the function preserves the original list and returns the modified list:

```
>>> print aList
['first entry']
>>> print result
['first entry', 'some value']
```

If you are familiar with the concepts of passing 'by reference' versus 'by value' from other programming languages, this is related to that. However, the 'by value' vs. 'by reference' discussion regarding Python leads to some confusing nuances about semantics. Instead, we chose to use the concept of mutable vs. immutable which avoids this confusion. Table 15.1 lists several familiar immutable and mutable data types. Sets and dictionaries will be covered in an upcoming chapter.

Table 15.1 Examples of
mutable and immutable data
types.

Immutable	Mutable
Numbers	Lists
Strings	Dictionaries
Tuples	Sets

15.5.2   Pass in Outside Variables

The first line of code in Example 15.8 assigns the value of 5 to the variable named
x. This variable is created within the script but not within a class or another module—we'll refer to this as a *script level* variable. When a function uses a variable
such as this, the variable should be passed into the script as an argument. The consequences of failing to do so are different depending on mutability. If the data type
is immutable, the function can not alter the value of the variable. Attempting to do
so throws an exception. In Example 15.8, the numerical variable, x, is defined at the
beginning of the script, outside of the function definition. This numeric (immutable)
variable can then be used anywhere in the script outside of the function. E.g., it is
printed before the call to addOne. But x cannot be modified inside the definition of
addOne. An UnboundLocalError occurs when the function attempts to assign
a new value to x.

Example 15.8

```
# passVars.py
# Purpose: Demonstrate 'UnboundLocalError'.
# Usage: No script arguments needed.
x = 5
def addOne():
    x = x + 1
    print 'In here', x

print 'Out here', x
addOne()
```

Output in the Interactive Window:

```
>>> Out here 5
Traceback (most recent call last):
...
    x = x + 1
UnboundLocalError: local variable 'x' referenced before assignment
```

The problem can be corrected either by redefining the function so that it takes the
value of x as an argument or by using a global statement inside the function. The

second option is generally frowned upon because it can lead to bugs that are difficult uncover, so we won't discuss this option further.

The behavior shown in Example 15.8 protects programmers from changing the values of immutable variables unintentionally. However, if the data type is mutable, the function can alter the value (even if that was not the caller's intent). The following function finds the maximum number in a list, adds one to the number, and then appends this number to the list:

```
def appendNext():
    maxVal = max(myList)
    maxVal = maxVal + 1
    myList.append(maxVal)
myList = [1,2,3]
appendNext()
```

The new number has been added to the myList variable:

```
>>> myList
[1, 2, 3, 4]
```

Since no arguments are passed into the function, when the code (and function) becomes more complex, it may not be obvious that calling this function is going to change an outside variable. So best practice is to explicitly pass in script level variables to be used within functions.

> **Note** For consistency and transparency, don't rely on mutability—pass in script level variables to be used within functions as arguments. But remember that the script level value of a mutable argument can be altered by a function.

15.6   Key Terms

Custom functions   The return keyword
Default arguments   Script level variable
Docstrings

15.7   Exercises

1. **Test how it works:** Sample script 'circles.py' contains three custom functions and several calls to the functions. Run the code in debug mode following the given instructions and answer the related questions.

(a) Step through the code using the 'Step over' button twice to step over the import statement and then use only the 'Step (in)' button, counting the steps until you reach line 9. How many clicks were required to reach line 9?

(b) What is the value of 'mode' at this point?

(c) Continue stepping with the 'Step (in)' button until you step into the `return` statement. What is the name of the script that opens at this point? Explain how to quickly step out of this script and return to 'circles.py'.

(d) Run the script to completion with the 'Go' button and explain why the print statement on line 13 is not printed in the Interactive Window.

(e) Again, step through the code, this time only using the 'Step over' button until you exit the script. How many clicks were required to exit the script? Why is this number about three times less than the number of lines of executable code in the script?

(f) Which line of code printed the built-in constant, None and why?

2. **Use function-related terms:** Sample script 'circles.py' contains three custom functions and several calls to the functions. Identify the following code components for each function:

(a) Function name.

(b) Signature.

(c) Docstring.

(d) Required arguments.

(e) Optional arguments.

(f) Data type of the return value (string, int, list, Boolean, Constant, etc. or N/A if there is none).

(g) Line(s) where the functions is called.

3. **functionPractice.py** Practice using function syntax by replacing the pseudo-code in 'functionPractice.py' to define two functions and call them within a loop. The functions are described in parts a and b. The output prints information about each polygon feature class in the input workspace separated by fish punctuation art. Comments in the script demonstrate the sample output for a given input.

(a) Function `printFish` should print a fish like this (no function arguments needed):

```
    ' ' '
<')}>)={
    ` `
```

(b) Function `printDescription` should take one argument, the name of a GIS data file and then it should print information Describe object properties name, dataType, and catalogPath. A feature class named 'park' in 'tester.gdb' yields the following output:

```
Name: park
DataType: FeatureClass
CatalogPath: C:/gispy/data/ch15/tester.gdb\park
```

4. **boxes.py** Practice converting existing code to functions. Sample script 'boxes.py' contains several related blocks of code. Replace the existing code with three functions: findPerimeter, isSquare, and createBoundingBoxes and six function calls (two calls to each function). findPerimeter should require two arguments (length and width) and return the numeric perimeter of the box. isSquare should require two arguments (length and width) and return a Boolean value (True if the box is square and False otherwise). createBoundingBoxes should require two arguments, an input workspace and an output directory. It should set the arcpy workspace variable inside the function and then get the Polygon feature classes and the bounding box geoprocessing loop as in the given script but it should not return anything to the caller. Remember to group all the function definitions at the beginning of the script, just after the imports. Also, place the function calls so that the output is created in the same order as the given script.

5. **largePolys.py** Run the sample script 'largePolys.py' to observe how it works. Then modify the script so that it does NOT use a function. In other words, remove the countLargePolygons function definition and replace the function call with code inside the loop that achieves the same results.

6. **latLong.py** The sample script 'latLong.py' defines and calls a function (dd2dms) that converts numeric decimal degrees (dd) to string decimal/minutes/seconds (dms) format and a second function, dms2dd that does the converse. The script calls both the functions for two points, Point1 and Point 2 and prints the results. But there is an error in each function. Use the debugger to identify and repair the errors. Place a breakpoint on lines 21 and 31 and run to these points. Put variables degs, mins, secs, dms, and dd in the Watch window. Use code comments to record the errors you find in the script. The current output looks like this:

```
Point 1: (35.684072, -78.728027) -> (None, None)
Point 2: (-33 43 27.6234, 24 31 17.521484) -> (-32.2756601667,
24.5215337456)
```

When both the errors are corrected, the printed output should look like this:

```
Point 1: (35.684072, -78.728027) -> (35 41 2.6592, -78 43 40.8972)
Point 2: (-33 43 27.6234, 24 31 17.521484) -> (-33.7243398333,
24.5215337456)
```

7. **trapezoid.py** Sample script 'trapezoid.py' gives four measurements for three trapezoids, two base lengths (b1 and b2), an altitude (alt), and an angle (angle). Add the following four functions to the script:

   (a) calculateArea should take three arguments, two base side lengths and the altitude, and it should return the area.
   (b) isParallelogram should take two arguments, the two base lengths. It should return True if the base lengths are the same and False otherwise.

(c) `isRectangle` should take three arguments, the two base lengths and a corner angle in degrees. It should return `True` if both the angle is 90 and it is a parallelogram (call `isParallelogram` to determine this). It should return `False` otherwise.

(d) `isSquare` should take four arguments, two base lengths, an altitude length, and an angle. It should return `True` if it is a rectangle (call `isRectangle` to determine this) and the base length equals the altitude.

Then call the functions for the three quadrilaterals given in the script and print the return values. The output should look like this:

```
Quad1: b1 = 4, b2 = 4, alt = 6, angle = 90
Area = 24.0
Is parallelogram? True
Is rectangle? True
Is square? False

Quad2: b1 = 5, b2 = 5, alt = 5, angle = 30
Area = 25.0
Is parallelogram? True
Is rectangle? False
Is square? False

Quad3: b1 = 8, b2 = 8, alt = 6, angle = 60
Area = 48.0
Is parallelogram? True
Is rectangle? False
Is square? False
```

8. **triangles.py** Write a script 'triangles.py' that takes three arguments, the lengths of three sides of a triangle. Then define the following three functions:

(a) `perimeter` takes three side lengths as argument and returns the perimeter length.

(b) `triangleType` takes three side lengths as arguments, determines if it is an equilateral triangle, an isosceles triangle, or neither. The result is returned.

(c) `area` takes three sides lengths as arguments and returns the area. Compute this as the square root of $(p*(p-a)*(p-b)*(p-c))$, where p is half the perimeter and a, b, and c are the side lengths. Use a `math` module function to compute the square root.

Call the functions and print the results as shown in these examples:

Example input1: 1 1 1

Example output1:
```
>>> Triangle sides: 1.0, 1.0, 1.0
Perimeter = 3.0
Type = Equilateral
Area = 0.433012701892
```

Example input2: 5 5 6

Example output2:
```
>>> Triangle sides: 5.0, 5.0, 6.0
Perimeter = 16.0
Type = Isosceles
Area = 12.0
```

Example input3: 3 4 5

Example output3:
```
>>> Triangle sides: 3.0, 4.0, 5.0
Perimeter = 12.0
Type = Neither equilateral nor isosceles
Area = 6.0
```

9. **segProc.py** Write a script 'segProc.py' that defines and calls a function named 'segLength' that computes the length of line segment between two points, $(x1, y1)$ and $(x2, y2)$. Use the Pythagorean theorem for Euclidean distance. Design 'segProc.py' like 'setEnv3.py' in Example 15.3 to use both script arguments and function arguments. The script should take four arguments and place these in variables, $x1$, $y1$, $x2$, and $y2$. The function should also be defined to take four arguments, the x and y values of two points. Call the function using the script arguments. Use a `return` statement in the function to return the distance. Print the returned values, formatting output as shown in the example.

Example script input: 3.2 1 6 8

Example script print output:
```
Segment (3.20,1.00) to (6.00,8.00) has length: 7.54
```

> **Tip** The following code formats a floating point value to two decimal places:
> ```
> >>> a = 4.33732984
> >>> print 'The number {0:.2f}'.format(a)
> The number 4.34
> ```

10. **fieldFunc.py** Write a script 'fieldProc.py'. The script should define a function named `printFields` that prints the name of the input file and the names of the fields in the input file. The script should take two arguments, a full path input file name and an output workspace. The function should take one argument, an input file name. Add code to the script to do the following: Call `printFields` on the script argument, the input file given by the user. Use the Copy (Data Management) tool to copy the file to the output directory. Add a FLOAT field named AREA to the new file. Finally, call `printFields` again on the new file.

Example script input: C:/gispy/data/ch15/park.shp C:/gispy/scratch/

Example script output:
```
>>> Fields in C:/gispy/data/ch15/park.shp:
FID
Shape
COVER
RECNO
C:/gispy/data/ch15/park.shp copied to C:/gispy/scratch/park.shp.
Fields in C:/gispy/scratch/park.shp:
FID
Shape
COVER
RECNO
AREA
```

11. **outNameProc.py** An output file name often needs to be dynamically created based on an input file name. For example, from input file 'C:/Data/NC.txt', we may want to create 'C:/Data/out/NCBuff.shp' and from input file 'C:/Data/VA.txt', we may want to create 'C:/Data/out/VABuff.shp', and so forth. We can use the same string and os module operations to generate each output name based on each input named.

Sample script 'outNameProc.py' calls a function named outName which has yet to be defined. Add code to this script to define function outName that will create an output file name based on an input file name and three additional arguments. Specifically, the function should be defined to take one required argument and three optional arguments:

A required input file name (possibly containing the full file path)
An optional string to insert in the name just before the extension. Set the default value for this as 'Out'.
An optional file extension. Set the default value for this to 'shp'.
An optional output directory. Set the default for this to an empty string ('').

In the function, use os.path commands (basename, splitext, and join) to get the basename of the input file, strip the file extension (if any) from the input file name, append the string to be inserted, append the new file extension, and prepend the output workspace.

Example 1:
```
>>> name1 = outName('C:/Data/NC.txt', 'Points', 'dbf',
'C:/gispy/outData/')
>>> name1
'C:/gispy/outData/NCPoints.dbf'
```

Example 2:
```
>>> name2 = outName('C:/states//PA.prj')
>>> name2
'PAOut.shp'
```

Define the function (including a docstring) and compare the results against the output shown in the ### script comments. This script only needs to perform string manipulation, so no data files are needed for testing.

12. **centralPoint.py** Add code to the script 'centralPoint.py' so that it takes two arguments, a full path polygon shapefile name and an output directory. Then have the script perform several geoprocessing operations to investigate the central trend of the polygons. Specifically, find the centroid of each polygon using the Feature to Point (Data Management) tool, and then find the central feature of the polygon centroids using the Central Feature (Spatial Statistics) tool and the mean center of the polygon centroids using the Mean Center (Spatial Statistics) tool. Finally, find the distance between the central feature and the mean center using the Point Distance (Analysis) tool. Use the functions `get-Time` and `timeDifference` given in the script to report the amount of time each of these geoprocessing tool call requires. When run without debugging, the computation times may all be so small that they are rounded to zero seconds. Step through with the debugger and to see non-zero times.

Example script input: C:/gispy/data/ch15/park.shp C:/gispy/scratch/

Example script output in *debug* mode:
```
[Dbg] >>> Time for FeatureToPoint to create parkPoints.shp:
0 weeks, 0 days, 0:0:2
[Dbg] >>> Time for CentralFeature to create parkCentral.shp:
0 weeks, 0 days, 0:0:3
[Dbg] >>> Time for MeanCenter to create parkMean.shp:
0 weeks, 0 days, 0:0:2
[Dbg] >>> Time for PointDistance to create parkMean2Central.dbf:
0 weeks, 0 days, 0:0:1
```

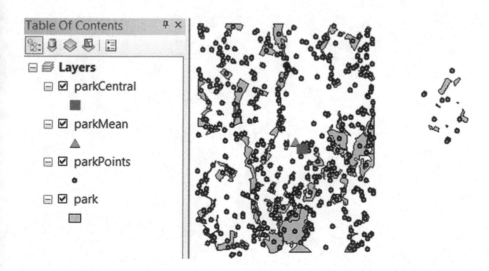

Chapter 16
User-Defined Modules

Abstract User-defined functions enable code reuse within a script; they can also be called from other scripts. To amplify code reusability, functions can be defined in a supporting script. Scripts that house sets of related function definitions and other related code are referred to as modules. This chapter focuses on defining and importing user-modules. It also discusses how to structure development code within a module. Finally, the chapter concludes with a practical example for managing GIS temporal data and time attributes.

Chapter Objectives
After reading this chapter, you'll be able to do the following:

- Articulate the purpose of a user-defined module.
- Create a user-defined module containing related functions.
- Import distributed user-defined modules.
- Write absolute and relative versions of a path.
- Call a function in a user-defined module.
- Use a conditional construct to exclude code in a user-defined module.

16.1 Importing User-Defined Modules

Functions within a user-defined module can be called from other scripts. A *module* is a Python script containing related function definitions and Python statements. Every Python script is a module, but we refer to it as a module particularly when we are importing it to use its functions in another script. You're already familiar with importing Python built-in modules and the `arcpy` package installed with ArcGIS software. Our geoprocessing scripts often begin with an import statement like this:

```
import arcpy, os, sys
```

The name of a module is simply the Python file name without the '.py' extension. For example, the `os` module refers to the 'os.py' script. Once you've imported the `os` module, you can use the `os` functions by referring to them with dot notation.

For example, you can use os.getcwd() to determine the current working directory and os.listdir(path) to get a list of the files in the path directory.

Importing user-defined modules and accessing the functions therein works in a similar way to using built-in modules. The main difference is that the location of a user-defined module may need to be specified explicitly before importing it, as discussed next.

Example 16.1: A user-defined module.

```
# Excerpt from 'listManager.py' in
#    C:\gispy\sample_scripts\ch16\supportCode
# Purpose: Provide list and delimited string manipulation functions.
def list2String(delimiter, L):
    '''Take a list and return a delimited string.'''
    # Join fails for non-string elements, so use list
    # comprehension to cast each element to string.
    stringL = [str(i) for i in L]
    # Join the string elements of stringL
    s = delimiter.join(stringL)
    return s

def string2List(delimiter, s):
    '''Take a delimited string and return a list'''
    L = s.split(delimiter)
    return L
```

The user-defined module, 'listManager.py', shown in Example 16.1, resides in a supportCode subdirectory of Chapter 16 sample scripts and contains two function definitions for manipulating Python lists. (The script also contains some commented code that will be used later in this chapter. For now, leave those lines commented. In other words, leave ## at the front of the lines). To use these functions outside of 'listManager.py', we need to first import the module, 'listManager'. The format for importing user-defined modules is the same as importing a built-in module. Try the following code:

```
>>> import listManager
```

Did it work? When you try this code, Python might throw the following exception:

```
Traceback (most recent call last):
File '<interactive input>', line 1, in <module>
ImportError: No module named listManager
```

If you didn't get this exception, try to generate it by restarting your IDE and entering this statement again. You should get an ImportError. Why should it work sometimes and not others? The answer lies in how Python imports modules. Python keeps a list of directory paths and searches the directories in this list for the files and packages that the user imports. If the file is located in one of the listed directories, the import succeeds. Otherwise, Python throws an ImportError exception. The directory path list is stored in the sys module as a property named path that gets initialized when the IDE is opened. You can see the list by importing sys and printing the path variable:

```
>>> import sys
>>> sys.path
['C:\\Windows\\SYSTEM32\\python27.zip','C:\\Python27\\ArcGIS10.3\\DLLs',...]
```

Only a few of the paths are shown here. The full list includes paths that are loaded automatically based on PythonWin installation dependent defaults, the PYTHONPATH environment variable, and the os current working directory. You can also modify the sys.path variable in your script. This way, you can define your own module and add its path to the list, so that it can be imported. Since sys.path is a Python list, you can just use the list append method. To add a path to the list. For instance, to add the directory 'C:/gispy/sample_scripts/ch16/support-Code' to the path, append this value. Then print the sys.path list again to see that the additional directory appears at the end of the list, once it has been appended:

```
>>> sys.path.append('C:/gispy/sample_scripts/ch16/supportCode')
>>> sys.path
['C:\\Windows\\SYSTEM32\\python27.zip', 'C:\\Python27\\ArcGIS10.3\\
DLLs',...,'C:/gispy/sample_scripts/ch16/supportCode']
```

When you open and run a script in PythonWin, the script's directory is automatically appended to the path. So if your supporting modules are in the same directory, as the script they support, their path does not need to be appended. For example, 'script1.py' can import the script2 module without any modifications to the system path because both scripts reside in 'C:/gispy/sample_scripts/ch16'. By contrast, 'script3.py' throws an exception when attempting to import script4 from the 'otherCode' subdirectory. Since 'script4.py' is not within the same directory, 'script3. py' fails to determine its location. The directory where module script4 resides must be appended to the system path first. To do so, you must use two separate import statements, one to import the sys module (and this must come first) and one to import the user-defined script. Appending the path comes after sys has been imported and before the user-defined module is imported, as in the following example:

```
import sys
sys.path.append('C:/gispy/sample_scripts/ch16/otherCode')
import script4
```

Try this by uncommenting the first two lines of code in 'script3.py' and running it again. This example hard-codes the file path to the supporting module. To make your code portable, you can dynamically specify the path. Dynamically specifying the path to append requires some understanding of absolute and relative paths. To illustrate these concepts, we'll use files from the multi-layered 'pics' directory ('C:\gispy\data\ch16\pics'). Back in Chapter 12, there is a drawing of the hierarchical structure of the same data (Figure 12.1). The *absolute path* of a file is the full path file name, starting with the name of the drive. The absolute path to 'backSeat.jpg' on the 'C:' drive is as follows:

```
filename = 'C:/gispy/data/ch16/pics/italy/backSeat.jpg'
```

The *relative path* of a file gives its position relative to another file. Hence, it doesn't need to start with a drive name. From the point of view of 'backSeat.jpg', there are several types of relative positions:

1. *Same directory*: For files in the same directory as 'backSeat.jpg', the relative path is simply the name of the other file. For example, 'bridge.jpg' is the entire relative path to 'bridge.jpg' from the point of view of the 'backSeat.jpg' file.
2. *Subdirectory*: 'backSeat.jpg' resides in the 'C:/gispy/data/ch16/pics/italy/' directory. To refer to files in subdirectories, the relative path lists the directories that fall below this one. For example, 'canal.jpg' is relatively referenced as 'venice/canal.jpg'.
3. *Up* or *up-and-over directories:* To climb upward in the hierarchy, you can use two dots. For example, 'istanbul.jpg' is relatively referenced as '../istanbul.jpg'. The double-dots take the path up out of the 'italy' directory to the next level up, which is the 'pics' directory. Relative paths have to follow the lines connecting the file in the tree. To reference 'old_city.jpg' from 'backSeat.jpg', you have to climb out of 'italy' using double-dots and then step down into the 'jerusalem' directory. So the relative path would look like this: '../jerusalem/old_city.jpg'.

You can test relative paths using Windows Explorer. For example, browse to 'C:/gispy/data/ch16/pics/italy/'. Then enter '../jerusalem/' in the window to confirm that it takes you to the directory that contains 'old_city.jpg'.

Supporting user-defined modules can be placed in another directory with a fixed position relative to the script that imports it. The goal is that you can move the entire workspace and the code won't break due to hard-coded paths. Figure 16.1 shows the relevant portion of the sample scripts directory tree. As an example, consider looking at the relative positions of 'script1.py' and 'script4.py'. The following four steps can be used to import the `script4` module dynamically from 'script1.py'.

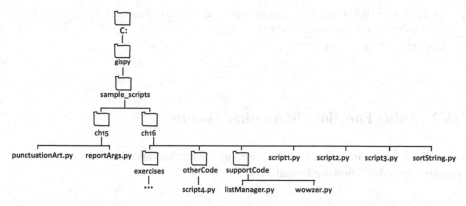

Figure 16.1 The directory tree structure for the examples discussed in this chapter.

1. Retrieve the full path of 'script1.py' using the `os` module:

```
import sys, os
scriptPath = os.path.abspath(__file__)
```

2. Combine the script path with the relative path of the user-defined module using the `os.path.join` method.

```
scriptDir = os.path.dirname(scriptPath)
relativePath = 'otherCode'
modulePath = os.path.join(scriptDir, relativePath)
```

3. Append `sys.path`.

```
sys.path.append(modulePath)
```

4. And finally, import the module.

```
import script4
```

These steps will usually be placed together near the beginning of the main script. As another example, to import the `punctuationArt` module (a Chapter 15 sample script) from 'script4.py', the following code could be placed at the beginning of 'script4.py':

```
import sys, os
scriptPath = os.path.abspath(__file__)
scriptDir = os.path.dirname(scriptPath)
relativePath = '../../ch15'
```

```
modulePath = os.path.join(scriptDir, relativePath)
sys.path.append(modulePath)
import punctuationArt
```

16.2   Using Functions in Another Module

To call a function in another module, import the module and then you can use dot notation with the following format:

```
moduleName.functionName (arg1, arg2, ...)
```

The two functions in Example 16.1 are designed for working with lists. One converts a delimited string (such as a string of comma separated values) to a list. The other converts a list to a delimited string. These functions are saved in the module 'listManager.py', so instead of just using the function names to call them, you need to use the dot notation to indicate where Python should look for them. The following code imports the module and calls the `string2List` function using dot notation:

```
>>> import listManager
>>> theString = 'z;x;y'
>>> theList = listManager.string2List(';' , theString)
>>> theList
['z', 'y', 'x']
```

This Interactive Window code works because the path to this module was appended to `sys.path` earlier in the same IDE session. The user-defined module functions can also be called from within another script, as in Example 16.2. The script 'sortString.py' resides in the directory above 'listManager.py'. This script first appends the path to 'listManager.py', then imports the `listManager` module and calls its functions. The purpose of 'sortString.py' is to sort the items in a delimited string. To do so, it converts the string to a list, sorts the list, and then converts the list to a string. To see how it works, run 'sortString.py'. The printed output should look like this:

```
>>> sortString.py results:
fireID~unit~shelter~alt~campgr -> alt~campgr~fireID~shelter~unit
```

Two of the functions it calls, `string2List` and `list2String`, are defined in 'listManager.py'. The third function called in this example is the list `sort` method discussed in Chapter 4. This is an object method, so instead of module-dot-functionName, `mylist.sort()` is called on the object, with the object-dot-method syntax. The purpose of the double hash comments, the reload statement, and the `delimStrLen` function are discussed next.

Example 16.2: Using a user-defined module from a script.

```
# excerpt from sortString.py
# Purpose: Sort the items in a delimited string

import sys, os
scriptPath = os.path.abspath(__file__)
scriptDir = os.path.dirname(scriptPath)
relativePath = 'supportCode'
modulePath = os.path.join(scriptDir, relativePath)
sys.path.append(modulePath)
import listManager

delimiter = '~'# set delimiter
# Set delimited string.
theString = 'fireID~unit~shelter~alt~campgr'

# Get a list from the string.
# module.function(arg1, arg2,...) format.
theList = listManager.string2List(delimiter, theString)

# Sort the list. Sort is a native list method
# (not a method defined in listManager.
# Notice the difference in how it is called.
# It's called with object.method format,
# where the object is mylist.
theList.sort()

# Get a string from the list.
# module.function(arg1, arg2,...) format.
newString = listManager.list2String(delimiter, theList)

print '\nsortString.py results:'
print '{0} -> {1}'.format(theString,newString)
```

16.3   Modifying User-Defined Modules (Reload!)

When you import a user-defined module, it loads the module and stores the name in
an internal list. As you step through a script in the debugger, you may notice that the
first time you step past an import statement for a large package, such as `arcpy`, the
import statement takes a few seconds; whereas, if you run the same script again
within the same PythonWin session, you can step past this import instantly. This is
because during the second run, Python quickly looks up the module in its internal
list of loaded modules, finds that module name, and does not reload the module
contents. This makes Python more efficient, but it can be baffling when you first
work with user-defined modules, because it can seem as if your updates to a user-
defined module are not being heeded. As an example, run sample script 'sortString.
py' (if you haven't already done so). Next, make a change to 'listManager.py' by
adding the `delimStrLen` function in 'listManager.py'. To do so, you can select
the following lines of code which are commented out in the script and use the block
uncomment shortcut (Alt + 4):

```
def delimStrLen(delimiter, s):
    '''Return the number of items in a delimited string.'''
    theList = string2List(delimiter, s)
    return len(theList)
```

Save 'listManager.py'. Then uncomment the following code in 'sortString.py',
to call the new function:

```
num = listManager.delimStrLen(delimiter,theString)
print 'Number of items: {0}'.format(num)
```

Save and run 'sortString.py' again. Despite the fact that you've just added the
function to listManager, Python throws the following traceback exception:

```
AttributeError: 'module' object has no attribute 'delimStrLen'
```

As far as Python is concerned, there is no such function in the `listManager`
module, because when the module was loaded, that function was not there. When
you modify a supporting module, the changes are not automatically recognized by
the file that imports it. To force new content in a module to be loaded, you can
reload it with the following steps:

1. Shift the focus to the user-defined module, by clicking on that module's window.
   In our case, click on 'listManager.py'.
2. Then, click the 'Import/Reload' button ⊟ on the PythonWin standard toolbar
   next to the 'Run' button,

The module has now been reloaded. To see the results, run 'sortString.py' again. The results should appear as follows:

```
>>> sortString.py results:
fireID~unit~shelter~alt~campgr -> alt~campgr~fireID~shelter~unit
Number of items: 5
```

The 5 is returned by the new function call, which counts the number of items in the delimited string.

There is also a built-in `reload` function which has the same effect as clicking the 'Import/Reload' button. This function reloads a previously imported module passed to it as an argument. The argument must be a `module` object, so it must be successfully imported before it can be reloaded. In other words, you must put an import statement before a reload statement, as in the following example:

```
import listManager
reload(listManager)
```

Temporarily placing a reload statement inside the main script can be useful while making repeated modifications to the code in user-defined modules. However, usually the reload statement should not be left in code to be shared, as it can slow the performance. You may have noticed that the first time a module is imported, Python creates a corresponding file with a '.pyc' extension. For example, look in the sample scripts Chapter 16 'supportCode' directory to see the 'listManager.pyc' file. This is an intermediate file that encodes the module in a format called 'byte code'. That file gets created based on the user-defined 'listManager.py' module the first time you import `listManager` and it gets rebuilt each time that module is reloaded. Once it has been created, this file is executed when the current IDE session uses `listManager`.

16.4   Am I the Main Module? What's My Name?

When you reload a module or import it for the first time, it not only stores the names of the functions, but it also executes any statements in that script that are outside of the functions. As an example, uncomment the reload statement in 'sortString.py' and then uncomment the following code at the end of 'listManager.py':

```
print '\nIn listManager.py, test string2List: '
theString = 'z;x;y'
theList = string2List(';', theString)
print '{0} -> {1}'.format(theString,theList)
```

Save both scripts and run 'sortString.py'. You'll see the following output:

```
In listManager.py, test string2List:
z;x;y -> ['z', 'x', 'y']

sortString.py results:
fireID~unit~shelter~alt~campgr -> alt~campgr~fireID~shelter~unit
Number of items: 5
```

The first two lines are output from the 'listManager.py' file. Usually, you don't want side-effects like this when you import a file. However, as you develop functions for user-defined modules, you need to test them. To include test code in your module and still avoid this undesired artifact, you can programmatically determine whether you're running the module directly or importing it from another module. The built-in variable, __name__, gets assigned the value '__main__' when the module is loaded as the main module. When the module is imported, the value of __name__ is set to the name of the module. To see this in action, we'll take a brief detour to another script, since you can only see this print the first time a script is imported during a PythonWin session. The script named 'wowzer.py' contains a line of code to print the value of __name__:

```
print '**** Name = ', __name__
```

Run 'wowzer.py' with the 'Run' button and it prints the following:

```
>>> **** Name = __main__
```

But import 'wowzer.py' in the Interactive Window and it prints wowzer:

```
>>> import wowzer
**** Name = wowzer
```

Now returning to 'listManager.py', you use this difference in the value of __name__ to restrict code in a user-defined module from being run when it is imported, by using a conditional expression that compares the value of __name__ to '__main__'. To try this, insert the following line of code into 'listManager.py' just after the last function definition:

```
if __name__ == '__main__':
```

Then indent the four lines of code that you uncommented at the start of Section 6.4 and add a dedented print statement to the end of the script, so that it appears as follows:

```
if __name__ == '__main__':

    print '\nIn listManager.py, test string2List: '
    theString = 'z;x;y'
```

```
    theList = string2List(';' , theString)
    print '{0} -> {1}'.format(theString,theList)
print 'Zowee!'
```

Save and run 'listManager.py' once more to see that it prints output. Then restart PythonWin, open 'sortString.py', and run it again to confirm that the code inside the conditional construct is not executed, and so it only activates the dedented print statement (Zowee!).

16.5   Time Handling Example

User-defined modules group related functions for convenient re-use. Python has two built-in time related modules, `time` and `datetime`. The user-defined `report-Time` module in Example 16.3 (located in 'C:\gispy\solutions\ch16\exercises\ exerSupportCode') contains some convenient functions that leverage these built-in modules. It uses the `ctime` and `sleep` methods from the `time` module. The `time` module can access the current time and convert time across time zones. The `ctime` function returns a string representing the current date and time. The following example shows what it would have printed if called just before midnight on New Year's Eve back in 1999:

```
>>> import time
>>> time.ctime()
'Fri Dec 31 23:59:59 1999'
```

The `sleep` function suspends processing for the given amount of seconds. This is used in 'timeReport.py' to demonstrate a time lapse. If you try the following code, you'll have to wait 10 seconds before the next command prompt appears:

```
>>> time.sleep(10)
```

The `datetime` module has data types for dealing with dates, times, and time intervals. The 'datetime' type stores both date and time information. The following code creates a `datetime` object for that moment before the start of the twenty-first century and then prints the hour, year, and weekday of the object:

```
>>> import datetime
>>> dt = datetime.datetime(1999, 12 ,31, 23, 59)
>>> dt
datetime.datetime(1999, 12, 31, 23, 59)
>>> dt.hour
23
```

```
>>> dt.year
1999
>>> dt.weekday()
4
```

The datetime type has a year, month, day, hour, minutes, and seconds properties. The weekday function returns an index, 0 for Monday, 1 for Tuesday, and so forth. Example 16.3 uses the weekday function and a dictionary to return a weekday name instead of an index.

Dates can be differenced. When a subtraction sign is used between two datetime objects, a timedelta type object is returned. The following code creates a datetime object for noon on the last day of 1999 and demonstrates ways to use the timedelta object:

```
>>> dt2 = datetime.datetime(1999,12,31, 11, 59)
>>> timeDiff = dt - dt2
>>> timeDiff
datetime.timedelta(0, 43200)
>>> timeDiff.days
0
>>> timeDiff.total_seconds()
43200.0
>>> hrs = timeDiff.total_seconds()/3600
>>> hrs
12.0
```

The timeDiff object stores the time difference. The total_seconds function returns the total time difference in seconds. These moments are 12 h apart (at noon and midnight of the same day). datetime objects can also be compared for temporal ordering, by using Python comparison operators.

```
>>> dt2 < dt
True
>>> datetime.datetime.now() < dt
False
```

The time.ctime() function call returns a string describing the current date and time; Whereas, the datetime.datetime.now() function returns a datetime object for the current moment—which is greater than dt. Example 16.3 uses datetime.datetime.now() in the getTime function to get a datetime object. In the test code at the bottom of the script this is called before and after a 5 second pause in execution. Then these two objects are passed into the reportDiffTime method to print how much time elapsed between the two.

Example 16.3: A user-defined module.

```python
# timeReport.py
import datetime, time

def reportTime(message='The current date and time is'):
    '''Print the current time'''
    now = time.ctime()
    print '{0}: {1}'.format(message, now)

def getDay(theDate):
    '''Given a date, return the day of the week'''
    index = theDate.weekday()
    wDict = { 0 : 'Monday', 1 : 'Tuesday', 2 : 'Wednesday',
              3 : 'Thursday', 4 : 'Friday', 5 : 'Saturday',
              6 : 'Sunday'}
    return wDict[index]

def getTime():
    '''Report the current time'''
    t = datetime.datetime.now()
    return t

def reportDiffTime(start, end, message= 'Time elapsed'):
    '''Print the number of seconds that passed
    between 'start' and 'end'.'''
    difference = end - start
    seconds = difference.total_seconds()
    print '{0}: {1} seconds.'.format(message, seconds)

if __name__ == '__main__':
    reportTime('Script began running at')
    # Get current time.
    beforeSleep = getTime()
    time.sleep(5)
    # Get current time.
    afterSleep = getTime()
    message = 'Time elapsed for sleeping'
    # Print the time difference.
    reportDiffTime(beforeSleep, afterSleep, message)
    reportTime('Script completed at')
    print 'Hurray! I like {0}s.'.format(getDay(afterSleep))
```

16.6   Summary

We'll be discussing a Python structure called a `class` in an upcoming chapter. User-defined modules are often used as containers for classes, so several of these concepts will come up again when we reach that topic. The following list summarizes the key concepts in this chapter:

- When calling a function from within the module where it is defined, you just need to use the name. When calling it from another module, you need to start with the name of the module and use dot notation, so that Python knows where to find the function.
- Importing a user-defined module in the same directory is the simplest scenario. The system path does not need to be appended. In all other cases, the system path needs to be appended.
- You must use a separate statement to import `sys` and a user-defined module when the system path needs to be appended. First import `sys` before appending the path, then use a second import statement for the user-defined module.
- You can append a hard-coded path, but in most applications this approach hinders portability. Generally, scripts dynamically determine their own location and then append a relative path to imported modules.
- When you make modifications to a module, remember to reload it so that the changes are recognized by the script that imports the module.
- You can use a conditional expression that checks if you're in the main script to exclude select portions of user-defined module code from imports. This allows you to use the script both as a main script and as a supporting module.

16.7   Key Terms

User-defined module
`sys.path` variable
Built-in `__file__` variable
Absolute vs. relative paths

Built-in `reload` function
Built-in `__name__` variable
Built-in `time` and `datetime` modules

16.8   Exercises

1. **owlCall.py** Write a script which imports sample script 'owl.py' and calls `printOwl`. Assume 'owlCall.py' and 'owl.py' reside in the same directory.

2. **wazzup.py** Sample script 'wazzup.py' (in 'C:\gispy\sample_scripts\ch16\exercises\favScripts') already imports scriptA. Modify the script so that it also

imports scriptB, scriptC, scriptD, and scriptE. These modules are scattered throughout 'C:/gispy/sample_scripts/ch16/exercises'. Append their relative paths to the system path variable, so that they can be imported. Check that the imports are working three ways and note your thoughts as comments within the script:

(a) First check the output. Each of the A-E scripts contains print statements to help you confirm successful import. The printed output should look like this:

```
>>> I am in scriptA.py
I am in scriptB.py
I am in scriptC.py
I am in scriptD.py
I am in scriptE.py
```

Can you explain why your output might only print the last line ('I am in scriptE.py'), when you finally have it working?

(b) Next, restart PythonWin, copy 'wazzup.py' to the 'C:/gispy/sample_scripts/ch16/exercises' directory, and rename it to 'wazzup2.py' and run it. When placed in this directory, only the first statement ('I am in scriptA.py') is printed. Can you say why this does print but the second statement doesn't? If the script used absolute paths, instead of relative paths would all the imports have worked? Why is it necessary to restart PythonWin before running this test?

(c) Finally, restart PythonWin and temporarily rename the 'ch16' directory containing these scripts to 'ch16_temp'. Open and run the original 'wazzup.py' script, again. If relative paths were used correctly, why should the imports still succeed? Rename the directory to ch16 when this test is complete.

3. **abc.py** Create a script named 'abc.py' and place it in the Chapter 16 exercises directory, 'C:/gispy/sample_scripts/ch16/exercises'. Add code to 'abc.py' that imports Chapter 15 sample script 'reportArgs.py' and calls `printArgs`. Make the script portable, assuming the scripts remain in the same relative positions. (For example, if the 'gispy' directory is renamed as 'gypsy', the import should still work.) When you run your initial solution, the argument list should be printed twice. Can you explain why?

Example input:
a b c

Example output:
```
Argument 0:  C:\gispy\sample_scripts\ch16\exercises\abc.py
Argument 1:  a
Argument 2:  b
Argument 3:  c
Argument 0:  C:\gispy\sample_scripts\ch16\exercises\abc.py
Argument 1:  a
Argument 2:  b
Argument 3:  c
```

Modify 'reportArgs.py' so that it only executes `printArgs` if it is run as the main script. Then run 'abc.py' again and confirm that the argument list is only printed once.

4. **mathMod.py** Write a user-defined module that contains three functions. One named `add` adds two arguments and returns the result, one named `mult` multiplies two arguments and returns the results, and one named `sub` subtracts two arguments and returns the results. Add a conditional expression that checks if `__name__` is equal to `'__main__'` and add the following test code inside this conditional block:

```
A = float(sys.argv[1])
B = float(sys.argv[2])
sumV = add(A, B)
product = mult(A, B)
difference = sub(A, B)
print sumV, product, difference
```

Test your code against the following sample input and output:

Example input:
5 2

Example output:
```
7.0 10.0 3.0
```

5. **listUniqueValues.py** The sample script named 'listManager2.py' which resides in the 'listManage' subdirectory is similar to 'listManager.py' in Example 16.1, but it has an additional function named 'uniqueList'. Run listManager2.py to see what the additional function does. Then create a script, 'listUniqueValues.py', in the 'C:/gispy/sample_scripts/ch16/exercises' directory, which imports the `listManager2` module (using a relative path) and uses this additional function. The script should use a search cursor to get a list of the values for a given field. Then it should get a list of unique field values. Next, it should sort the unique field value list. Finally, it should print the list. The script should take two arguments: a full path file name of a shapefile and the name of a text field.

Example input: C:/gispy/data/ch16/park.shp COVER

Example output:
```
[u'orch', u'other', u'woods']
```

6. **csvStrings.py** The sample script, 'csvString.py', which can be found in the 'C:/gispy/sample_scripts/ch16/exercises' directory, contains a list of annual wheat yields between 2000 and 2007 for five farms. These are stored as comma separated value (csv) strings. Add code to 'csvString.py' to import the `listManager3` module, using a relative path to 'listManager3.py' which resides in

the 'listManage' subdirectory. Then add code to 'csvString.py' to use functions from the listManager3 module to analyze the wheat yield data: First, use `delimStrLen` to find and print the number of wheat yield sample time steps recorded for each farm. Run the script and print the output as follows:

```
Site 1 has 8 samples: 1,4.07,4.21,4.15,4.64,4.03,3.74,4.56
Site 2 has 8 samples: 2,,4.29,4.4,4.69,3.77,4.46,4.76
Site 3 has 8 samples: 3,3.9,4.64,4.05,4.04,3.49,3.91,4.52
Site 4 has 8 samples: 4,3.63,,4.92,4.64,3.75,4.1,4.4
Site 5 has 8 samples: 5,3.15,,4.08,4.73,3.61,3.66,
```

Second, uncomment `delimStrLen2` in 'listManager3.py' and modify 'csvStrings.py' to call this function instead of `delimStrLen`. You'll see that the results are more accurate. In fact, wheat yield readings are not available for some farms during some years; Therefore, some of the entries are empty strings. The simpler function `delimStrLen` counted these, but `delimStrLen2` avoids counting empty strings:

```
Site 1 has 8 samples: 1,4.07,4.21,4.15,4.64,4.03,3.74,4.56
Site 2 has 7 samples: 2,,4.29,4.4,4.69,3.77,4.46,4.76
Site 3 has 8 samples: 3,3.9,4.64,4.05,4.04,3.49,3.91,4.52
Site 4 has 7 samples: 4,3.63,,4.92,4.64,3.75,4.1,4.4
Site 5 has 6 samples: 5,3.15,,4.08,4.73,3.61,3.66,
```

7. **timeTrig.py** Create and save a script named 'timeTrig.py' in the directory named 'C:/gispy/sample_scripts/ch16/exercises/timeTrigDir'. The 'timeTrig.py' script will import the user-defined module `timeReport` using a relative path. Find 'timeReport.py' in the 'excrSupportCode' subdirectory. Run 'timeReport.py' as the main script to observe how the functions work. Then use the time functions in your 'timeTrig.py' script to report how long it takes to calculate the sine of the cells of each raster in the given workspace using the Sine (Spatial Analyst) tool. Be sure to check out the spatial analyst extension in the script before calling the Sine tool. The script should take two input arguments using `sys.argv` to set an input geodatabase workspace and an output directory. To name the output files, append '.tif' to each raster name. Print the name of each successful sine output and the overall time elapsed while performing the sine calculations.

Example input: C:/gispy/data/ch16/rastTester.gdb C:/gispy/data/ch16

Example output:
```
>>> C:/gispy/scratch\elev.tif created.
C:/gispy/scratch\landcov.tif created.
C:/gispy/scratch\soilsid.tif created.
. . . (and so forth)
Calculating sine took: 11.695 seconds.
```

8. **bufferArgs.py** Write a portable script named 'bufferArgs.py' in the 'C:/gispy/ sample_scripts/ch16/exercises' directory that performs a buffer based on two user input arguments: the full path file name to a shapefile and a numerical distance measure (use the default units). Use the functions from the user-defined module, 'argHandler.py' in the `exerSupportCode` subdirectory to check that the first user argument is a shapefile and the second user argument is a number. If either fails, exit the script, else create 'buffOut.shp' in the 'C:/gispy/scratch/' directory. The code can assume this directory already exists.

Example input 1: C:/gispy/data/ch16/park.shp 5

Example output 1:
```
>>> buffOut.shp created in C:/gispy/scratch
```

Example input 2: C:/gispy/data/ch16/jack.jpg 4.5

Example output 2:
```
>>> Expecting a shapefile for arg 1. Got RasterDataset instead.
No buffer can be computed.
```

Example input 3: C:/gispy/data/ch16/park.shp ba

Example output 3:
```
>>> Input ba is not a floating point number.
No buffer can be computed.
```

Chapter 17
Reading and Writing with Cursors

Abstract Batch processing can be extended to work with individual records within GIS data tables. Previous chapters listed fields in multiple files within nested directories; this chapter adds another dimension to data exploration. Each data table is made of *records*, the rows of values in the attribute table, which can also be listed within a script. Though the term 'records' may be more familiar to database users, we'll use the term 'rows' to refer to data records for consistency with the arcpy documentation. Cursors and file input/output techniques, enable reading and writing individual rows within data tables. The focus of this chapter is arcpy cursor functions, which work with the rows in Esri attribute tables, such as feature class tables.

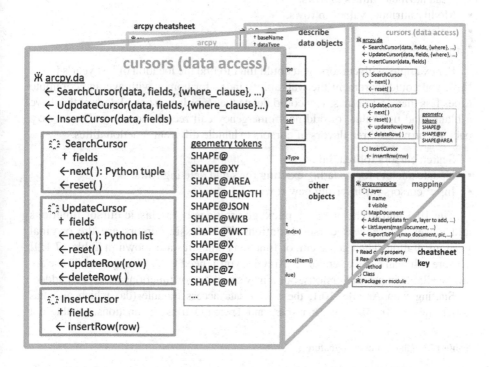

© Springer International Publishing Switzerland 2015
L. Tateosian, *Python For ArcGIS*, DOI 10.1007/978-3-319-18398-5_17

Chapter Objectives
After reading this chapter, you'll be able to do the following:

- Read and modify GIS attribute table files.
- Query the data for a select set of table rows.
- Use error handling in cursor scripts.
- Prevent data locking errors.
- Update or insert field values.
- Insert point and polygon data.

17.1   Introduction to Cursors

GIS attribute tables are in proprietary formats not accessible with generic Python text file reading methods. `arcpy` cursors are objects designed specifically to accommodate GIS tables. You can create a cursor by calling an arcpy function and specifying a table. Then, depending on the type of cursor you requested, it can do one or more of the following:

- Read attribute values in rows.
- Modify attribute values in rows.
- Delete rows.
- Insert new rows.

For example, with cursors, you could filter to find the location of last year's forest fires, you could increment the walkability index for homes with average commute times less than 15 minutes, you could delete the records in a parcel dataset that have a blank land use value, or add new emergency call records to a log. The `arcpy` package provides three flavors of cursors to handle different functionalities:

- Search cursors for reading.
- Update cursors for updating existing rows and deleting rows.
- Insert cursors for inserting new rows.

The 10.1 and higher `arcpy` package provides both classic cursor functions and newer data access cursor functions. The classic cursors are the original implementation. The classic cursor function signatures are shown in Table 17.1. If you are using an earlier version of ArcGIS, you'll need to use this syntax. These are still available in newer versions, but a new set of cursor functions have been added.

Starting with ArcGIS 10.1, there is a data access module (da) which also has `SearchCursor`, `UpdateCursor`, and `InsertCursor` functions. Double dot

Table 17.1 Classic cursor signatures.

`arcpy.SearchCursor(dataset, {where_clause}, {spatial_reference}, {fields}, {sort_fields})`
`arcpy.UpdateCursor(dataset, {where_clause}, {spatial_reference}, {fields}, {sort_fields})`
`arcpy.InsertCursor(dataset, {spatial_reference})`

Table 17.2 Data access cursor signatures.

```
arcpy.da.SearchCursor(in_table, field_names, {where_clause},
{spatial_reference}, {explode_to_points}, {sql_clause})
```

```
arcpy.da.UpdateCursor(in_table, field_names, {where_clause},
{spatial_reference}, {explode_to_points}, {sql_clause})
```

```
arcpy.da.InsertCursor(in_table, field_names)
```

notation is used to call these functions within the da module. Table 17.2 shows their signatures. Notice the differences between the classic and data access SearchCursor signatures; Single dot versus double dot and one required argument versus two are some of the obvious differences. Both sets of functions return a cursor object which allows you to iterate through the rows. However, the data access cursors are significantly faster than the classic cursors, so this chapter focuses on the data access cursors.

Cursors are quite useful, but proceed with caution; Cursors lock the data while they are operating, meaning other ArcGIS Desktop processes cannot simultaneously delete or otherwise modify the data. Data cannot be previewed while edits are being made with update or insert cursors. If necessary, make a copy of the data and view this table instead of the original during cursor operations. Scripts need to delete cursor objects for locks to be released. Also, update and insert cursors can modify your data, so backup data before testing cursor code. This chapter begins with search cursors, to illustrate basic cursor object syntax.

FID	Shape*	FireId	CalendarYe	FireNumber	FireName	FireType_P	SizeClass	StartTime	Authoriz_1
0	Point	239008	1997	9702	MEADOW	11	A	6/9/1997	Fire Technician
1	Point	239009	1997	9703	LITTLE CRK	11	A	7/20/1997	ND Fire Techn
2	Point	239016	1997	9710	T.Calvin	11	B	10/9/1997	Fire Technician
3	Point	239017	1997	9711	VISITORC	11	A	10/11/1997	Park Ranger
4	Point	239031	1998	9810	PILGRIM HT	12	B	9/6/1998	fmo
5	Point	239036	1999	9905	DUMP	13	A	6/24/1999	Engine Foreman
6	Point	239039	1999	9908	PETRELEIF	13	A	7/11/1999	Engine Foreman
7	Point	239042	1999	9911	COCONUT	11	B	7/18/1999	FMO
8	Point	239060	2000	10	HIGHHEAD	11	B	12/19/2000	FMO
9	Point	239127	2004	424	HRCOVEDUNE	11	A	7/26/2004	CACO FMO
10	Point	513169	2005	506	Beech Fost	13	A	3/22/2005	CACO FMO
11	Point	513179	2005	507	Pilg. Hgts	13	B	3/23/2005	CACO FMO

Figure 17.1 fires.shp table contains 11 wildfire records.

17.2 Reading Rows

The data access SearchCursor function has two required arguments: the input table and a list of field names. This function returns a search cursor object. The cursor object is used to access the rows. The following code creates a search cursor for the 'FID', 'FireId', and 'FireName' fields in the table shown in Figure 17.1:

```
>>> import arcpy
>>> # Create a cursor
>>> fc = 'C:/gispy/data/ch17/fires.shp'
>>> cursor = arcpy.da.SearchCursor fc, ['FID', 'FireId', 'FireName'])
```

The cursor object has two methods next and reset and one property fields. Rows can be extracted from the cursor object using the next method or using FOR-loops. The next method gets the next row in the table. This provides a convenient way to examine row contents interactively as you become familiar with cursors. The following code uses the next method to get a row and prints the row:

```
>>> row = cursor.next() # Get an individual row.
>>> row
(0, 239008.0, u'MEADOW')
```

0, 239008.0, and MEADOW are the values for 'FID', 'FireId', and 'FireName' in the first row of the table in Figure 17.1. The row variable is a Python tuple that can be indexed to get individual field values. These indices correspond to the order in which the field names are passed into the cursor. For example, the 'FID' index is zero, but the 'FireId' index is 1, even though it is the third column in the attribute table, and the 'FireName' index is 2, even though it is the sixth column in the table. The indices are 0, 1, and 2 because a list of three field names was used to create the cursor.

```
>>> row[0] # FID
0
>>> row[1] # FireId
239008.0
>>> row[2] # FireName
u'MEADOW'
```

The next method gets the first row the first time it is called; Calling it again gets the second row and calling it again gets the third row:

```
>>> row = cursor.next()      # Get the second row
>>> row
(1, 239009.0, u'LITTLE CRK')
>>> row = cursor.next()      # Get the third row
>>> row
(2, 239016.0, u'T.Calvin')
```

The cursor reset method brings the pointer back to the first row.

```
>>> cursor.reset()
>>> row = cursor.next()
>>> row
(0, 239008.0, u'MEADOW')
```

The `fields` property returns a tuple containing the field names, as shown in the following code:

```
>>> fds = cursor.fields
>>> fds
(u'FID', u'FireId', u'FireName')
```

Notice that `fields` is not followed by parentheses because it's a cursor property; Whereas, `next` and `reset` require parentheses because they are methods. Omitting the parentheses will not throw an exception, but it can lead to other perplexing errors as shown in the following code:

```
>>> row = sc.next     # Missing parentheses!
>>> row
<method-wrapper 'next' of da.SearchCursor object at 0x10CF8700>
>>> row[0]
Traceback (most recent call last):
File '<interactive input>', line 1, in <module>
TypeError: 'method-wrapper' object is not subscriptable
```

17.3   The Field Names Parameter

The field names can be specified in any order within the `field_names` parameter list. As mentioned earlier, the row index for that field depends on the order in which they are specified, not the order of the fields in the attribute table. Compare the field name ordering in the following code to the order of the fields in the attribute table shown in Figure 17.1:

```
>>> # Create a cursor
>>> fc = 'C:/gispy/data/ch17/fires.shp'
>>> cursor = arcpy.da.SearchCursor(fc, ['FireName', 'FID'])
>>> row = cursor.next()
>>> row
(u'MEADOW', 0)
>>> row[0]
u'MEADOW'
>>> row[1]
0
```

The classic `arcpy` cursors do not require a field name list; These cursors always contain all fields. Data access cursors can be made to simulate this behavior. Using a string asterisk (`'*'`) for the field name parameter obtains all fields. The drawback is that this may affect performance for large files.

```
>>> # Create a cursor with access to ALL the fields.
>>> cursor = arcpy.da.SearchCursor(fc, '*')
>>> cursor.fields
(u'FID', u'Shape', u'FireId', u'CalendarYe', u'FireNumber',
u'FireName', u'FireType_P', u'SizeClass', u'StartTime',
u'Authoriz_1')
```

In case the field names are not hard-coded in the script, the field names can be derived using the ListFields method. Example 17.1 gets a subset of the fields based on the field type. This script lists all the fields, then calls a user-defined function that returns a list of field names for all fields except those that are 'Geometry' or 'OID' type data. OID stands for object identifier, the unique key for each record created automatically when a table is created. Example 17.1 uses a data access cursor without hard-coding the field names. The last line of code in Example 17.1 uses the del key word to delete the cursor. We'll look at what this does and why this is important shortly.

Example 17.1: Print data values with unknown field names.

```
# printTableExclude.py
# Purpose: Print the names of the non-geometry non-OID type
#          fields in the input file and the value of these
#          fields for each record.
# Usage: No script arguments required.
import arcpy

def excludeFields(table, types=[]):
    '''Return a list of fields minus those with
       specified field types'''
    fieldNames = []
    fds = arcpy.ListFields(table)
    for f in fds:
        if f.type not in types:
            fieldNames.append(f.name)
    return fieldNames

fc = 'C:/gispy/data/ch17/fires.shp'
excludeTypes = ['Geometry', 'OID']
fields = excludeFields(fc, excludeTypes)

with arcpy.da.SearchCursor(fc, fields) as cursor:
    print cursor.fields
    for row in cursor:
        print row
del cursor
```

17.4   The Shape Field and Geometry Tokens

Cursors not only provide access to the row values you see in the attribute table of a file, they can also access each feature's geometric information. When a feature class attribute table is viewed in ArcGIS, the 'Shape' column shows a shape type, such as 'Polygon'. However, more information about each feature is stored internally for visualization and geoprocessing operations. This information can be accessed through `arcpy Geometry` objects. Notice that the field 'type' of the 'Shape' field is a 'Geometry'.

```
>>> fds = arcpy.ListFields('C:/gispy/data/ch17/park.shp')
>>> [f.name for f in fds]
[u'FID', u'Shape', u'COVER', u'RECNO']
>>> [f.type for f in fds]
[u'OID', u'Geometry', u'String', u'Double']
```

The `Geometry` objects have properties to do with geometric features (Figure 17.2). Some properties are intuitive, such as `area`, `length`, `isMultipart`

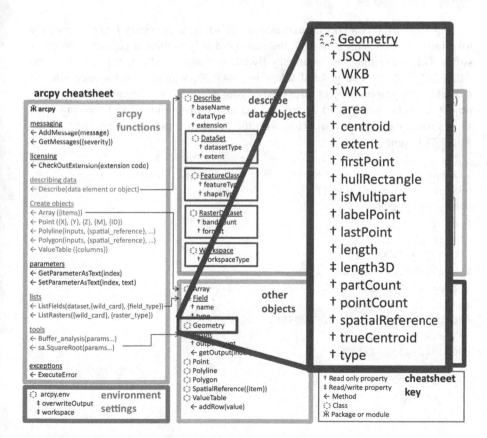

Figure 17.2   Selected `Geometry` object properties.

and so forth; Others, such as WKT (for well-known text, a portable geometry format) may be less familiar. Search the ArcGIS Resources site for 'Geometry (arcpy)' to see an explanation of each property. Previously, we used the Geometry object in field calculation expressions, as in the following code:

```
>>> data = 'C:/gispy/data/ch17/special_regions.shp'
>>> fieldName = 'PolyArea'
>>> expr = '!shape.area!'
>>> arcpy.CalculateField_management(data, fieldName, expr, 'PYTHON')
```

Cursor notation for Geometry objects differs from the field calculator notation. The following code creates a cursor for polygon areas and prints the area of the first polygon feature:

```
>>> parkData = 'C:/gispy/data/ch17/parks.shp'
>>> cursor = arcpy.da.SearchCursor (parkData, ['SHAPE@AREA'])
>>> row = cursor.next()
>>> row[0]
600937.0921638225
```

SHAPE@AREA is a special expression known as a *geometry token*. Geometry token strings start with SHAPE@ and can optionally include a geometry property suffix. Other examples of geometry tokens include SHAPE@, SHAPE@XY, and SHAPE@LENGTH. The general token is SHAPE@, which provides access to all available geographic object properties. The general token returns a geometry object; Use dot notation to access the geometry properties. The following code shows some examples of working with the geometric properties returned by the SHAPE@ token:

```
>>> cursor = arcpy.da.SearchCursor (parkData, ['SHAPE@'])
>>> row = cursor.next()
>>> row[0].type
'polygon'
>>> row[0].area
600937.0921638225
>>> row[0].centroid
<Point (2131483.24062, 191220.538898, #, #)>
>>> row[0].firstPoint
<Point (2131312.35816, 190450.120062, #, #)>
>>> row[0].area
600937.0921638225
```

These are called tokens because they represent the Geometry object. Using the 'Shape' field without the @ symbol, only produces a tuple with x and y coordinates of the feature centroids as shown in the following code:

```
>>>> cursor = arcpy.da.SearchCursor (parkData, ['SHAPE'])
>>> row = cursor.next()
>>> row[0].centroid
Traceback (most recent call last):
File '<interactive input>', line 1, in <module>
AttributeError: 'tuple' object has no attribute 'centroid'
>>> row[0]
(2131483.240622718, 191220.5388983136)
```

17.5   Looping with Cursors

Repeatedly calling the next method yields each subsequent row until the rows are exhausted. However, most search and update cursor scripts use looping. If the intention of the script is to use all rows, a FOR-loop should be used instead of the next method. The cursor is not a Python list, but is an *iterable* object—an object which is capable of returning its members one at a time. In other words, you can use it as the sequence in a FOR-loop just like looping on a list. The iterating variable gets the row tuples as it loops. The following code loops through the rows of the fire table and prints each FireName:

```
import arcpy
fc = 'C:/gispy/data/ch17/fires.shp'
fields = ['FireName']

cursor = arcpy.da.SearchCursor(fc, fields)
for row in cursor:
    print row[0]

del cursor
```

The del keyword deletes the cursor after the loop. This is necessary to prevent the dataset from becoming locked by the cursor object.

17.6   Locking

Locking is a mechanism to ensure that two processes don't modify the same dataset at once. You may have noticed 'LOCK' files in your data directories. Computer processes such as ArcCatalog or PythonWin create these locks so that they can perform operations on the data without risking data corruption. The operations that can be performed on the data depend on the type of locks placed on it.

Computers use two types of locks, exclusive and shared locks. These types of locks can be defined by how they restrict other processes while they are in place. If a process has an *exclusive lock* on the data, it won't allow another process to lock the data. If a process has a *shared lock* on data, it will allow other processes to obtain shared locks on the data, but it won't allow exclusive locks to be obtained for this data.

You can think of the data as a shared resource like a kitchen in a hostel. If someone is making a large meal for dinner party guests, he may ask that no one else use the kitchen that evening and the other house members would have to find elsewhere to eat dinner (an exclusive lock). Whereas, everyone will come and go in the mornings, so that some people may be preparing breakfast at the same time (a shared lock). It is against house rules to eject others from the kitchen and take over the room while others are preparing bowls of oatmeal (no one can claim an exclusive lock on the kitchen while a shared lock exists).

IDEs need to get an exclusive lock on data to iterate through it with update or insert cursors. Imagine a script in PythonWin iterating through a large file with an update cursor. A copy of the same script run simultaneously in PyScripter won't be able to use an insert cursor on this file, because PythonWin has an exclusive lock on the data.

Search cursors require shared locks. As an example, data can be viewed in ArcMap while a search cursor is iterating through its rows, but the lock will not allow the data to be deleted in ArcCatalog while the search cursor is active because deletion requires an exclusive lock. The following screen shot of Windows Explorer shows two 'LOCK' files associated with the 'fire' shapefile:

fires.dbf	DBF File
fires.prj	PRJ File
fires.sbn	SBN File
fires.sbx	Adobe Illustrator ...
fires.shp	SHP File
fires.shp.TUZIGOOT.2876.4588.sr.lock	LOCK File
fires.shp.TUZIGOOT.2876.rd.lock	LOCK File
fires.shp.xml	XML Document
fires.shx	SHX File

Use the search cursor `next` method in the Interactive Window to see these locks appear in the data list. `arcpy` cursors can create schema, read, and write locks. The schema and read locks are shared locks and write locks are exclusive locks. Once a cursor obtains a lock, the lock persists until the cursor is released. Lock release behavior depends where the code is being run. The best way to ensure that locks are released is to delete the cursor object after use.

17.6.1 The del Statement

The del Python keyword deletes objects from memory. The following code deletes the cursor object and releases all locks:

```
>>> del cursor
```

Run the following code at the prompt while watching the 'fires*' files in the 'ch17' data directory in Windows Explorer to see a lock file appear:

```
>>> fc = 'C:/gispy/data/ch17/fires.shp'
>>> cursor = arcpy.da.SearchCursor (fc, ['FireName'])
>>> for row in cursor:
...     print row[0]
```

Next, run the following code and observe the lock file disappear from the 'ch17' directory:

```
>>> del cursor
```

fires.dbf	DBF File
fires.prj	PRJ File
fires.sbn	SBN File
fires.sbx	Adobe Illustrator ...
fires.shp	SHP File
fires.shp.xml	XML Document
fires.shx	SHX File

Deleting the cursor explicitly deletes the schema, read, or write locks created by the cursor. Delete each cursor you create in a script to make sure that locks don't linger. The deletion needs to be done carefully. If something causes a script to crash, the locks might not be released. For example, the following code throws an IndexError before reaching the deletion statement:

```
cursor = arcpy.da.SearchCursor (fc, ['FireName'])
for row in cursor:
    # Try to index a second field, but there is none.
    print row[1]
del cursor
```

Using the deletion statement in combination with error handling ensures the cursor is deleted if an error occurs in the script. Example 17.2 illustrates this, placing a cursor deletion statement in the except block.

Example 17.2: Using try/except blocks with cursor looping.

```
# searchNprint.py
# Purpose: Print each fire name in a file.
import arcpy, traceback
try:
    cursor = arcpy.da.SearchCursor (fc, ['FireName'])
    for row in cursor:
        print row[0]
    del cursor
except:
    print 'An error occurred'
    traceback.print_exc()
    del cursor
```

17.6.2 The with Statement

The ESRI help describes a second technique for writing cursor scripts to release locks using the Python `with` and `as` keywords. When scripts are run within the ArcGIS Python Window, the `with` block automatically deletes data locks when the code exits the `with` block, even if there is an exception. Though this technique also works for `arcpy` cursors when code is run inside the ArcGIS Python window, this does not work for stand-alone scripts. The format for using a `with` block for an `arcpy` cursor is as follows:

```
with arcpy_cursor_function as cursor:
    code statement(s) using the cursor
```

If you use this technique to run stand-alone script in IDEs, such as PythonWin or PyScripter, locks are not guaranteed to be released this way. Since this doesn't work for stand-alone scripts, most `arcpy` cursor code should use error handling with try/except blocks and delete the cursor explicitly, instead of using the `with` block. We'll revisit this structure in Chapter 19, as it is useful for text file reading and writing.

Note Though the cursor examples in the `arcpy` documentation use the `with` statement, stand-alone scripts should use error handling and delete the cursor, as in Example 17.2.

17.7   Update Cursors

Update cursors are not only able to read field values, but can also modify them. Update cursors share some commonalities with search cursors. They have the same required and optional parameters (See Table 17.2). They can both be used in a loop and in a with block. Like the search cursor, the update cursor has next and reset methods (See Table 17.3). The next method returns a list instead of a tuple. Recall that Python lists are mutable, unlike Python tuples. Using a Python list during update operations makes it possible to modify individual elements of the row. Update cursors additionally have updateRow and deleteRow methods for modifying records. Examples 17.3 and 17.4 demonstrate these methods.

Example 17.3 increments the FireType_P by 2 and standardizes the FireName capitalization scheme. It performs multiple updates within the same loop by modifying the corresponding row elements. Notice that there are two steps to updating a table row. The first step changes the value of the row list. This just changes the value of a Python variable, but this does not change the value in the table. The second step changes the table value with the updateRow method.

Example 17.3: Perform more than one update.

```
# updateValues.py
# Purpose: Modify the fire type value and the fire
#          name in each record.
# Usage: No script arguments needed.
import arcpy, traceback
fc = 'C:/gispy/data/ch17/firesCopy.shp'
fields = ['FireType_P', 'FireName']
cursor = arcpy.da.UpdateCursor(fc, fields)
try:
    for row in cursor:
        # Make changes to the list of values in 'row'
        # Example: 13->15
        row[0] = row[0] + 2
        # Example: LITTLE CRK->Little Crk
        row[1] = row[1].title()
        # Update the table (otherwise changes won't be saved)
        cursor.updateRow(row)
        print 'Updated {0} and {1}'.format(row[0], row[1])
except:
    print 'An error occurred'
    traceback.print_exc()
del cursor
```

Table 17.3 Data access cursor methods and properties.

Cursor type	Methods	Properties
Search cursor	`next() # Returns a tuple` `reset()`	Fields

Update cursor	`next() # Returns a list`
	`reset()`
	`updateRow(row)`
	`deleteRow()`
		Fields

Insert cursor	`insertRow()`	Fields

Example 17.4, deletes the first seven rows. The file's object identifier field (FID) is updated automatically in the remaining rows. When the deletions occur, the FID values are automatically renumbered to be contiguous starting with zero.

Example 17.4: Delete rows.

```
# deleteRows.py
# Purpose: Delete the first x rows.
# Usage: No script arguments required.
import arcpy, traceback
fc = 'C:/gispy/data/ch17/firesCopy.shp'
field = 'FID'
x = 7
try:
    cursor = arcpy.da.UpdateCursor(fc, [field])
    # Delete the first x rows
    for row in cursor:
        if row[0] < x:
            # Delete this row
            cursor.deleteRow()
            print 'Deleted row {0}'.format(row[0])
    del cursor
except:
    print 'An error occurred.'
    traceback.print_exc()
    del cursor
```

17.8   Insert Cursors

Use insert cursors to insert new rows. Like the other cursor functions, the InsertCursor method requires two arguments, an in_table and a field_names list. The code should create the cursor with field_names set to list the fields that you want to specify in the new row. Not all fields need to be specified. Fields with no values specified will be assigned a default value. Then there are two

FID	Shap	FireId	Calen	FireNu	FireName	FireTy	Si	StartTime
0	Point	239042	1999	9911	Coconut	13	B	7/18/1999
1	Point	239060	2000	10	Highhead	13	B	12/19/2000
2	Point	239127	2004	424	Hrcovedune	13	A	7/26/2004
3	Point	513169	2005	506	Beech Fost	15	A	3/22/2005
4	Point	513179	2005	507	Pilg. Hgts	15	B	3/23/2005
5	Point	513180	2009	0		0		<Null>

Figure 17.3 New row added by 'insertRow.py'.

steps to inserting a new row. First, create a Python list of the values to insert in the row. The list should be the same length as the field_names parameter list and specify the value for each field in the order of this list. Second, to insert the row, call the insertRow method. After the insertion, the cursor must be deleted to release locks. The code in Example 17.5 creates the last row in Figure 17.3.

Example 17.5: Insert a row.

```
# insertRow.py
# Purpose: Insert a new row without geometry.
# Usage: No script arguments needed.
import arcpy, traceback

# Create an insert cursor
fc = 'C:/gispy/data/ch17/firesCopy.shp'
fields = ['FireId','CalendarYe']

try:
    cursor = arcpy.da.InsertCursor(fc, fields)
    # Create a list with FireId=513180 & CalendarYr=2009
    newRecord =[513180, 2009]
    # Insert the row (otherwise no change would occur)
    cursor.insertRow(newRecord)
    print 'Record inserted.'
    del cursor
except:
    print 'An error occurred.'
    traceback.print_exc()
    del cursor
```

More than one row can be inserted at a time. Example 17.6 first uses a search cursor and the built-in max function to find the maximum fire number in the table. Then it uses an insert cursor to insert five additional fires for calendar year 2015 with successive fire numbers starting after the previous maximum.

Example 17.6: Find the maximum value of an attribute and insert multiple rows using this information.

```
# insertRows.py
# Purpose: Insert multiple rows without geometry.
# Usage: No script arguments needed.
import arcpy, traceback

fc = 'C:/gispy/data/ch17/firesCopy.shp'

# Find the current fire numbers.
try:
    fields = ['FireNumber']
    cursor = arcpy.da.SearchCursor(fc, fields)
    fireNumbers = [row[0] for row in cursor]
    print '{0} fire numbers found.'.format(len(fireNumbers))
    del cursor
except:
    print 'An error occurred in the search.'
    traceback.print_exc()
    del cursor

# Insert 5 new fires for year 2015.
try:
    fields.append('CalendarYe')
    cursor = arcpy.da.InsertCursor(fc, fields)
    # Find the max value in list and increment by 1
    fireNum = max(fireNumbers) + 1
    for i in range(5):
        # Create a row with unique fire number & year=2015
        newRow = [fireNum, 2015]
        fireNum = fireNum + 1
        # Insert the row.
        cursor.insertRow(newRow)
        print '''New row created with \
        fire {0} and year = {1}.'''.format(
        newRow[0], newRow[1])
    del cursor
except:
    print 'An error occurred in the insertion.'
    traceback.print_exc()
    del cursor
```

17.8.1   Inserting Geometric Objects

The rows added in Examples 17.6 and 17.7 appear in the tables, but no new points appear because we have not specified locations. To set the 'Shape' field, the script must create the appropriate arcpy Geometry objects. The arcpy package has functions named Point, Multipoint, Polyline, and Polygon for creating Geometry objects. The following code creates an arcpy Point object with x = -70.1 and y = 42.07 and stores it in the variable named myPoint:

```
>>> myPoint = arcpy.Point(-70.1, 42.07)
>>> myPoint
<Point (-70.1, 42.07, #, #)>
```

Multipoint, Polyline, and Polygon features are composed of multiple points. These more complex shape types require the user to create an Array object and then pass this to the Geometry object to create the feature. The following code creates a Polyline connecting points a and b:

```
# Create 2 polyline endpoints.
>>> x = 2134000
>>> y = 179643
>>> a = arcpy.Point(x,y)
>>> x = 2147000
>>> y = 163267
>>> b = arcpy.Point(x,y)
>>> a
<Point (2134000.0, 179643.0, #, #)>
>>> b
<Point (2147000.0, 163267.0, #, #)>

>>> # Create an array with a Python list of Point objects.
>>> myArray = arcpy.Array([a,b])

>>> # Create a line with an Array object that has points.
>>> line = arcpy.Polyline(myArray)
>>> line.length
20908.691398554813
```

Examples 17.7 and 17.8 use Geometry objects, along with insert cursors, to insert features. Example 17.7 inserts a point. It uses the SHAPE@XY token since it is only specifying a point to be added. The script creates a Point object and uses the point to define the new row. The new row is a Python list variable, as usual. The order of the values in the new row list corresponds to the order of the fields in the

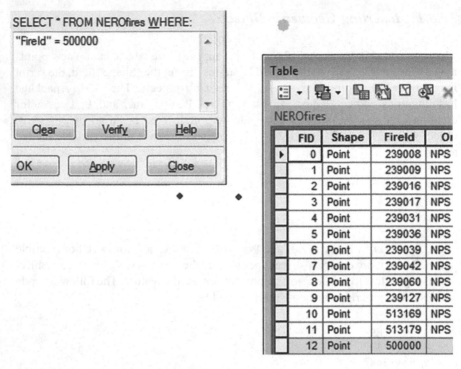

Figure 17.4 New point selected, highlighted in blue in the top right corner of this image.

field_names parameter list when the cursor was created. In other words, the FireId value, of 500000, is placed first and the Shape object is placed second. There is nothing special about this ordering except that these two lists must use the same ordering. Finally, the new row is added to the table. Prior to the call to the insertRow method, no changes were made to the attribute table. Once this call has succeeded, the new point is added. Figure 17.4 shows this point highlighted when the new row is selected in ArcMap.

Example 17.7: Insert a point.

```
# insertPoint.py
# Purpose: Insert a point with a Geometry object.
# Usage: No script arguments needed.

import arcpy, traceback

fc = 'C:/gispy/data/ch17/firesCopy.shp'
try:
    ic = arcpy.da.InsertCursor(fc, ['FireId', 'SHAPE@XY'])
```

```
    # Create a point with x = -70.1 and y = 42.07
    #     to be used for the Shape field.
    myPoint = arcpy.Point(-70.1, 42.07)

    # Create a row list with FireId=500000 and the new point.
    newRow =[500000, myPoint]

    # Insert the new row.
    ic.insertRow(newRow)
    print 'New row inserted.'

    del ic
except:
    print 'An error occurred.'
    traceback.print_exc()
    del cursor
```

Example 17.8 inserts a polygon. It uses the SHAPE@ token to get the entire Geometry object, since it is inserting a polygon instead of a line. Inserting any feature more complex than a point requires this because SHAPE@XY is the shape's centroid. The script creates three points and then creates an Array object by passing it a list of the points (myArray = arcpy.Array([a,b,c])). Then the polygon is created from the Array object (the polygon is created with this line of code: poly = arcpy.Polygon(myArray)) and the Shape field is set using that polygon. In this example, the geometry token is given first in the field_names parameter list, so it is listed first in the new row. The new row is inserted, which adds the triangle highlighted in Figure 17.5.

Figure 17.5 New triangular polygon created with the code in Example 17.8.

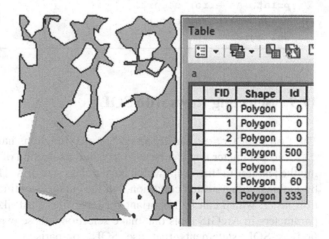

FID	Shape	Id
0	Polygon	0
1	Polygon	0
2	Polygon	0
3	Polygon	500
4	Polygon	0
5	Polygon	60
6	Polygon	333

Example 17.8: Insert a polygon.

```
# insertPolygon.py
# Insert cursor polygon example.

import arcpy, traceback

# Create 3 point objects for the triangle
a = arcpy.Point(2134000, 179643)
b = arcpy.Point(2147000, 163267)
c = arcpy.Point(2131327, 167339)

# Create an array, needed for creating a polygon
myArray = arcpy.Array([a,b,c])

# Create a polygon.
poly = arcpy.Polygon(myArray)
fc = 'C:/gispy/data/ch17/aggCopy.shp'

try:
    # Create an insert cursor
    cursor = arcpy.da.InsertCursor(fc, ['SHAPE@', 'Id'])

    # Create row list
    newRow = [poly, 333]
    # Insert the new row.
    # It's automatically given an FID one greater
    # than the largest existing one.
    cursor.insertRow(newRow)
    print 'Polygon inserted.'
    del cursor
except:
    print 'An error occurred.'
    traceback.print_exc()
    del cursor
```

17.9   Selecting Rows with SQL

The SearchCursor and UpdateCursor functions have an optional where_
clause parameter that can be used to refine a selection of features, such as selecting the features in 'fires.shp' which have a class size of A. This can be accomplished
by looping with a nested condition, but the cursor is optimized to make this selection faster, which can be important for large files. Just like the where_clause
parameters in ArcGIS tools that make selections, these expressions must be specified as SQL statements that use SQL comparison operators. The following

expressions are examples of `where_clause` values that could be used to select rows in the 'fires' shapefile data:

```
>>> query1 = "SizeClass = 'A'" # Fires of size class A.
>>> query2 = "FireName <> 'MEADOW'" # Fires not named MEADOW.
>>> query3 = 'FID > 6' # Fires with ID greater than 6.
>>> query4 = "StartTime = date '2000-01-06'" # After Jan.6,2000
```

The same formatting described in Section 9.2, 'ArcGIS Tools that make selections', applies here. We can make a few observations from these examples. The first two expressions involve text fields so they use the single quotations around the field value. The SQL inequality operator (<>) is different from the Python one we use (!=). Only one pair of quotation marks is used for numeric fields. Fields other than text and numeric types, such as the date field used in `query4`, require specialized syntax. Read more in the ArcGIS Desktop help topic, 'SQL reference for query expressions'.

To select a subset of rows, create the cursor and use a `where_clause` query as the third parameter. Example 17.9 uses `query1` to print the years for the records when the fire size class is 'A':

Example 17.9: Use a `where_clause` with a cursor.

```
# sqlQueryCursor.py
# Purpose: Use a SQL query to select specific records.
# Usage: No script arguments needed.
import arcpy, traceback
fc = 'C:/gispy/data/ch17/fires.shp'

# Create the where_clause
query = "SizeClass = 'A'"
try:
    sc = arcpy.da.SearchCursor(fc, ['CalendarYe'], query)

    for row in sc:

        print row[0],
    del sc
except:
    print 'An error occurred.'
    traceback.print_exc()
    del cursor
```

The `SizeClass` field is used to specify a selection (those rated 'A') and the `CalendarYe` is printed for these seven records. The printed output looks like this:

```
>>> 1997.0 1997.0 1997.0 1999.0 1999.0 2004.0 2005.0
```

Queries can be formed without hard-coded field names, by using string formatting. The query used in Example 17.10 will have the same effect as query3 if the user passes FID as the second script argument. This query would select the row with FID greater than 6. A list comprehension is used to create a list of the values of fieldToPrint. When this script is run with the example input, the results look like this:

```
[u'COCONUT', u'HIGHHEAD', u'HRCOVEDUNE', u'Beech Fost',
u'Pilg. Hgts']
```

Example 17.10: Use a cursor where_clause with a variable.

```python
# whereClauseWithVar.py
# Purpose: Use a SQL query to select specific
#     records based on user arguments.
# Example: C:/gispy/data/ch17/fires.shp FID FireName
import arcpy, sys, traceback

fc = sys.argv[1]
numericField = sys.argv[2]
fieldToPrint = sys.argv[3]

query = '{0} > 6'.format(numericField) # string formatting

try:
    with arcpy.da.SearchCursor(fc, [fieldToPrint],
                               query) as cursor:
        recordList = [row[0] for row in cursor]
    del cursor
    print recordList
except:
    print 'An error occurred.'
    traceback.print_exc()
    del cursor
```

17.10   Key Terms

Cursor

Record

arcpy DataAccess (da) module search, update, and insert cursors

Search and update cursor next and reset methods

Cursor fields property

Update cursor updateRow and deleteRow methods

Insert cursor insertRow method

arcpy geometry tokens

arcpy Geometry objects

arcpy Point, Array, Polyline, and Polygon methods

17.11 Exercises

1. **updateUPPER.py** Use the Copy (Data Management) tool to make a copy of the input file in 'C:/gispy/scratch/' and then use an update cursor to modify the string field input by the user so that all characters in that field are uppercased. Use two required arguments, an input table and a field name. In the example below, the output 'COVER' field values become WOODS, ORCH, and OTHER.

   Example input:
   ```
   C:/gispy/data/ch17/park.shp COVER
   ```

   Example output:
   ```
   Copied C:/gispy/data/ch17/park.shp to C:/gispy/scratch/park.shp
   Modified C:/gispy/scratch/park.shp
   ```

2. **cursorBasics.py** Practice using search and update cursors. Write a script that copies 'C:/gispy/data/ch17/park.shp' to 'C:/gispy/scratch/' and then does each of the tasks in a-c for the file in 'scratch'. Since the table and FID values are hard-coded, this script takes no arguments. Find the specified rows using the where_clause parameter.

   (a) Using a search cursor, find the row with an FID value of 45 and print the COVER field value in that row.

   Printed output:
   ```
   Row with FID = 45 has cover orch
   ```

   (b) Using an update cursor, find the row with an FID value of 120 and change the cover name to 'park'.

   Printed output:
   ```
   Row with FID = 120 has been updated to use cover: park
   ```

   (c) Using an update cursor, find the row with an FID value of 22 and delete it. Note that when you delete a row, it updates the FIDs so that there will be a new entry with the FID of 22 even after successful deletion. To confirm that the row was deleted, call the Get Count (Data Management) tool before and after deleting this row and print the before and after counts.

   Printed output:
   ```
   Count before deletion: 426
   Count after deletion: 425
   ```

3. **buggyCursor.py:** Sample script 'buggyCursor.py' is supposed to count the number of records in 'parkCopy3.shp' with a 'COVER' field value that is neither 'woods' and nor 'orch'. Correct the six errors in the script. Insert comments describing each error. The correctly working script should print these results:

   ```
   Number of records with other cover types: 62
   ```

4. **printStringFields.py** Modify sample script 'printTableExclude.py' so that it prints the field names and row tuples of only the string type fields. Rename the `excludeFields` function as `includeFields` and modify it to *include* (instead of exclude) the field types being passed in. The printed output should look like this:

```
(u'FireName', u'SizeClass', u'Authoriz_1')
((u'MEADOW', u'A', u'Fire Technician')
(u'LITTLE CRK', u'A', u'ND Fire Techn')
(u'T.Calvin', u'B', u'Fire Technician')
```
. . . and so forth.

5. **deleteHigh.py** Write a script that copies the input data to 'C:/gispy/scratch/' and then uses cursor functionality to remove every record of a table where a given field is higher than a given value. Use three required arguments, the full path file name of an input file, the name of a numeric field in that file, and a search value.

Example input: C:/gispy/data/ch17/firesCopy2.shp FID 8

Example output:
```
Remove row: [9]
Remove row: [10]
Remove row: [11]
```

6. **deleteSpaceRows.py** Write a script that copies the input data to 'C:/gispy/scratch/' and then uses cursor functionality to remove every row of a table in which the value of a given field contains a space. Use two required arguments, the full path file name of an input file and the name of a text field in that file.

Example: C:/gispy/data/ch17/firesCopy3.shp FireName

Example output:
```
Removed row containing: [u'LITTLE CRK']
Removed row containing: [u'PILGRIM HT']
Removed row containing: [u'Beech Fost']
Removed row containing: [u'Pilg. Hgts']
```

7. **insertLine.py** Sample script 'insertLine.py' finds the centroids (`point1` and `point2`) of the first two records in 'C:/gispy/data/ch17/park.shp'. Add code to this script to insert a line from `point1` to `point2` in the polyline shapefile 'C:/gispy/scratch/parkLines.shp' and set the value of 'LEFT_FID' to 50 for this record. The new line should look like the highlighted line segment in the image below:

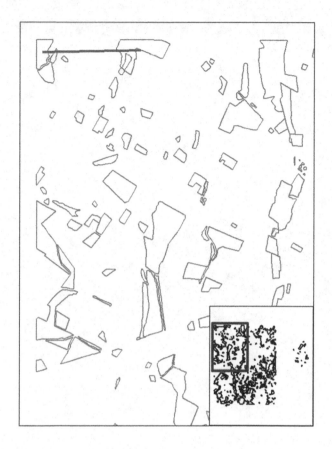

Chapter 18
Dictionaries

Abstract In addition to integers, floats, strings, and lists, Python has a powerful built-in Python data type called a dictionary. Dictionaries are useful for storing tables of information with a unique identifier for each record. Dictionaries provide a mapping from a set of keys to a set of values. In other words, given a key, a dictionary can look up the value associated with it. These have many applications in GIS, including reading GIS attribute tables or text data files and modifying them within a script. Also, they are often used to store pairs of items that go together. For example, soil science uses standard classifications for soil, abbreviated with terms such as 'Ap' and 'Cg'; However, more explicit names such as 'Plinthic Acrisol' (for 'Ap') and 'Gleyic Chernozem' (for 'Cg') are needed for some analysis. A dictionary can be used to store these terms so that the abbreviations are associated with the full names. This chapter shows the dictionary syntax for creating associations like this and then it shows how to access, update, and modify dictionaries.

Chapter Objectives
After reading this chapter, you'll be able to do the following:

- Create Python dictionaries.
- Add an item to a dictionary.
- Modify or delete an item from a dictionary.
- Check if a dictionary has a key.
- Explain `KeyError` exceptions.
- Replace `IF`/`ELSE IF` structures with dictionaries.
- List the keys, values, and items in a dictionary.
- Loop through a dictionary.
- Populate a dictionary based on user input.
- Populate a dictionary using `arcpy` cursors.
- Embed Python lists as values.

© Springer International Publishing Switzerland 2015   335
L. Tateosian, *Python For ArcGIS*, DOI 10.1007/978-3-319-18398-5_18

18.1   Dictionary Terms and Syntax

The terms used to refer to the elements in a Python dictionary are 'items', 'keys', and 'values'. Python dictionaries store a collection of *items;* Each item consists of a *key* and a *value*. An item is like a GIS attribute table record. Each record has a unique identifying attribute, such as its FID value, which allows you to access the rest of the values in that record. A Python dictionary item is a unique identifier, a key, along with the related value. The following code creates a Python dictionary containing five items:

```
>>> zipcodeDictionary = {27522 : 'Granville', 28736 : 'Jackson',
27953: 'Dare', 27511: 'Wake', 27607: 'Wake'}
```

The items in the dictionary are zipcode-county pairs; Each key is a North Carolina zip code and each value is the enclosing county name. Keys must be unique, but values can be duplicated. For example, two zip codes fall within Wake County. This is consistent with English language dictionaries; The words defined in an English dictionary must have only one entry, but synonyms have the same definition.

Dictionary keys are most often strings or numbers. Mutable types such as lists or dictionaries can not be used as keys. However, the item value can be of any data type, including a mixture of numbers, strings, lists, and so forth:

```
>>> flickDict = {8.5:1963, 'oceans':[11,12], 'up' : 'dog'}
```

The definition of a dictionary can be distributed across multiple lines. When multiple lines are used to assign a dictionary in the Interactive Window, three dots appear at the beginning of the lines:

```
>>> newVocab = {'mouse potato': 'a frequent computer user',
... 'wasband' : 'former husband',
... 'himbo': 'an attractive but vacuous man'}
```

The triple dots appear automatically if the dictionary extends across multiple lines in the Interactive Window to indicate that the lines are grouped together until the closing curly brace.

Python reports the data type of Python dictionaries as type 'dict':

```
>>> type(newVocab)
<type 'dict'>
```

Dictionaries can have zero or more items. The syntax for an assignment statement for a dictionary that has three items looks like this:

```
dictionaryName = {key1 : value1, key2 : value2, key3 : value3}
```

Spacing doesn't matter, but the punctuation is required. Each key is followed by a colon and then its corresponding value. The pairs are separated by commas. The curly braces indicate that it is a dictionary, just as square braces are used to create lists. Also similar to lists is the syntax for creating an empty dictionary. If dictionary items are not known ahead of time, you can create an empty dictionary and then add items to it. To create an empty dictionary, set a variable equal to a pair of empty curly braces:

```
dictionaryName = {} # Create an empty dictionary
```

18.1.1   Access by Key, Not by Index

The similarities between list and dictionary syntax can be misleading. Dictionaries store sequences of items, so you might assume you can try to index into a dictionary like a list or a string. But dictionaries are not designed to work that way. Instead, values need to be accessed using keys. The following code assigns a dictionary of student ids and dormitory room numbers ('emforste' and 'pgwodeho' are roommates):

```
>>> roomDict = {'cslewis': 4139, 'emforste': 4118,
... 'jkrowlin': 4098, 'jrtolkie': 4259, 'pgwodeho': 4118,
... 'jrtolkie': 4259, 'vmhugo': 2121, 'vwoolf': 3145}
```

The following code uses the key, 'emforste', to retrieve this student's room number:

```
>>> roomDict['emforste']
4118
```

The syntax format for accessing an item is as follows:

```
theValue = dictionaryName[key1] # Get value paired with key1
```

The syntax looks similar to indexing into a list, but the numeric index is replaced by a key. Attempting to use zero-based indexing to access the second item in the dictionary throws an exception:

```
>>> roomDict[1]
Traceback (most recent call last):
File '<interactive input>', line 1, in <module>
KeyError: 1
```

Since roomDict is a dictionary, the number (1) is interpreted as a key. But the dictionary has no key of 1, so it reports a KeyError, an exception that is thrown whenever a key is not found in the set of existing keys.

18.1.2   Conditional Construct vs. Dictionary

A dictionary can sometimes be used to replace a multi-way conditional construct, one with multiple serially checked conditions. When a selection statement is controlling flow based on the value of some variable and actions taken inside each branch are based on the value of a second variable, the pairing of these values can be replaced with a dictionary. Some programming languages use a switch statement for this purpose; Python uses dictionaries instead. Take the following code as an example:

```
if num == 0:
    day = 'Monday'
    print 'Weekday: {0}'.format(day)
elif num == 1:
    day = 'Tuesday'
    print 'Weekday: {0}'.format(day)
elif num == 2:
    day = 'Wednesday'
    print 'Weekday: {0}'.format(day)
elif num == 3:
    day = 'Thursday'
    print 'Weekday: {0}'.format(day)
elif num == 4:
    day = 'Friday'
    print 'Weekday: {0}'.format(day)
```

The code prints the day of the week based on the value of num. The flow is controlled by the num variable and the code inside each branch is identical, except for the value of the day variable. num and day are paired by the branching. As a sleeker alternative, a dictionary can create this mapping from num values to day values, as in Example 18.1.

Example 18.1

```
# weekdays.py
weekdayDict = {0: 'Monday', 1: 'Tuesday', 2: 'Wednesday',
               3: 'Thursday', 4: 'Friday'}
day = weekdayDict[num]
print 'Weekday: {0}'.format(day)
```

These four lines of code replace the entire multi-way conditional construct. Accessing the dictionary with num as the key, returns the value of day used in the print statement. The key access statement replaces the hard-coded weekday name assignment statements in each block of the conditional construct.

18.1.3 How to Modify: Update/Add/Delete Items

Dictionaries can be modified by updating existing items, adding new items, and deleting items. Each of these three actions access the dictionary using a key in square braces. Modifying mutable and immutable item values is handled slightly differently. When the item values are immutable data types like strings or numbers, the syntax for modifying an item and adding a new item is identical:

```
# update or add an item with an immutable value
dictionaryName[ key ] = value
```

If the key exists in the dictionary, the item is modified. If not, a new item is added. As an example of modifying an existing item, suppose 'emforste' and 'pgwodeho' had a falling out, causing 'emforste' to move into room 4139 with 'cslewis'. The following code updates an existing item, modifying the entry for 'emforste' with an assignment statement to change the value associated with key 'emforste':

```
>>> # Modify the room number for 'emforste'.
>>> roomDict['emforste'] = 4139
>>> roomDict # The dictionary is updated.
{'emforste': 4139, 'cslewis': 4139, 'jrtolkie': 4259,
'vmhugo': 2121, 'vwoolf': 3145, 'pgwodeho': 4118,
'jkrowlin': 4098}
>>> len(roomDict)
7
```

Because the student, 'emforste' was already listed in the dictionary, this code modified that item. The number of items in the dictionary remains 7. However, if a new student, 'lcarroll' moves into the dorm, this is not an existing key. The following code adds a new item and assigns 'lcarroll' to room 2121 with 'vmhugo':

```
>>> # Add 'lcarroll' in room 2121 to the dictionary.
>>> roomDict['lcarroll'] = 2121
>>> roomDict # A new item is added to the dictionary.
{'emforste': 4139, 'cslewis': 4139, 'jrtolkie': 4259,
'lcarroll': 2121, 'vmhugo': 2121, 'vwoolf': 3145,
'pgwodeho': 4118, 'jkrowlin': 4098}
>>> len(roomDict) # The length of the dictionary becomes 8.
8
```

When the item values are mutable data types, they may need to be updated differently. Consider the following dictionary containing two weather gauge sites and a list of temperature readings at each site.

```
>>> weather = {'S1' : [15,20],'S2' : [25,30,40]}
```

Adding another site to the dictionary or replacing one of the items, has the same format as the numerical and string examples. The value for new site 'S3' is a list, so the statement has two sets of square braces, one around the key and one around the value (list) being assigned on the right:

```
>>> weather['S3'] = [33,40]
>>> weather
{'S3': [33, 40], 'S2': [25, 30, 40], 'S1': [15, 20]}
```

Other operations, such as adding another temperature reading to a site list or incrementing each temperature in a site list, require techniques that are specific to list handling. To add a temperature to an existing item list, you must use the append method on that item. No assignment statement is used with the append method since it is an in-place method that modifies the mutable object. For example, the following statement adds 36 to the 'S2' temperature list:

```
>>> weather['S2'].append(36)
>>> weather
{'S3': [33, 40], 'S2': [25, 30, 40, 36], 'S1': [15, 20]}
```

A list comprehension can be used to increment each temperature in a site list. The following code uses list comprehension to creates a list of incremented temperatures and then assigns the new list to site 'S1':

```
>>> update = [i + 1 for i in weather['S1']]
>>> weather['S1'] = update
>>> weather
{'S3': [33, 40], 'S2': [25, 27, 40, 36], 'S1': [16, 21]}
```

Deleting a dictionary item works the same for both mutable and immutable values. Access the item with its key and use the Python keyword `del` with the following syntax:

```
# Delete the item with the key named key1
del dictionaryName[key1]
```

Returning to the dormitory room example, suppose `'jrtolkie'` moves out of the dormitory so that he can focus on his next novel, tentatively titled 'Everybody loves rings'. The following code deletes the item with key `'jrtolkie'`, reducing the number of items to 7:

```
>>> del roomDict['jrtolkie']
>>> roomDict
{'emforste': 4139, 'cslewis': 4139, 'lcarroll': 2121, 'vmhugo':
2121, 'vwoolf': 3145, 'pgwodeho': 4118, 'jkrowlin': 4098}
>>> len(roomDict)
7
```

If the key used in a deletion statement is missing, a `KeyError` exception is thrown:

```
>>> del roomDict['cslouis']
Traceback (most recent call last):
File '<interactive input>', line 1, in <module>
KeyError: 'cslouis'
```

18.2   Dictionary Operations and Methods

Like strings and lists, dictionaries have specialized operation and methods for dictionary operations. This section focuses on four frequently used operations and methods: the `in` keyword, `keys`, `values`, and `items`.

18.2.1   Does It Have That Key?

The `in` method can check if a dictionary has a particular key to avoid overwriting an existing key or to avoid throwing a `KeyError`. The `in` keyword can be used on keys in a manner similar to how it works on lists. It returns `True` or `False`. To check for the keys 'cslewis' and 'cslouis' in the dormitory room dictionary, use the following code:

```
>>> 'cslewis' in roomDict
True
>>> 'cslouis' in roomDict
False
```

The `in` keyword can be used to avoid `KeyError` exceptions that occur when a statement attempts to access or delete an item with a nonexistent key:

```
>>> key = 'cslouis'
>>> if key in roomDict
...      print roomDict[key]
... else:
...      'No {0} key found.'.format(key)
...
'No cslouis key found.'
```

18.2.2 Listing Keys, Values, and Items

Like other strings and lists, dictionaries have specialized methods for dictionary operations. Certain dictionary methods allow you to list the keys, values, and items. `keys` returns a list of the keys in the dictionary in arbitrary order:

```
>>> roomDict.keys()
['emforste', 'cslewis', 'lcarroll', 'vmhugo', 'vwoolf', 'pgwodeho',
'jkrowlin']
```

`values` returns a list of the values in the dictionary:

```
>>> roomDict.values()
[4139, 4139, 2121, 2121, 3145, 4118, 4098]
```

`items` returns a list of the items in the dictionary as key-value tuples:

```
>>> roomDict.items()
[('emforste', 4139), ('cslewis', 4139), ('lcarroll', 2121), ('vmhugo', 2121),
('vwoolf', 3145), ('pgwodeho', 4118), ('jkrowlin', 4098)]
```

18.2.3 Looping Through Dictionaries

The listing methods can be used to loop through a dictionary. Typically, you won't need to store the lists returned by these methods, since they are already stored in the dictionary; In this case, you can just use the method call as the sequence that comes after the `in` keyword in a FOR-loop. For the `keys` and `values` method, place a single variable between the `for` and `in` keywords. The variable iterates through the keys or values:

```
>>> for k in newVocab.keys():
...     print k
...
himbo
wasband
mouse potato

>>> for v in newVocab.values():
...     print v
...
an attractive but vacuous man
former husband
a frequent computer user
```

For the `items` method, place two variables between the `for` and `in` keywords. One iterates through the keys and the other one iterates through the values:

```
>>> for k, v in newVocab.items():
...     print k, ':', v
...
himbo : an attractive but vacuous man
wasband : former husband
mouse potato : a frequent computer user
```

The `keys` method can be used to update each item's value. The following example uses the `keys` method to append zero to every weather gauge temperature list:

```
>>> for k in weather.keys():
...     weather[k].append(0)
...
>>> weather
{'S3': [33, 40, 0], 'S2': [25, 27, 40, 36, 0], 'S1': [16, 21, 0]}
```

The `values` method can be used to derive information from each value. The following example uses the `values` method to find the average temperature at each site:

```
>>> for v in weather.values():
...      print sum(v)/len(v),
...
24 25 12
```

The `items` method is useful when both key and value are needed in the loop. The following example uses the `items` method to print the fourth floor residents in the dormitory room dictionary:

```
>>> for k, v in roomDict.items():
...      room = str(v)
...      if room.startswith('4'):
...          print '{0}, room {1}'.format(k,v)
...
emforste, room 4139
cslewis, room 4139
pgwodeho, room 4118
jkrowlin, room 4098
```

18.3   Populating a Dictionary

The examples thus far have hard-coded dictionaries to introduce dictionary syntax, but you'll often want to generate dictionaries dynamically. This section shows how to populate a dictionary to store data from user input. As a starting point, the first examples use a hard-coded set of keys and the built-in `raw_input` function to gather values. The remaining examples generate the entire dictionary dynamically, using directory listings or `arcpy` cursors to retrieve data from attribute tables.

The script 'healthyLiving.py' in Example 18.2 surveys user preferences. The script starts with an empty dictionary and uses an assignment statement to add new values. An empty dictionary `favDict` is created, then the script loops through the topics, asks about each topic in turn and populates the dictionary inside the loop.

Example 18.2

```
# healthyLiving.py
# Purpose: Collect user preferences; only keep most recent responses.
topics = ['fruit','fruit','fruit',
          'veg','veg','veg','exercise','park']
favDict = {}     # Create empty dictionary
for topic in topics:
    question = 'What is your favorite {0}?'.format(topic)
    answer = raw_input(question)
    favDict[topic] = answer     # Add or update item
print favDict
```

The following assignment statement from Example 18.2 populates the dictionary:

```
favDict[topic] = answer # Add or update item
```

The script doesn't check if the dictionary has a key already before assigning a value. The scripts asks for a favorite fruit and vegetable more than once. When the user responses are `grape`, `banana`, `mango`, `fennel`, `lettuce`, `carrots`, `rollerblading`, and `Central` in that order, the dictionary records `grape` but overwrites it when `banana` is given. In the end, the dictionary shows `mango` as a favorite, since this was the latest favorite fruit response:

```
>>> favDict
{'veg': 'carrots', 'fruit': 'mango', 'park': 'Central', 'exercise':
'rollerblading'}
```

By using only the latest response, Example 18.2 assigns only one answer to each question. This overwriting could be prevented by checking for the key. Replacing the loop in Example 18.2 with the following code would record only first responses and discard all other answers:

```
for topic in topics:
    question = 'What is your favorite {0}?'.format(topic)
    answer = raw_input(question)
    if key not in roomDict:
        favDict[topic] = answer   # Add item (only 1st responses recorded)
```

The script 'healthLivingV2.py', in Example 18.3 takes an alternative approach that preserves multiple responses by using Python lists as item values.

Example 18.3

```
# healthyLivingV2.py
# Purpose: Collect user preferences; keep all responses.
topicList = ['fruit','fruit','fruit',
             'veg','veg','veg','exercise','park']
topDict = {}     # Create empty dictionary
for topic in choiceList:
    question = "What is your favorite {0}?".format(topic)
    answer = raw_input(question)
    if topic not in topDict:
        # Add a new item to the dictionary.
        topDict[topic] = [answer]
    else:
        # Update an item by adding to an item's list.
        topDict[topic].append(answer)
print 'topDict {0}'.format(topDict)
```

The `topDict` dictionary includes every answer, listing it with the corresponding question:

```
>>> topDict
{'veg': ['fennel', 'lettuce', 'carrots'], 'fruit': ['grape', 'banana',
'mango'], 'park': ['Central'], 'exercise': ['rollerblading']}
```

Each key has exactly one value, a list. But we are storing one or more answer for each question. For example, `topDict` includes three favorite vegetables and three favorite fruits. The script collects the list type values by using decision-making blocks. If the dictionary does not yet have a topic as a key, it adds a new item to the dictionary. The values need to be lists, so that more than one answer can be associated with each question. The value of the new item is the current answer surrounded by square braces. In other words, the new item is a list with one element, the current answer. The ELSE block appends the new answer to the existing list for that topic. Table 18.1 juxtaposes the pseudocode for Example 18.2 (collecting string values) and Example 18.3 (collecting list values).

Variations of the pseudocode in Table 18.1 can be used in many applications. As another example of using dictionaries to collect information, consider 'fileDates.py' in Example 18.4, which uses a dictionary to collect file names and modification time stamps. The script lists the files in a given directory and stores a file name in a dictionary along with a modification time-stamp. Key conflicts might arise if the script were to walk into subdirectories. But since this script only looks in one directory, there is no chance of a duplicate file name causing a dictionary item to be overwritten. Therefore, the dictionary population code was modeled after P1 in Table 18.1.

Table 18.1 Pseudocode for Examples 18.1 and 18.2 collecting data in a dictionary.

	Dictionary population pseudocode	Add and append syntax
P1	# String values	
	CREATE empty dictionary, D	
	GET list, L	
	FOR each v_i in L	
	GET v_j	
	ADD to new item to D	D[v_i] = v_j
	with key of v_i & value of v_j	
	ENDFOR	
P2	# List values	
	CREATE empty dictionary, D	
	GET list, L	
	FOR each v_i in L	
	GET v_j	
	IF D does not have key v_i THEN	
	ADD new item to D with key of v_i	D[v_i] = [v_j]
	and value of a list	
	containing v_j	
	ELSE	D[v_i].append(v_j)
	APPEND v_j to D item with key v_i	
	ENDIF	
	ENDFOR	

fileDict contains file name-time pairs. The first few items in the dictionary from Example 18.4 look something like this following:

```
{'a.shx': 'Thu Jan 01 13:34:24 2009', 'a.dbf': 'Thu Jan 01 13:34:24
2009', 'dog.JPG': 'Thu Jan 01 13:34:24 2009',...
```

Example 18.4

```
# fileDates.py
# Purpose: Collect filenames and modification dates in a dictionary.
import os, time
inputDir = 'C:/gispy/data/ch18/smallDir'

fileList = os.listdir(inputDir)
fileDict = {}
for f in fileList:
    epochNum = os.path.getmtime(inputDir + '/' + f)
    modTime = time.ctime(epochNum)
    fileDict[f] = modTime

print fileDict
```

18.3.1 Dictionaries and Cursors

Another common use of dictionaries is to store GIS data gathered by `arcpy` cursors. Example 18.5 finds the median area of the polygons in a shapefile and then determine the IDs of any polygons whose areas are close to the median area. The script collects the polygon areas of a shapefile in a dictionary. Then the built-in `numpy` module is used to calculate the median of the dictionary value list. A final loop through the dictionary items identifies the polygons with area measurements in the range of plus or minus 400 square feet of the median area.

Example 18.5

```
# areaMedian.py
import arcpy, numpy

fc = 'C:/gispy/data/ch18/parkAreas.shp'
idField = 'FID'
areaField = 'F_AREA'
areasDict = {}

# Populate dictionary with id:area items.
sc = arcpy.da.SearchCursor(fc, [idField, areaField])
for row in sc:
    fid = row[0]
    area = row[1]
    areasDict[fid] = area
del sc

# Find the median area.
areas = areasDict.values()
medianArea = numpy.median(areas)
print 'Median area: {0}'.format(medianArea)

# Find the polygons with values close to median.
sqft = 400
print 'Polygons close to median:'
for k, v in areasDict.items():
    if medianArea - sqft < v < medianArea + sqft:
        print '{0}: {1}, {2}: {3}'.format(idField, k, areaField, v)
```

Populating the dictionary using a cursor looks similar to the previous examples. The only difference is that, in this case, the values are collected by looping with a search cursor. In Example 18.5, no conditional constructs are used when the dictionary is populated because the keys are FID values, which are unique. The values are simply numbers, so Example 18.1 follows the Table 18.1 P1 pseudocode.

If you want to collect information about a field that has duplicate values, you may want to follow the Table 18.1 P2 pseudocode instead. Example 18.5 finds the median area of each land cover type. Multiple polygons have the same land cover types, so the script needs to collect a list of areas for each type. This script encounters three cover types ('woods', 'other', and 'orch'), so the final dictionary has three items, one key for each cover type and one list of areas for each cover type. The final loop finds the median of each area list in the dictionary. The output from Example 18.5 is as follows:

```
Polygons with cover 'woods' have median area 83095.3479504
Polygons with cover 'other' have median area 55491.6260843
Polygons with cover 'orch' have median area 83477.7527484
```

Example 18.6

```
# coverMedianArea.py
import arcpy, numpy
arcpy.env.workspace = 'C:/gispy/data/ch18'
fc = 'parkAreas.shp'

# Populate the dictionary,
# accumulate a list of areas for each cover type.
d = {}
sc = arcpy.da.SearchCursor(fc, ['COVER', 'F_AREA'])
for row in sc:
    cover = row[0]
    area = row[1]
    if d.has_key(cover):
        d[cover].append(area)
    else:
        d[cover] = [area]
del sc

# Calculate the median area for each cover type.
for k, v in d.items():
    median = numpy.median(v)
    print '''Polygons with cover '{0}' have \
            median area {1}'''.format(k, median)
```

These examples used the numpy modules. numpy is a scientific computing module optimized for efficiently performing operations on n-dimensional arrays. See the online documentation and examples for more information on the numpy module.

18.4   Discussion

Dictionaries provide a way to store a mapping between one set of values and another set of values. Like strings, lists, and tuples, they are a built-in Python sequence data type that has methods and special operations. Items are accessed via a key, not by indexing. The 'dict' dictionary data type is not ordered (though Python does have other variations of dictionaries that are ordered). A dictionary can be hard-coded or populated dynamically. The syntax for adding items uses an assignment statement with the dictionary name and key in square braces on the left and a value on the right. If the key is not already in the dictionary, a new item is added. If, on the other hand, the key is already in the dictionary, the item with this key is overwritten. Dictionaries with value lists are often useful. In this case, list syntax and methods must be used to modify the lists.

Common applications of lists include replacing multi-way conditional constructs (if/elif/elif...) and storing data collected from user input or with arcpy cursors. Care must be taken to use the in keyword as appropriate when populating a dictionary, since overwriting can occur. Handling these situations depends on the application. If the input already has unique keys (such as FIDs) no checking may be needed. But if collisions can occurs, thought should be given to how to handle these. Should the script keep the first instance of that key? Should it keep the last one? Should it use a list and store every occurrence? The examples in this chapter can be used to help with these decisions.

18.5   Key Terms

'dict' data type   Access by key vs. indexing multi-way
Key   conditional construct
Value   Dictionary methods (keys,
Item   values, items)
                                              numpy module

18.6   Exercises

1. **landUse.py** Sample script, 'landUse.py' has a land usage dictionary landUse = {'res': 1, 'com': 2, 'ind': 3, 'other' : [4,5,6,7]}. Add code to the script to use common dictionary operations and methods to perform the following tasks:

   (a) Print the value of the item with key 'com'.
   (b) Check if the dictionary has key 'res' and print the result.
   (c) Increment the value of the item with key 'ind' and then print the result.

(d) Use access by key format to add an item with land use of `'agr'` and a value of 0. (Then print the dictionary.)

(e) Change the land use `'res'` value to 10. (Then print the dictionary.)

(f) Print a list of the dictionary keys.

(g) Print the dictionary values, one per line, using a FOR-loop.

(h) Print a list of the dictionary items.

(i) Delete the item with key `'ind'`. (Then print the dictionary.)

(j) Check for membership of the key `'ind'`, using the keyword `in`. Print the result.

Printed output should appear as follows:

```
>>> 1. Print the value of the item with key 'com': 2
2. Check if the dictionary has key 'res': True
3. Increment the value of the item with key 'ind': 3
4. Add an item with land use 'agr' a value of 0:
   {'ind': 3, 'res': 1, 'other': [4, 5, 6, 7], 'com': 2,
   'agr': 0}
5. Change land use 'res' value to 10.
   {'ind': 3, 'res': 10, 'other': [4, 5, 6, 7], 'com': 2,
   'agr': 0}
6. Print a list of the dictionary keys:
   ['ind', 'res', 'other', 'com', 'agr']
7. Print the dictionary values:
3
10
[4, 5, 6, 7]
2
0
8. Print a list of the dictionary items:
   [('ind', 3), ('res', 10), ('other', [4, 5, 6, 7]), ('com',
   2), ('agr', 0)]
9. Delete the item with key 'ind':
   {'res': 10, 'other': [4, 5, 6, 7], 'com': 2, 'agr': 0}
10. Check for membership of the key 'ind': False
```

2. **tideRecords.py** Sample script, 'tideRecords.py' has a tide gauge dictionary `tides={'G1': [1,6], 'G2': [2], 'G3': [3,8,9]}`. The dictionary currently stores tide gauge readings for three gauges, G1, G2, and G3. Notice that each value in the dictionary is a Python list, so that one or more readings can be recorded for each gauge. Whenever a new item is added to the dictionary, the value must be a Python list, so that the item can store multiple readings for that gauge. Use dictionary operations to reflect each of the following events and print the dictionary after each step:

(a) A new gauge, G5 has been installed and the first reading was 2.

(b) Record an additional reading for gauge G5, the number 6.

(c) The latest reading from gauge G3 is invalid. Discard this information using the list `pop` method.

(d) A new gauge G6 has been installed but no readings are recorded yet. Use an empty list as a placeholder.

(e) Gauge G3 is no longer collecting data. Remove the item that represents this gauge.

(f) Gauge G1 recorded measurements two times higher than they should be. Use list comprehension to correct this error.

Printed output should appear as follows:
```
>>> 1. {'G5': [2], 'G3': [3, 8, 9], 'G2': [2], 'G1': [1, 6]}
2. {'G5': [2, 6], 'G3': [3, 8, 9], 'G2': [2], 'G1': [1, 6]}
3. {'G5': [2, 6], 'G3': [3, 8], 'G2': [2], 'G1': [1, 6]}
4. {'G6': [], 'G5': [2, 6], 'G3': [3, 8], 'G2': [2], 'G1':
   [1, 6]}
5. {'G6': [], 'G5': [2, 6], 'G2': [2], 'G1': [1, 6]}
6. {'G6': [], 'G5': [2, 6], 'G2': [2], 'G1': [0.5, 3.0]}
```

3. **tideRecordsLoop.py** Sample script, 'tideRecordsLoop.py' has a dictionary `tides={'G1': [1,6], 'G2': [2], 'G3': [3,8,9]}`. These represent tide gauge readings for three gauges, `'G1'`, `'G2'`, and `'G3'`. Use the `keys`, `values`, and `items` dictionary methods to perform the following updates and calculations:

(a) Append a reading of 7 for each gauge (Then print the dictionary.)

(b) Print the number of readings for each gauge.

(c) Square the value of every reading for every gauge. (Then print the dictionary.)

(d) Find the minimum reading for each gauge in the dictionary.

Printed output should appear as follows:
```
>>> 1. {'G3': [3, 8, 9, 7], 'G2': [2, 7], 'G1': [1, 6, 7]}
2. Number of readings: 4
2. Number of readings: 2
2. Number of readings: 3
3. {'G3': [9, 64, 81, 49], 'G2': [4, 49], 'G1': [1, 36, 49]}
4. Min reading at gauge G3 = 9
4. Min reading at gauge G2 = 4
4. Min reading at gauge G1 = 1
```

4. **check4keys.py** The sample script 'check4keys.py' contains a dictionary. Add code to the script so that it takes 1 or more numerical arguments, checks if the dictionary has those numbers as keys, and reports the results as shown in the following example:

Example input: 22 11 50 75

Example output:
```
>>> Key 22.0 has value: 4
Key 11.0 not found.
Key 50.0 not found.
Key 75.0 has value: 8
```

5. **sizeDict.py** Given a directory path as an argument, collect the file names and file sizes in a dictionary and print the dictionary.

Example input: C:/gispy/data/ch18/smallDir

Example output:
```
>>> Size dictionary:
{'a.shx': 124L, 'a.dbf': 2540L, 'dog.JPG': 579880L, 'pines.JPG':
588644L, 'a.shp': 620L, 'a.prj': 145L, 'a1.lyr': 10752L, 'cluck.
wav': 33970L, 'a.kml': 4684L, 'xyData2.txt': 124L, 'a.shp.xml':
1644L, 'abc.dbf': 250L}
Average file size: 98466.0769231
```

6. **typeDict.py** Given a directory path as an argument, collect the file data types and names in a dictionary. The dictionary should have one item for each data type, as determined by the `arcpy Describe` object. Each item should have a list of files with the key data type. Print the resulting dictionary and compare it to the given example.

Example input: C:/gispy/data/ch18/smallDir

Example output:
```
>>> Type dictionary:
{u'Layer': ['a1.lyr'], u'DbaseTable': ['abc.dbf'], u'ShapeFile':
['a.dbf', 'a.shp'], u'File': ['a.kml', 'a.prj', 'a.shp.xml', 'a.
shx', 'cluck.wav'], u'TextFile': ['xyData2.txt'], u'RasterDataset':
['dog.JPG', 'pines.JPG']}
```

7. **multiwayDict.py** The sample script 'multiwayDict.py' computes a list of values based on a given sea level rise (SLR) scenario and a time interval. It uses one branch for each of six SLR scenarios (B1, A1T, B2, A1B, A2, and A1F1). Each branch applies a rise rate that corresponds to the scenario. Rewrite this script to use the following dictionary of SLR scenarios and rates:

```
rateDict = {'B1': 0.0038, 'A1T': 0.0045, 'B2': 0.0043, 'A1B':
0.0048, 'A2': 0.0051, 'A1F1':0.0059}
```

Remove the multi-way conditional construct. Replace the repeated code in each block with a single set of statements that uses the dictionary keys and values to achieve the same affect. Include code to check that the dictionary has the input key. The revised script should be less than 25 lines.

Example input1: A1B 50

Example output1:
```
>>> Running Sea Level Rise Model for A1B [0.24, 0.48, 0.72]
```

Example input2: AZ 5

Example output2:
```
>>> Warning: Invalid resolution. Choose B1, A1T, B2, A1B, A2,
or A1F1.
```

8. **trees.py** Write a script that will read a dBASE file and collect information in a dictionary about two of the dBASE fields. The script should require three arguments, the full path name of a dBASE file, the name of a classification field in the dBASE file (e.g., species), and the name of a second field in the dBASE file (e.g., DBH). Create a dictionary with a key for each distinct classification. Use the value of the second field to set the dictionary item's value. Use only the *first occurrence* of a classification encountered by the cursor (e.g., record only the first 'LOB', which has a DBH of 24). Print the dictionary.

Example input: C:/gispy/data/ch18/rdu_forest1.dbf SPECIES DBH

Example output:
```
>>> Tree dictionary:
{u'LOB': 24, u'BE': 17, u'WD': 17, u'WO': 14, u'HK': 9, u'YP': 17,
u'POP': 21, u'RB': 15, u'RM': 7, u'ASH': 11, u'LP': 23, u'SLP': 14,
u'VP': 9, u'SRW': 5, u'SHL': 16, u'CV': 18, u'RO': 8, u'MPL':
5, u'SP': 13, u'SW': 5, u'MW': 4, u'SL': 11, u'SG': 14}
```

9. **trails.py** Practice using cursors with dictionaries by writing a script to use a dictionary to calculate some statistics for a shapefile. The script should require three arguments, the full path name of a shapefile, a unique identifying field name, and a numerical field. As an example, we'll use a shapefile containing trail widths for a park in Narnia. First, populate and print a dictionary to pair the input fields using the ID field for the keys and the numeric field as the value. The 'numeric values' in the sample data, the trail width field, are stored in a text type field, so they should be cast to floats before being inserted into the dictionary. Next, use the dictionary values to find and print the minimum and maximum values of the trail width field. Also, determine and print the dictionary keys whose trail width values are minimal and the keys whose trail width values are maximal as shown in the example (The example output only shows part of the dictionary).

Example input: C:/gispy/data/ch18/narniaHike.shp FID Tra_Width

Example output:
```
>>> FID_Tra_Width_Dict = {{0: 12.0, 1: 30.0, 2: 24.0, 3: 18.0,...
Minimum Tra_Width = 6.0
Features(s) with minimum Tra_Width: 6 65 82 114
Maximum Tra_Width = 48.0
Features(s) with maximum Tra_Width: 56 164
```

10. **trailWidths.py** Write a script to use a dictionary and find the average trail width for each classification in the Narnia trail shapefile. The script should require three arguments, the full path name of a shapefile, a classification field name, and a trail width field. Populate and print a dictionary with classifications as keys and lists of trail widths as values. The trails widths in the sample data are stored as strings, so they must be cast to floats. Then find and print the average trail width for each classification as shown in the example (The example output only shows part of the dictionary).

Example input: C:/gispy/data/ch18/narniaHike.shp Classifica Tra_Width

Example output:
```
>>> Classifica_Tra_Width_Dict = {u'Barren': [18.0, 24.0, 30.0, 18.0,
18.0, 12.0, 30.0, 18.0, 24.0, 30.0, 18.0], u'Some Bare Ground':
[18.0, 12.0, 12.0,...
Classifica: Barren, average Tra_Width: 21.82
Classifica: Some Bare Ground, average Tra_Width: 20.68
Classifica: Stunted Vegetation, average Tra_Width: 18.39
```

Tip The following code rounds one number (Pi) to two decimal places and another number (e) to three decimal places:

```
>>> import math
>>> print 'The numbers {0:.2f} and {1:.3f}'.format(math.pi, math.e)
The numbers 3.14 and 2.718
```

Chapter 19
Reading and Writing Text Files

Abstract GIS data often comes in formats that can't be imported into ArcGIS or read with `arcpy` cursors without modification. Automating these modifications can be a significant time saver if you have numerous files that require the same changes. The `arcpy` cursors discussed in Chapter 17 are specialized for reading and writing attribute tables that conform to a prescribed format, such as dBASE files, shapefiles, and GRID rasters. This chapter discusses Python file handling functions that can read and write text files, regardless of format. The chapter begins with Python file handling syntax and then uses GIS examples to demonstrate common tabular text file modifications.

Chapter Objectives

After reading this chapter, you'll be able to do the following:

- Write and read text files.
- Use the Python `with` structure for reading and writing.
- Separate the parts in a line of a tabular data file.
- Modify a tabular data file.
- Differentiate between the Python working directory and the `arcpy` workspace.
- Handle `IOError` exceptions.
- Avoid file locking.
- Write a file that preserves Python data structures.

19.1 Working with `file` Objects

The first step to reading or writing a file is to call the built-in `open` function which returns a `file` object. The `'file'` data type has a number of methods for reading and writing a file. The following line of code creates a `file` object named `'f'` that allows you to read a file:

```
>>> f = open('C:/gispy/data/ch19/poem.txt', 'r')
```

The `open` function requires one argument, the name of the file. The `file` object can be created for reading, writing, or both. The second argument is an optional

Table 19.1 Approaches to reading a text file.

with open(<filename>, 'r') **as** f:	
f open in 'r' mode	**What it does**
f.read()	Read the entire contents of the file and returns a string
f.read(8)	Read the next 8 bytes of the file and return a string
f.readline()	Read the next line and return a string
f.readlines()	Read every line and return a list of strings (one string per line)
for line **in** f:	Iterates through each remaining line in the file

argument for setting the mode. The mode describes how the file will be used: 'r' for reading text, 'w' for writing text, 'r+' for both reading and writing ASCII text files. Other modes can be used to handle binary (non-text) file formats, but the examples in this chapter focus on reading and writing text files ('r' and 'w'), as these are the most common usage scenarios. Depending on the mode, the file object returned by the open function has either read or write methods. In the code above, 'poem.txt' was opened in read mode. There are several methods for reading all or part of a file (See Table 19.1). The read method reads the entire file and returns it as a string:

```
>>> f = open('C:/gispy/data/ch19/poem.txt', 'r')
>>> f.read()
'\nScripterwocky\n`Twas brillig, and the Python lists\nDid join
and pop-le in the loop:\nAll-splitsy were the string literals,\nAnd
the boolean values were true.\t \n'
```

Run the code above, then browse to 'poem.txt' and open it to view the contents in a text editor, such as 'Wordpad'. The file contains five lines of text, a Pythonic verse inspired by Lewis Carroll's poem, 'Jabberwocky'. The string printed by Python contains a carriage return escape sequence ('\n') at the end of each line of text.

Add another line to the poem and try to save the file. If you have run the file opening code, you will receive a message that the file is in use by another application and it cannot be saved. Like arcpy cursors, Python file objects lock the file. To release the file, you must use the close method. There should be one close statement for each open statement to prevent locking issues. The close method does not require any arguments, but parentheses must be used, since it is a method. Use the following statement to close the file object and release the lock:

```
>>> f.close()
```

Similar to the process of reading a book (open->read->close), when reading or writing a file, the code should open the file, read or write, and then close the file. If a book is left open, it can be splattered with soup. If a file is left open, it will be locked and can cause errors or data corruption.

19.1.1   The WITH Statement

The WITH statement is an alternative to closing the file explicitly. The usage of WITH statements for file objects is similar to the usage described in Chapter 17 for arcpy cursors. Place the open function call between the with and as keywords; Place the file object variable name after the as keyword; Complete the statement with a colon. Indent the code that deals with the file object. The file object is only usable inside this indented code block. Dedented code signals the end of the block. The following two lines of code,

```
>>> with open('C:/gispy/data/ch19/poem.txt', 'r') as f:
...     print f.read()
```

are equivalent to the following three lines of code,

```
>>> f = open('C:/gispy/data/ch19/poem.txt', 'r')
>>> f.read()
>>> f.close()
```

except that the with block ensures that locks are released even if an exception is thrown while attempting to read the file. When a with statement is used, the file close method does not need to be called. Upon exiting the with block, the file is automatically closed. Attempting to call the read method outside of this block returns a value error, since the file is automatically closed when the code exits the with code block:

```
>>> with open('C:/gispy/data/ch19/poem.txt', 'r') as f:
...     print f.read()
...
>>> f.read()
Traceback (most recent call last):
File "<interactive input>", line 1, in <module>
ValueError: I/O operation on closed file
```

The open->read->close approach is handy for Interactive Window examples since the read feedback is immediate and the explicit close statements make the scope of the file object obvious. Otherwise, the with statement is a good technique. This chapter uses both approaches.

19.1.2   Writing Text Files

There are other methods for reading files which we'll return to momentarily. Writing to file is less nuanced, so we'll address that first. To write a text file, call the open function in 'w' mode. Then the file object has a write method which requires a string argument, the text to be written to the file:

```
>>> f = open('C:/gispy/data/ch19/sneeze.txt', 'w')
>>> f.write('haa')
>>> f.write('choo')
>>> f.close()
```

Separate calls to the `write` method do not automatically place the strings on separate lines. Though two separate write statements are used, the output is all on the same line in the file:

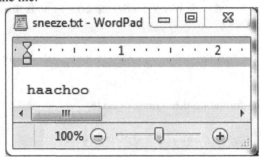

Carriage returns (`'\n'`) must be placed in the strings where the new lines are desired. Notice the new line can be placed anywhere in a string (e.g., the middle as well as the end):

```
>>> f = open('C:/gispy/data/ch19/sneeze.txt', 'w')
>>> f.write('snork\nsniffle\n')
>>> f.write('haaack\n')
>>> f.write('*sigh*')
>>> f.close()
```

The output from the following code contains four lines:

Notice that the original text in 'sneeze.txt.' has been overwritten. When a file is opened for writing, any existing text is erased, unless an append mode ('a') is specified. The following code creates an empty file:

```
>>> f = open('C:/gispy/data/ch19/sneeze.txt', 'w')
>>> f.close()
```

The string join function can be used to write a list of strings as comma separated values on the same line or to write a list of strings on separate lines.

```
>>> f = open('C:/gispy/data/ch19/sneeze.txt', 'w')
>>> csv = ','.join(['glug', 'gulp', 'erp'])
>>> f.write(csv)
>>> f.write('\n')
>>> lines = '\n'.join(['ack', 'hmmm', 'sniff'])
>>> f.write(lines)
>>> f.close()
```

The write method only takes string arguments. Attempting to write a number results in a TypeError:

```
>>> f = open('C:/gispy/data/ch19/sneeze.txt', 'w')
>>> f.write(5000)
Traceback (most recent call last):
File "<interactive input>", line 1, in <module>
TypeError: expected a character buffer object
```

Casting or string formatting can be used to write numeric values to a file:

```
>>> f.write(str(5000))
>>> lament = 'I sneezed {0} times today.'.format(5000)
>>> f.write(lament)
>>> f.close()
```

19.1.3   Safe File Reading

When a file is opened in write mode, if the file already exists, the contents will be overwritten. If the file doesn't exist yet, it will be created. For a file to be opened in read mode, however, it must exist. If you attempt to open a nonexistent or corrupted file in read mode, an IOError is thrown:

```
>>> f = open('C:/gispy/data/ch19/bogus.txt', 'r')
Traceback (most recent call last):
File "<interactive input>", line 1, in ?
IOError: [Errno 2] No such file or directory: 'bogus.txt'
```

When the file might not exist, you can use code like the following to catch an IOError exception and handle the possibility of failure gracefully:

```
# safeFileRead.py
import os, sys

infile = sys.argv[1]

try:
    f = open(infile, 'r'
    print f.read()
    f.close()
except IOError:
    print "{0} doesn't exist or can't be opened.".format(infile)
```

19.1.4   The os Working Directory vs. the arcpy
Workspace

When you're writing geoprocessing scripts and setting the arcpy workspace, it's easy to make the mistake of assuming this will work for Python file objects as well. The built-in open function is not part of the arcpy package, so the full path file name may need to be used even if the arcpy workspace is set. For example, the following code sets the arcpy workspace and then attempts to open 'poem.txt' which resides in the workspace:

```
>>> import arcpy
>>> arcpy.env.workspace = 'C:/gispy/data/ch19'
>>> f = open('poem.txt', 'r')
Traceback (most recent call last):
File "<interactive input>", line 1, in <module>
IOError: [Errno 2] No such file or directory: 'poem.txt'
```

An IOError exception is thrown even though the os.path method isfile proves that this file does exist in that workspace:

```
>>> import os
>>> os.path.isfile('C:/gispy/data/ch19/poem.txt')
True
```

In fact, Python has its own native workspace which is stored by the os module. If the full path file name is not specified, Python searches for the file in the os module working directory. The getcwd method returns the current working directory (the CWD). The CWD depends on a number of factors, including the IDE. If PythonWin is opened by browsing to a script, right-clicking, and selecting 'Edit with PythonWin', the CWD is the path to the script. If, instead, PythonWin is opened first (e.g., by clicking on a shortcut to PythonWin), the CWD is the path to the PythonWin executable, as in the following example:

```
>>> os.getcwd()
'C:\\Python27\\ArcGIS10.3\\Lib\\site-packages\\pythonwin'
```

The chdir method changes the current working directory. The following code changes the os current working directory and then the file is opened for reading:

```
>>> os.chdir('C:/gispy/data/ch19')
>>> os.getcwd()
'C:\\gispy\\data'
>>> f = open('poem.txt', 'r')
>>> f.readline()
'Scripterwocky\n'
>>> f.close()
```

The arcpy workspace and the os current working directory are mutually independent. Changing the os module working directory does not modify the arcpy workspace.

19.1.5 Reading Text Files

The read method returns the entire contents of the file as a string with lines separated only by carriage return characters. For many applications, you may want to read each line, one at a time, using the readline method or a FOR-loop. The readline method reads the next line in the file. When the file is first opened, the "next" line is the first line of the file:

```
>>> f = open('C:/gispy/data/ch19/poem.txt', 'r')
>>> f.readline()
'Scripterwocky\n'
```

The file object keeps track of where it is in the file. You can think of the file object as maintaining an internal position cursor. When readline is called again, it reads the next line:

```
>>> f.readline()
'Twas brillig, and the Python lists\n'
```

The readline method returns each line with a trailing carriage return sequence. This sequence is interpreted and printed as a blank line when a print function is used.

```
>>> f.readline()
Did join and pop-le in the loop:

>>>
>>> f.readline()
'All-splitsy were the string literals,\n'
>>>
```

The readline method returns an empty string when the end of the file is reached:

```
>>> f.readline()
'And the boolean values were true.\t \n'
>>> f.readline()
''
>>> f.close()
```

The file object can also be used as an iterator in a FOR-loop. This approach is frequently used.

```
>>> for line in f:
...         print line
...
```

```
Scripterwocky

`Twas brillig, and the Python lists

Did join and pop-le in the loop:

All-splitsy were the string literals,

And the boolean values were true.
```

The loop iterates over each line until the end of the file is reached.

```
>>> f.readline()
''
>>> f.close()
```

Looping over a file object can be used together with the readline method. The iteration in the FOR-loop will start wherever the readline stopped. The following code reads the first line of the file, the poem title, with the readline method, then reads the rest of the file by looping over the file object.

```
>>> f = open('C:/gispy/data/ch19/poem.txt', 'r')
>>> f.readline()
'Scripterwocky\n'
>>> for line in f:
...     print line
...
Twas brillig, and the Python lists

Did join and pop-le in the loop:

All-splitsy were the string literals,

And the boolean values were true.

>>> f.close()
```

Calling readline followed by a FOR-loop is often used to handle header lines separately from table records. This approach is used again in upcoming examples on handling file content. This section used read, readline, and FOR-loops to read a file. Though other approaches exist, these are sufficient for the most common applications. Table 19.1 summarizes these methods and variations.

19.2   Parsing Line Contents

Once you read a line of a text file with a script, you need to *parse* the contents, in other words, break them down into usable parts. Each line is returned as a single string with a trailing carriage return. Depending on the text file contents, parsing the text may involve stripping trailing whitespace, splitting the string into a list of items, casting the string to a numeric value and so forth. Several common parsing operations are demonstrated here.

Suppose, for example, you want to find the sum of a line in 'report.txt'. Though the file contains tab delimited numbers, the `readline` function always returns a single string. In the following example, readline returns a string containing numbers and escape sequences:

```
>>> f = open('C:/gispy/data/ch19/report.txt', 'r')
>>> f.readline()
'1\t2.07\t5.21\t4.05\t3.64\t2.03\t3.74\n'\'
```

To use the contents, you need to store them in a variable. The code above read line one, but did not store it. For example's sake, we'll move on to line two. The following code stores and prints line two:

```
>>> line = f.readline()
>>> line
'2\t3.51\t7.29\t4.2\t4.44\t3.67\t4.46\n'
>>> print line
2    4.51    7.29    4.2    4.44    3.67    4.46
```

The tabs appear as whitespace when the print statement is used because it interprets them, but the `line` variable is a string data type containing all the numbers and tabs within one string. Now that the line contents are stored in a variable, you can use the string `split` method to separate the numbers. By default, the `split` method separates the string contents at the white spaces, so no argument is needed. The following code splits the values into a list:

```
>>> lineList = line.split()
>>> lineList
['2', '3.51', '7.29', '4.2', '4.44', '3.67', '4.46']
```

The `lineList` variable is not quite ready for numeric calculations, since the list elements are strings. You can use a list comprehension to convert each one to a number:

```
>>> nums = [float(i) for i in lineList]
>>> nums
[2.0, 3.51, 7.29, 4.2, 4.44, 3.67, 4.46]
```

Suppose the first column in 'report.txt' is an ID number, which you don't want to include in the sum. The following code uses slicing to exclude the first column and then recomputes the sum:

```
>>> data = nums[1:]
>>> data
```

```
[3.51, 7.29, 4.2, 4.44, 3.67, 4.46]
>>> sum(data)
27.57
```

If other operations need to be performed on individual elements of a line, list comprehension or FOR-loops can be used. Suppose, for example, the values in the report need to be capped at 5. Example 19.1 uses a nested FOR-loop to cap the data values of 'report.txt'. This script also finds the sum and count of data elements for each line. The third line is missing an element, as the output shows:

```
ID: 1.0 Sum: 20.53 Count 6
Data: [2.07, 5, 4.05, 3.64, 2.03, 3.74]
ID: 2.0 Sum: 25.28 Count 6
Data: [3.51, 5, 4.2, 4.44, 3.67, 4.46]
ID: 3.0 Sum: 19.72 Count 5
Data: [3.9, 4.24, 4.05, 4.04, 3.49]
ID: 4.0 Sum: 22.64 Count 6
Data: [3.18, 3.5, 4.73, 4.39, 3.28, 3.56]
```

Example 19.1

```python
# parseTable.py
# Purpose: Parse numeric values in a tabular text file.
cap = 5
infile = 'C:/gispy/data/ch19/report.txt'
try:
    with open(infile, 'r') as f:
        for line in f:
            # String to list of strings.
            lineList = line.split()
            # String items to float items.
            nums = [float(i) for i in lineList]
            # First col is ID, rest are data values.
            ID = nums[0]
            data = nums[1:]
            # Cap the data values at 5.
            for index, val in enumerate(data):
                if val > cap:
                    data[index] = cap
            # Count and sum the values and report the results.
            count = len(data)
            total = sum(data)
            print 'ID: {0} Sum: {1} Count {2}'.format(ID, total, count)
            print 'Data: {0}'.format(data)
except IOError:
    print "{0} doesn't exist or can't be opened.".format(infile)
```

The basics steps outlined above are used frequently for reading tabular data: Store the line in a variable, split the contents, cast each element to a number, and slice the columns.

The split method is needed whenever you're reading tabular data. The argument for the split method varies depending on the delimiter (comma, space, and so forth). There is a specialized Python module `csv` for handling comma separated values, but these can also be handled using the generic file handling described here. Example 19.2 reads 'cfactors.txt', which contains a set of numerical erosion factors and their labels, delimited by an equals sign (=). The factors and labels are placed in a dictionary:

```
{1: 'stable', 2: 'low deposition', 3: 'moderate deposition', 4:
'high deposition', 5: 'severe deposition'}
```

The script uses both a `readline` call and a FOR-loop. The first line of the file, a header, is not needed. The code reads past the first line, (not saving it), then reads the rest into the dictionary. The erosion factor numbers are cast to integer and the labels are right-stripped of whitespace using the following statements:

```
factor = int(line[0])
label = line[1].rstrip()
```

If these statements were replaced by simpler statements:

```
factor = line[0]
label = line[1]
```

the resulting dictionary would have string keys and the values would have trailing carriage returns:

```
{'1': 'stable\n', '3': 'moderate deposition\n', '2': 'low deposition\n',
'5': 'severe deposition\n', '4': 'high deposition\n'}
```

Example 19.2

```python
# cfactor.py
# Purpose: Read a text file contents into a dictionary.
factorDict = {}
infile = 'C:/gispy/data/ch19/cfactors.txt'
try:
    with open(infile, 'r') as f:
        f.readline()
        for line in f:
            line = line.split('=')
            factor = int(line[0])
```

```
            label = line[1].rstrip()
            factorDict[factor] = label
    print factorDict
except IOError:
    print "{0} doesn't exist.".format(infile)
```

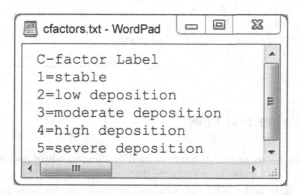

```
cfactors.txt - WordPad

C-factor Label
1=stable
2=low deposition
3=moderate deposition
4=high deposition
5=severe deposition
```

19.2.1 Parsing Field Names

Example 19.2 hard-codes the index of the fields (0 and 1) in the split line containing those fields. If the field name is known, but the position of the column is not, you can use the list index method, which returns the index of the first occurrence of a value in the list:

```
>>> mylist = ['a','b','c','d']
>>> mylist.index('c')
2
```

The 'fieldIndex.py' script in Example 19.3 calls a getIndex function that strips the trailing whitespace from the field names string, splits the string into a list, and then finds the index of the field based on its name 'Label'. Since 'Label' is the second field in the 'cfactors.txt' data file, the script finds that the field named 'Label' has index 1.

Example 19.3

```
# fieldIndex.py
# Purpose: Find the index of a field name in a text
#          file with space separated fields in the first row.
infile = 'C:/gispy/data/ch19/cfactors.txt'
fieldName = 'Label'
```

```
def getIndex(delimString, delimiter, name):
    '''Get position of item in a delimited string'''
    delimString = delimString.strip()
    lineList = delimString.split(delimiter)
    index = lineList.index(name)
    return index

with open(infile, 'r') as f:
    line = f.readline()
    ind = getIndex(line, ' ', fieldName)
    print '{0} has index {1}'.format(fieldName, ind)
```

19.3   Modifying Text Files

For many applications you will need to modify files. Reading and writing to the
same file using ' r+ ' mode is not usually the best solution for this, because managing existing content and modifications simultaneously can be complicated. A simpler approach is to read the original file, make modifications to the content within
the script, then write the modified content to another file. The script can then replace
the original file with the new one, if desired. This flow would be as follows:

open->read->parse->modify->write->close->replace

As an example, suppose you want to use the c-factor table from 'cfactor.txt' in
ArcGIS. Because of the spacing in the file, ArcGIS parses the field values
incorrectly:

C_factor	Label
▶ 1=stable	
2=low	deposition
3=moderate	deposition
4=high	deposition
5=severe	deposition

To correct this, 'cfactorModify.py' in Example 19.4 places a comma between the
field names and replaces the equals signs with commas. When there is a space on
the first line of a text file, ArcGIS interprets the space separated items on this line as
field names and then it interprets spaces on additional lines as the delimiter.

In Example 19.4, the script names the output file based on the input file name,
opens the input file for reading and the output file for writing. Each line is read and

modified and then written to the output file. In this case, the first line is handled differently than the rest, so a `readline` call is used for the first line and a FOR-loop is used for the rest. The input and output files are then closed implicitly when the code exits the WITH statement. The output table displays correctly in ArcGIS:

	C_factor	Label
▶	1	stable
	2	low deposition
	3	moderate deposition
	4	high deposition
	5	severe deposition

The examples and exercises in this book forgo the step of replacing the original file with the new one so that the input and output can be easily compared. You can add this functionality using the built-in Python shell utilities module named `shutil`. The `move` method takes two file names as arguments, a source and a destination. It moves the first file to the location of the second file. Moving the modified version to the location of the original version overwrites the original with the modified file. Add the following code to the end of Example 19.3 and 'cfactorsv2.txt' will be moved to 'cfactors.txt', replacing the original and deleting version 2 at the same time:

```
import shutil
shutil.move(outfile, infile)
```

Example 19.4

```
# cfactorModify.py
# Purpose: Demonstrate reading and writing files.

import os

infile = 'C:/gispy/data/ch19/cfactors.txt'
baseN = os.path.basename(infile)
outfile = 'C:/gispy/scratch/' + os.path.splitext(baseN)[0] + 'v2.txt'
try:
    #OPEN the input and output files.
    with open(infile, 'r') as fin:
        with open(outfile, 'w') as fout:
            #READ/MODIFY/WRITE the first line.
            line = fin.readline()
            line = line.replace(' ',',')
            fout.write(line)
            #FOR the remaining lines.
            for line in fin:
```

```
              #MODIFY the line.
              line = line.replace('=', ',')
              #WRITE to output.
              fout.write(line)
          print '{0} created.'.format(outfile)
except IOError:
    print "{0} doesn't exist.".format(infile)
```

19.3.1   Pseudocode for Modifying Text Files

The flow for modifying text files can usually be described by the following
pseudocode:

```
SET the input and output file name
IF the input file exists:
  OPEN the input and output files
  FOR EACH line in the input file
    MODIFY the line.
    WRITE the modified line to the output file.
  END FOR
  CLOSE the input and output files (automatic if WITH block is used).
  REPLACE original
ENDIF
```

Placing this pseudocode in a script and then filling in the details can be a good
place to start any file modification task.

19.3.2   Working with Tabular Text

Some common modification tasks, such as removing lines or columns, take a
slightly different approach. To demonstrate these functions, we'll use the raw output
from a device that tracks eye movement and records the points on a map where the
eye fixates. The raw eye tracking data is not designed for use in ArcGIS, so it needs
to be modified for import into ArcGIS and visualized as a shapefile. Some entries
need to be deleted. The arcpy update cursor function has a deleteRow method
for this purpose, but when the data is not ArcGIS compatible this method can't be
used. The Python native file objects do not have a delete function. However, deletion can be achieved by omission. The usual flow is to read a line of the input file,
modify the line, and then write the line to the output file. For lines that are to be
deleted, you can read the line from the input, but not write it to the output. The

internal position cursor needs to read each line in the input file to step through it methodically, but not all the lines that are read need to be written to the output!

ArcGIS expects the first line of a table to be a set of field names; However, tables are often preceded by metadata, so removing header lines is a common task. The first two lines of the 'eyeTrack.csv' file contain metadata about the conditions under which eye movements were tracked. These lines must be deleted so that the MakeXYEventLayer management tool can be used to bring the points into ArcGIS. To strip these lines from the modified file, 'removeHeader.py', in Example 19.5 simply reads these lines, but does not write them to output.

Example 19.5

```
# removeHeader.py
# Purpose: Remove header rows.

import os
headers = 2
infile = 'C:/gispy/data/ch19/eyeTrack.csv'
baseN = os.path.basename(infile)
outfile = 'C:/gispy/scratch/' + os.path.splitext(baseN)[0] \
                    + 'v2.txt'
try:
    with open(infile, 'r') as fin:
        with open(outfile, 'w') as fout:
            #READ header lines, but don't write them.
            for i in range(headers):
                fin.readline()
            #READ and WRITE the remaining lines.
            for line in fin:
                fout.write(line)
            print '{0} created.'.format(outfile)
except IOError:
    print "{0} doesn't exist.".format(infile)
```

Deleting header lines boils down to reading but not writing the first few lines to the output file. Sometimes other lines need to be deleted. For example, we also need to remove lines in 'eyeTrack.csv' where the fixation point-of-gaze x or y values (FPOGX and FPOGY fields) are not positive; These represent invalid readings. The pseudocode for removing delimited records based on a condition is as follows:

```
FOR each line in the input file:
    SPLIT the line (string to list)
    IF condition is TRUE THEN
        WRITE the string line to the output file.
    ENDIF
ENDFOR
```

In short, you check the condition and only write the line when the condition is
true. Example 19.6 shows a code snippet from sample script 'removeRecords.py'.
This code only writes table lines to the output file that have values greater than zero
for the fields FPOGX and FPOGY. This has the effect of deleting the lines with non-
positive values for either of these fields. This script calls the getIndex function
defined in Example 19.3 and then when it encounters each line, it splits the line, and
gets the float values of the columns of interest using the field indices. It then checks
the required condition (positive for both columns) and writes the line only if the
condition is met.

Example 19.6

```
# Excerpt from sample script 'removeRecords.py'
# REMOVE selected lines (only write acceptable lines)
# READ and WRITE field names
fieldNameLine = fin.readline()
fout.write(fieldNameLine)
# FIND field indices
sep = ','
findex1 = getIndex(fieldNameLine, sep, 'FPOGX')
findex2 = getIndex(fieldNameLine, sep, 'FPOGY')

# FOR the remaining lines:
for line in fin:
    lineList = line.split(sep)
    v1 = float(lineList[findex1])
    v2 = float(lineList[findex2])
    # IF condition is TRUE, write this record.
    if v1 > 0 and v2 > 0:
        fout.write(line)
```

Several columns in the eye tracking data contain non-numeric values. These columns need to be removed entirely. Since the file object reads each line individually
and the tabular line contents can be split into lists, dealing with columns involves
list operations. To remove columns, you need to remove the corresponding item in
each line's list. The getIndex function defined in Example 19.3 can be used again
to determine the index of a column. Then the string pop method can be used to
remove that index:

```
>>> fields = ['FireId', 'Org', 'State', 'FireType', 'Protection']
>>> index = 2
>>> fields.pop(index)
'State'
>>> fields
['FireId', 'Org', 'FireType', 'Protection']
```

The `pop` removes the item with the specified index. It returns the value of the item it removed, but we're not using that value in this application, so we can just discard it. Instead, we are using the resulting list which has one less item. Care must be taken to remove the intended items. When `pop` is used, the indices of the rest of the items coming after that one in the list are decremented by one. Trying to `pop` indices 2 and 4 in that order results in an index error:

```
>>> fields = ['FireId', 'Org', 'State', 'FireType', 'Protection']
>>> indexA = 2
>>> indexB = 4
>>> fields.pop(indexA)
'State'
>>> fields.pop(indexB)
Traceback (most recent call last):
File "<interactive input>", line 1, in <module>
IndexError: pop index out of range
```

To avoid this problem, pop the indices in descending order:

```
>>> fields = ['FireId', 'Org', 'State', 'FireType', 'Protection']
>>> fields.pop(indexB)
'Protection'
>>> fields
['FireId', 'Org', 'State', 'FireType']
>>> fields.pop(indexA)
'State'
>>> fields
['FireId', 'Org', 'FireType']
```

Eight columns in the 'eyeTrack.csv' file need to be removed. Example 19.7 lists these fields in the `removeFields` variable. This code snippet from sample script 'removeColumns.py' reads the line containing the field names, determines the indices of these unwanted columns, then calls a `removeItems` function. This function sorts the indices in reverse order and then pops the unwanted values off a list that is derived from a delimited string. The `removeItems` function is called once for the field names line and then it is called again for each record in the table. The function splits the line string into a list, removes the unwanted items in the list and then rejoins the modified list into a delimited string. This shortened string is returned to the caller and the script writes the modified line to the file.

Example 19.7

```
# Excerpt from sample script 'removeColumns.py'
# REMOVE named columns (only write valid columns)

removeFields = ['LPCX','LPCY', 'RPCX', 'RPCY',
                'LGX', 'LGY', 'RGX', 'RGY']
```

```
def removeItems(indexList, delimiter, delimString):
    '''Remove items at given indices in a delimited string'''
    lineList = delimString.split(delimiter)
    indexList.sort(reverse = True)
    for i in indexList:
        lineList.pop(i)
    lineString = delimiter.join(lineList)
    return lineString

###code to OPEN files and READ past header omitted###
        # READ field names.
        fieldNamesLine = fin.readline()
        # DETERMINE field indices.
        rfIndex = []
        for field in removeFields:
            index = getIndex(fieldNamesLine, sep, field)
            rfIndex.append(index)
        # REMOVE items with these indices and WRITE the line.
        line = removeItems(rfIndex, sep, fieldNamesLine)
        fout.write(line)
        # READ/MODIFY/WRITE the remaining lines.
        for line in fin:
            line = removeItems(rfIndex, sep, line)
            fout.write(line)
```

You can combine functionality such as removing lines and removing columns of a tabular text file within a single script by working from top to bottom in a text file, one line at a time, parsing and modifying data along the way. The 'eyeTrackSHP.py' sample script in the exercises deletes rows and columns and performs a transformation from an eye tracking reference system to a geographic reference system. In Figure 19.1, the eye fixation point shapefile is overlaid on the world map that the viewer was viewing as the eye movements were recorded.

19.4 Pickling

The standard file object write function requires a string argument. Suppose your scripts generate a number of dictionaries and lists that you want to save in a file for further analysis. This means converting other data types before writing them to file. For example, to write a number in a file, you need to first cast it to a string; To write a list to a file, you need to represent it as a string by joining the elements. If instead, you want to preserve the original data types within a file, you can use a built-in Python module named pickle.

Pickling allows you to write and read any Python data type. The file itself contains information that encodes the data type. When opened in Wordpad, the encoding

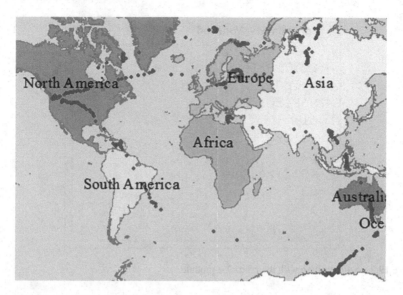

Figure 19.1 Eye fixation points drawn on the stimulus map.

doesn't look the same as the Python syntax for these types, but Python can read them using the `pickle` module.

The pickle methods for writing and reading are named `dump` and `load`. The `dump` method can be used when a file is opened for writing. It takes two arguments, the object to be written, which can be any data type, and a `file` object created in write mode. The following code dumps a float and then a list into the 'gherkin.txt' file:

```
>>> import pickle
>>> f = open('C:/gispy/data/ch19/gherkin.txt', 'w')
>>> pickle.dump(2.71828,f)
>>> pickle.dump(['FireId', 'Org', 'FireType'],f)
>>> f.close()
```

The resulting 'gherkin.txt' file is shown in Figure 19.2. The following code opens the pickle file in read mode and then uses the `load` method to read the first item:

```
>>> f2 = open('C:/gispy/data/ch19/gherkin.txt', 'r')
>>> thing1 = pickle.load(f2)
>>> thing1
2.71828
```

As expected, `thing1` has a `'float'` data type:

```
>>> type(thing1)
<type 'float'>
```

Figure 19.2 A file written with the `pickle` module.

and the second item loaded from the file into thing2 has a 'list' data type.

```
>>> thing2 = pickle.load(f2)
>>> thing2
['FireId', 'Org', 'FireType']
>>> type(thing2)
<type 'list'>
>>> f2.close()
```

Files written with pickling are only meant to be read with pickling. Standard `file` object reading will not preserve the data types. The following code reads the first line of the pickled file and instead of a float, it returns a string that starts with 'F':

```
f3 = open("C:/gispy/data/ch19/gherkin.txt", "r")
f3.readline()
'F2.71828\n'
f3.close()
```

19.5   Discussion

This chapter discussed `file` object methods and techniques for parsing and modifying files. `file` objects have additional methods, such as `seek` and `writelines`, but most file handling can be accomplished using only the `read`, `write`, `readline`, and FOR-loop methods discussed here. Using the WITH statement to open files provides a convenient protection against locking. If a WITH statement is not used to open the file, the `close` method must be called before the file is released. If an exception occurs before the `close` method is called, the file will remain locked and you may need to type a call to the `close` method in the Interactive Window. The `close` method can be used along with the WITH structure, but it is not necessary.

The built-in `file` object `read` functions always read file content as strings and file content must be written as strings. This means that most of the parsing and modification work in file handling involves string operations—casting when numbers are involved and converting to lists when the lines are part of a table. Pickling is a useful alternative for writing files that are only meant to be read by Python; Whereas, generic Python `file` objects provide techniques for manipulating the data into a format that can be imported into GIS software.

19.6 Key Terms

The `'file'` data type
The built-in `open` function
The file `read`, `readline`, `write`, and `close` methods
The WITH statement for opening files.
The **for** line **in** f reading approach
The `IOError` exception

Current working directory vs. `arcpy` workspace
Parse
The open/read/parse/modify/write/close/ replace workflow
The `shutil` module
The list `pop` method
The `pickle` module

19.7 Exercises

1. **writeBio.py** Create a Python script which writes a file in 'C:/gispy/scratch' called 'mybio.txt' containing your name on the first line, your favorite GIS tool on the next line, your home town on the third line, and a short autobiography on the last line.

2. **fileIO.py** Write a script to use common file handling operations and methods to perform the following tasks in the order given:

   (a) Open a file named 'C:/gispy/scratch/test.txt' for writing.
   (b) Write the number 365 on the first line of the file.
   (c) Write tab separated week days on the next line of the file.
   (d) Write the numbers 1, 2, and 3 separated by commas on the third line.
   (e) Close the file.
   (f) Open the same file for reading.
   (g) Read the first line and print the results.

(h) Read the second line, split the line on the tabs and print the results one day per line.

(i) Read the third line, split the line on the commas, sum the numbers, and print the results.

(j) Close the file.

The printed output should look like this:

```
>>> 365

Monday
Tuesday
Wednesday
Thursday
Friday

The sum is 6.
The output text file should look line this:
365
Monday  Tuesday  Wednesday  Thursday  Friday  1,2,3
```

3. **countLines.py** Write a script that counts the number of lines of an input text file and prints the number of lines. Take the full path of the input file name as an argument. Make sure to catch the IOError if the input file does not exist.

Example input1: C:/gispy/data/ch19/RDUforest.txt

Example output1:
```
>>> C:/gispy/data/ch19//RDUforest.txt has 1271 lines.
```

Example input2: C:/gispy/data/ch19/noSuchFile.txt

Example output2:
```
>>> C:/gispy/data/ch19/noSuchFile.txt doesn't exist or can't
be opened.
```

4. **eyeTrackSHP.py** Sample script 'eyeTrackSHP.py' is missing ten lines of code which need to be filled in. Practice file reading and writing methods by replacing the ### comments as instructed in the script. When complete, the script reads an eye tracking data file, creates a modified version of the text file, and creates a point shapefile of eye fixations. The fixations indicate where the viewer was looking on the screen. The script takes two arguments, the full path file name of an eye tracking data file and an output directory. The shapefile points are shown in Figure 19.1 for the following script input:

Example input: C:/gispy/data/ch19/eyeTrack.csv C:/gispy/scratch/

Example output:
```
>>> C:/gispy/scratch/eyeTrackv2.csv complete.
Shapefile complete. View C:/gispy/scratch/eyeTrack.shp in ArcCatalog.
```

5. **ipcc2ESRI.py** Sample script 'ipcc2ESRI.py' is missing eight lines of code which need to be filled in. Practice file reading and writing methods by replacing the ### comments as instructed in the script. When the code is complete, the script will modify a weather data file downloaded from the International Panel on Climate Change (IPCC) Web site. These IPCC files contain monthly weather conditions averaged over 10 or 30 year periods. However, the raw format downloaded from the site can not be directly imported into ArcGIS. Once you've filled in the missing lines of code, this script will read the raw IPCC data and convert it to ESRI ASCII format, which will then be imported into an ESRI GRID raster with the `arcpy` ASCIIToRaster (Conversion) tool. The script takes two arguments, a full path input file name and an output directory.

Example input: C:/gispy/data/ch19/precipitation6190.dat C:/gispy/scratch/

Example output:
```
>>> Converting IPCC format file
C:/gispy/data/ch19/precipitation6190.dat to ESRI ascii format.
Processing header:
Processing lines...
ASCII file created. View ASCIIout.txt in Windows Explorer.
Converting ASCII to Raster
Conversion complete. View precGRID in ArcCatalog.
```

The beginning of the output ASCII file appears as follows:

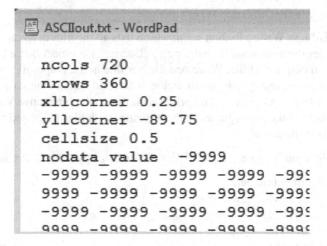

```
ASCIIout.txt - WordPad

ncols 720
nrows 360
xllcorner 0.25
yllcorner -89.75
cellsize 0.5
nodata_value    -9999
-9999 -9999 -9999 -9999 -999
9999 -9999 -9999 -9999 -999
-9999 -9999 -9999 -9999 -999
9999 -9999 -9999 -9999 -999
```

The output raster appears as follows:

6. **loblolly.py** Write a script that reads the tab separated sample data table, 'C:/ gispy/data/ch19/RDUforest.txt', and writes a modified version to 'C:/gispy/ scratch/loblollyPine.txt'. The modified version should contain the field names (the first line in 'RDUforest.txt') and the 696 records (one per line) with the species 'LP' (loblolly pine). Use the getIndex function to find the index of the SPECIES column. No arguments are required (hard-code the input full path file name, the output full path file name, the field name, the delimiter, and the selected species). Print feedback to the user as the following:

Example input: (No arguments needed.)

Example printed output:
```
>>> Reading file C:/gispy/data/ch19/RDUforest.txt...
696 records with SPECIES 'LP' written to
C:/gispy/scratch/loblollyPines.txt
```

7. **wrDirInfo.py** Write a Python script which lists all the files in an input directory and writes the names, sizes, and arcpy Describe object dataType of the files in an output text file. Write one file per line in the output file and separate the name, size, and type by a semicolon. Then add code to the script that reads the text file you just created and prints each line to the Interactive Window. The script should take two arguments, the directory with files to list and the full path output text file name.

Example input: C:/gispy/data/ch19/smallDir C:/gispy/scratch/dirContents.txt

Example printed output:
```
>>> a.dbf;2540;ShapeFile
a.kml;4684;File
a.prj;145;File
... (and so forth)
```

8. **avgNumbers.py** Write a script that reads a file with lines of tab separated numeric values, finds the average of each line, and prints these averages to an output file named 'out.txt'. Use two script arguments, the full path filename of the input file and an output directory. Use try and except to handle IOError exceptions as shown in the first example.

SiteNo.	6/2/2000	6/24/2000	7/14/2000	6/13/2001	6/28/2001	6/2/2002	6/27/2002	7/2/2002
1	4.07	4.21	4.15		4.03	3.74	4.56	4.5
2	4.51	4.29	4.4	-4.69	3.77	4.46	4.76	3.76
3	3.8		4.05	4.04	3.49	3.91	4.52	4.52
4	3.6		4.92	4.64	3.75	4.1	4.4	4.17
5	3.1		4.08	4.73	3.61	3.66	4.39	4.84

Figure 19.3   The first six lines of wheatYield.txt.

Example input1: C:/gispy/data/ch19/cop_yield.txt C:/gispy/scratch

Example output1:
```
>>> Warning: C:/gispy/data/ch19/cop_yield.txt could not be opened.
```

Example input2: C:/gispy/data/ch19/cop_yield.txt C:/gispy/scratch

Example output2:
```
>>> C:/gispy/scratch/out.txt created.
```

The first three lines of the 'out.txt' will appear as follows:
4.13777777778
6.27111111111
3.82333333333
. . . (and so forth)

9. **removeLines.py** Write a script to read the 'wheatYield.txt' sample dataset, remove invalid records, and write the results to an output text file. The output file should be named 'wheatYield_edited.txt' and created as described here. Figure 19.3 shows the first six rows of the dataset which contains wheat yield samples collected at various sites between June 2000 and July 2002. The first line of the file contains the dates. The values in each record are separated by spaces. Some records contain errors: all values should be positive and each record should contain eight sample wheat yields, but some entries are missing or negative. None of the site numbers are missing, so you can check for missing values by checking the length of the list of items in each record against the number of fields. Any sites with invalid data entries should be removed. For example, the errors shown in Figure 19.3 mean that sites 2 and 4 should be removed, as in Figure 19.4. Use two script arguments, the full path filename of the input file and an output directory.

Example input: C:/gispy/data/ch19/wheatYield.txt C:/gispy/scratch/

Example output:
```
>>> Number of records with errors is: 10
Corrected file is: C:/gispy/scratch/wheatYield_edited.txt
```

10. **remove2000.py** Write a script that removes the 'wheatYields.txt' samples that were taken in the year 2000. Use the removeItems function found in Example 19.7 to remove columns 2–4. Write the results to 'wheatYieldv2.txt'.

SiteNo.	6/2/2000	6/24/2000	7/14/2000	6/13/2001	6/28/2001	6/2/2002	6/27/2002	7/2/2002
1	4.07	4.21	4.15	4.64	4.03	3.74	4.56	4.5
3	3.9	4.64	4.05	4.04	3.49	3.91	4.52	4.52
5	3.15	3.55	4.08	4.73	3.61	3.66	4.39	4.84
7	3.42	3.35	4.07	4.66	3.72	3.84	4.44	3.4
8	3.97	3.61	4.67	4.49	3.75	4.11	4.64	2.99

Figure 19.4   The first six lines of wheatYield_edited.txt.

Use two script arguments, the full path filename of the input file and an output directory.

Example input: C:/gispy/data/ch19/wheatYield.txt C:/gispy/scratch/

Example output:
```
>>> Number of columns removed: 3
Corrected file is: C:/gispy/scratch/wheatYieldv2.txt
```

The first three lines of the modified file will appear as follows:

SiteNo. 6/13/2001 6/28/2001 6/2/2002 6/27/2002 7/2/2002
1 4.64 4.03 3.74 4.56 4.5
2 -4.69 3.77 4.46 4.76 3.76
. . . (and so forth)

11. **avgDBH.py** Write a script that reads a text file and collects information into a dictionary. The 'RDUforest.txt' sample data file has a table with four fields: Block, Plot, Species, and DBH. The last field, DBH (Diameter Breast Height) measures the outside bark diameter of a tree at breast height, an indicator of maturity. This script will determine the average DBH for each species in the RDU Forest by reading the file and populating a dictionary. Each dictionary item will consist of a species as the key and a list of DBH measurements for that species as a value. A portion of the dictionary appears as follows:

```
{ 'LOB': [24.0, 18.0, 25.0, 17.0, ...
'BE': [17.0],
'WD': [17.0, 15.0, 15.0, 11.0, 16.0, ...
...}
```

As the last step, the script should loop through the items in the completed dictionary and use the Python built-in sum and len functions to calculate the average DBH for each species. The script should take a full path file name as input and print the average DBH as follows:

Example input: C:/gispy/data/ch19/RDUforest.txt

Example output:
```
>>> Average DBH for SPECIES BE = 17.0
Average DBH for SPECIES WD = 12.2
Average DBH for SPECIES WO = 12.329787234
Average DBH for SPECIES HK = 11.4210526316
. . . (and so forth)
```

Chapter 20
Working with HTML and KML

Abstract GIS tasks often involve working with HTML and KML files. You may need to generate an HTML page showing the results of some GIS analysis or you may need to parse the data attributes in the description that pops up when you click on a KML element in Google Earth. HTML and KML files are simply text files that use tags to delineate elements within the data. Chapter 19 discussed how to read and write text files with Python; This chapter explains some additional techniques for working with HTML and KML formats.

Chapter Objectives
After reading this chapter, you'll be able to do the following:

- Identify basic HTML tags, such as links, images, and text formatting.
- Set HTML tag attributes.
- Create an HTML file with text formatting, links, and embedded images.
- Identify KML tags that define geographic features.
- Write HTML and KML with Python.
- Download and save Web site contents with Python.
- Extract files from Zip and KMZ archives.
- Parse HTML and KML with the Python.
- Convert KML files to ESRI shapefiles with Python.

20.1 Working with HTML

HTML, which stands for Hyper Text Markup Language is a popular format for Web pages. This section provides a brief introduction to the format of these files. Readers who are interested in learning more should search for interactive HTML tutorials online which allow you to enter HTML code and view the results.

HTML is a language for encoding web content. Files containing HTML code have '.html' or '.htm' file extensions. HTML is not a programming language like Python, but rather a language that uses tags to create web pages. A *tag* is a set of characters surrounded by angle brackets, such as <html> and <body>, as shown in Example 20.1. This simple HTML file is named 'elephant1.html'.

**Example 20.1: HTML code from the file
'C:\gispy\data\ch20\htmlExamplePages\elephant1.html'.**

```
<!DOCTYPE html>
<html>
    <body bgcolor='Aquamarine'>
            <!-- blablabla -->
            <h1>Elephants Say</h1>
            <p> We <b>like</b> HTML.</p>
            We <i>love</i> Python <br /> and GIS.
    </body>
</html>
```

If you open 'elephant1.html' (found in 'C:\gispy\data\ch20\htmlExamplePages\')
without specifying a program, by default it opens in a Web browser, such as Chrome
or Firefox. To view the HTML tags, open the file using a text editor. The freely
downloadable text editor named 'Notepad++' is good for viewing HTML because
it provides context highlighting for HTML code elements; That is, it colors the
HTML file contents to differentiate amongst various code components as IDE's do
for Python scripts.

The tags are visible when the file is viewed in a text editor, but not when the same
file is viewed in a Web browser. The 'elephant.html' file looks as pictured in
Figure 20.1 when viewed in a Web browser. The browser interprets the tags instead
of displaying them. For example, the tag makes the word 'like' display with bold
font. The tags in Example 20.1, serve the following purposes:

- The DOCTYPE tag defines the document type as HTML5. HTML5 defines a set
  of standards for HTML.
- The text between <html> and </html> are wrapped around the entire contents of the page so that it will be interpreted as HTML.
- The text between <body> and </body> are wrapped around all of the page
  content that is meant to be displayed. Other content, such code that controls the
  page style, will be between the <html> tags, but not between the <body> tags.
  The background color of the page is specified using bgcolor="Aquamarine".

Elephants Say

We like HTML.

We *love* Python
and GIS.

Figure 20.1   The HTML page created with the code in Example 20.1.

- The `<!--blablabla-->` tag is a code comment. Like Python comments, HTML comments are only there for the human reader. In Python, anything placed on a line following a # sign is a comment; However in HTML, comments are wrapped in a set of characters. Start comments with `<!--` and end them with `-->`. The text between these characters is a comment.
- The text between `<h1>` and `</h1>` is displayed as a heading.
- The text between `<p>` and `</p>` is displayed as a paragraph, hence the vertical space surrounding the 'we like HTML' paragraph.
- The text between `<b>` and `</b>` is displayed as bold.
- The `<br/>` tag inserts a line break, so that the phrase 'We love Python' and the phrase 'and GIS' appear on separate lines.
- Can you guess what the `<i>` and `</i>` tags do?

Many tags pair a *start tag* with an *end tag*. For example, `<b>` is a start tag, which starts the bold text and `</b>` is an end tag which ends the bold text. The end tag always starts with a forward slash. Only the word 'like' is bold because the `</b>` tag ends the bold font setting. We say pairs of tags are *wrapped* around the *tag content* because tag instructions are applied to whatever appears between them in the HTML text. The word 'like' is the tag content wrapped in tags that apply bold font. Some tags such as the line break tag and comment tags are not paired. Some tags have *tag attributes* that provide more information about the tag. Most tag attributes are optional and they are placed inside the tag after the tag name before the closing angle bracket. An attribute is specified with the attribute name equal to the attribute value in quotes (similar to a Python string assignment statement). For example, the `body` tag has an attribute for specifying the background color (`bgcolor`). Example 20.1 sets the background color to `"Aquamarine"`. Replace `"Aquamarine"` with `"Pink"` or another named color (there are over 100 allowable color names) to change the background. HTML5 files usually control the background with other code in cascading style sheets (css files). The background color is only used here as a simple example of a tag attribute. Tags and attributes for links, lists, and tables are discussed next.

20.1.1 Specifying Links

The text between `<a>` and `</a>` can be linked to a Web address. The `ref` attribute specifies the hypertext reference, the linked Web page. If the file resides in the same directory, specify the reference as the name of the file. For example, since 'elephant1.html' is in the same directory as 'elephant2.html', the following code is placed in 'elephant2.html' to link to the 'elephant1.html' file:

```
<a href="elephant1.html">A link</a>
```

If the file is within the same Web site, use a relative path for the hypertext reference (review relative paths in Section 16.1). If the file is not on the same Web site, the full Web address needs to be specified. The second link in 'elephant2.html' uses the following code to point to Google's home page:

```
<a href="https://www.google.com/">Another link</a>
```

20.1.2 Embedding Images

The `<img>` tag is used to embed images in a web page. The `src` attribute specifies the image source, a path to an image. The image source can be specified with a relative or complete path. Other attributes such as `width` and `height` can be used to format the image. When more than one attribute is used, they are separated by spaces, as in the following example:

```
<img src="../pics/lakshmi.jpg" width="32" height="32">
```

The file 'elephant2.html' specifies two links and an image, as shown in Figure 20.2. The code uses the width and height attributes to reduce the image size. Add the following line of code to 'elephant2.html' to embed a full-sized version of the image (and refresh your browser to view the change):

```
<img src="../pics/lakshmi.jpg">
```

Notice, the image source is specified by a relative path to the image. If you move the html file, you would need to place the image in the same relative position. If the relative path to an image is incorrect, it will not appear in the page. When you add the following code for a third image link to 'elephant2.html', you will still only see two pictures, since there is no 'lakshmi.jpg' image in the 'htmlExamplePages' directory:

```
<img src="lakshmi.jpg">
```

Figure 20.2 The HTML page created with the code in 'C:\gispy\data\ch20\htmlExamplePages\elephant2.html'.

20.1.3 HTML Lists

HTML lists and tables are important for reporting GIS results. These structures both have tags nested inside of tags. There are two types of HTML lists, ordered (ol) and unordered (ul) lists. A list is either wrapped in `<ol>` and `</ol>` tags or `<ul>` and `</ul>` tags. The items in the list are nested inside these tags and each item is wrapped in list item tag pairs (`<li>` and `</li>`). The following HTML code creates an ordered list:

```
<ol>
    <li>Savannah</li>
    <li>Bush</li>
    <li>African</li>
</ol>
```

The ordered list automatically generates a number or letter for each list item. By default, numbers are used. The HTML code above creates the following ordered list:

1. Savannah
2. Bush
3. African

The unordered list tag generates a bulleted list. The following code specifies an unordered list with two items:

```
<ul>
    <li>Savannah</li>
    <li>Bush</li>
</ul>
```

The HTML code above creates the following bulleted list with one bullet for each 'li' tag:

• Savannah
• Bush

Indentation is not required in HTML, but indented list items in the code are easier to read.

20.1.4 HTML Tables

HTML tables are wrapped in `<table>` and `</table>` tags and the contents are inserted in row major order. In other words, the first row is specified left to right, followed by the second, and so forth. Each row is wrapped in `<tr>` and `</tr>` tags.

Species	2011 Population
African	Less than 690000
Asian	Less than 32700

Figure 20.3 The table created by the HTML table code example in 'C:\gispy\data\ch20\htmlExamplePages\elephant3.html'.

The cells in each row are again individually wrapped in `<th>` and `</th>` or `<td>` and `</td>` tags ('th' stands for table heading and 'td' stands for table data). The following code sets table border thickness to 3 and the three `tr` tag pairs create three rows, while the pairs of nested `th` or `td` tags create two columns within each row (see the resulting table in Figure 20.3):

```
<table border="3">
    <tr>
        <th>Species</th>
        <th>2011 Population</th>
    </tr>
    <tr>
        <td>African</td>
        <td>Less than 690000</td>
    </tr>
    <tr>
        <td>Asian</td>
        <td>Less than 32700</td>
    </tr>
</table>
```

20.1.5 Writing HTML with Python

HTML can be used to present GIS analysis results. Since HTML is simply a text file containing HTML tags, you can use Python file handling to write HTML. There are specialized packages for writing HTML which are not covered in this book; Instead, the upcoming examples show how to do basic HTML file writing with standard Python file handling, familiar from Chapter 19. The Python script in Example 20.2 hard-codes a string variable `mystr` with HTML code, opens a file for writing, and writes the string to file. The triple quotes are wrapped around the string literal value to preserve the line breaks in the HTML code.

file:///C:/gispy/data/ch20/htmlOut/output2.html

Asian Elephant

Figure 20.4 HTML page generated by the Python code in Examples 20.2 and 20.3.

Example 20.2

```
# writeSimpleHTML.py
mystr = '''<!DOCTYPE html>
<html>
    <body>
        <h1>Asian Elephant</h1>
        <img src="../data/ch20/pics/lakshmi.jpg" alt="elephant">
    </body>
</html>
'''

htmlFile = 'C:/gispy/scratch/output.html'
outf = open(htmlFile, 'w')
outf.write(mystr)
outf.close()
print '{0} created.'.format(htmlfile)
```

As the HTML file becomes more complex, breaking the HTML code into three parts (beginning, middle, and end) can make it easier to manage the string contents. Example 20.3 breaks the HTML code from Example 20.2 into three parts and writes each part to file. Example 20.3 also uses string formatting to dynamically generate the HTML content based on user input (the resulting Web page is shown in Figure 20.4).

Example 20.3

```
# writeSimpleHTML2.py
# Purpose: Write HTML page in 3 parts.
# Usage: workspace title image_path
# Example input:
# C:/gispy/scratch "Asian Elephant" ../data/ch20/pics/lakshmi.jpg
```

```python
import os, sys
workspace = sys.argv[1]
title = sys.argv[2]
image = sys.argv[3]

beginning = '''<!DOCTYPE html>
<html>
    <body>'''

middle = '''
            <h1>{0}</h1>
            <img src='{1}' >\n'''.format(title, image)

end = '''    </body>
</html>
'''

htmlfile = workspace + '/output2.html'
with open(htmlfile,'w') as outf:
    outf.write(beginning)
    outf.write(middle)
    outf.write(end)
print '{0} created.'.format(htmlfile)
```

Compound structures can be built separately too. The `python2htmlList`
function in Example 20.4 converts a Python list to an HTML list. First each item in
the Python list is wrapped in list item tags. Then the items are joined into a single
string. Finally, the list tag (`ol` or `ul`) is wrapped around the string.

Example 20.4

```python
# excerpt from printHTMLList.py
def python2htmlList(myList, listType):
    '''Convert a Python list to HTML list.
    For example, convert [rast1,rast2] to:
    <ul>
        <li>rast1</li>
        <li>rast2</li>
    </ul>
    '''
    # Wrap items in item tags.
    htmlItems = ['<li>' + str(item) + '</li>' for item in myList]

    # Join the item list into a string with a line break
    # after each item.
    itemsString = '''\n    '''.join(htmlItems)
```

```
# Wrap the string of items in the list tag.
htmlList = '''
<{0}>
    {1}
</{0}>
'''.format(listType, itemsString)
return htmlList
```

The following code creates a Python list, calls the function, and prints the resulting unordered HTML list:

```
>>> rasts = [u'elev', u'landcov', u'soilsid', u'getty_cover']
>>> htmlList = python2htmlList(rasts, 'ul')
>>> print htmlList
<ul>
    <li>elev</li>
    <li>landcov</li>
    <li>soilsid</li>
    <li>getty_cover</li>
</ul>
```

By calling this function on Python lists, you can generate HTML lists to write in HTML files. In an upcoming chapter, you'll also learn how to generate screen captures of maps with Python, enabling you to automatically generate HTML output reports to show GIS analysis results, with graphical elements (using the 'img' tag).

20.1.6   Parsing HTML with BeautifulSoup

By reading and parsing markup language content with Python, you can access linked or embedded GIS data. The previous section showed that HTML pages can be generated using Python `file` objects in write `'w'` (write) mode. HTML can also be read with Python `file` objects in `'r'` (read) mode; However, the challenge comes in deciphering the HTML, i.e., parsing the content. To simplify this process, it's worth learning a few things about a module that supports markup language parsing. Python does have a built-in module for HTML parsing (named `HTMLParser`), but the non-built-in module named `BeautifulSoup`, in reference to a Lewis Carroll verse, is easier to use for high-level tasks such as finding all the links in a page.

The `BeautifulSoup` module has methods and objects for reading tags and their attributes and content. Online documentation explains how to download and install the latest version of `BeautifulSoup`, but for consistency, a stand-alone version which consists of just one Python file named 'BeautifulSoup.py' is included with the sample scripts for this chapter. Use this module while you're learning the basics, so that you don't get caught up in the installation procedures. Code may need minor changes to use the latest installable versions.

The given module doesn't require any special installation to be imported, but since it is not a built-in Python module, you have to make sure Python looks in the right directory. Handle this just as you handle importing user-defined modules. To import BeautifulSoup, first import the sys module and append the path for BeautifulSoup to the path list, then use a separate import statement to import BeautifulSoup:

```
import sys
sys.path.append('C:/gispy/sample_scripts/ch20')
import BeautifulSoup
```

Once you have imported it, you can make use of its methods and objects. To use the module, the first step is to create a soup object. We'll start by using a hard-coded HTML string to demonstrate how it works. The basic approach to using the module is to put the HTML content in a Python string variable (either by hard-coding or by reading it from a file), create a soup object, then call the soup methods on the soup object. The following statement hard-codes mystr with some HTML code:

```
>>> mystr = '<!DOCTYPE html><html><body><h1>Asian Elephant</h1><img
src="lakshmi.jpg" alt="elephant"></body></html>'
```

To create a soup object, call the BeautifulSoup class in the BeautifulSoup module with dot notation (the module and the class do have the same name) and pass in HTML content as an argument. The following code shows this use of dot notation with the HTML string variable passed in as content:

```
>>> soup = BeautifulSoup.BeautifulSoup(mystr)
```

The soup object has a function named prettify. Though HTML doesn't require each tag to be on a separate line or nested tags to be indented, this kind of spacing certainly makes the HTML easier to read. The prettify function, as the name implies, makes the HTML code pretty. Use dot notation on the soup object to call this object method and it returns a prettified HTML string:

```
>>> h = soup.prettify()
>>> print h
<!DOCTYPE html>
<html>
     <body>
       <h1>
            Asian Elephant
       </h1>
```

```
        <img src="lakshmi.jpg" alt="elephant" />
      </body>
</html>
```

The input HTML string was difficult to decipher when it was written on a single line. Now that it has been prettified, you can clearly see that this HTML sets a title and embeds an image, like the HTML page ('output.html') generated in Example 20.2.

The BeautifulSoup module makes it easy to find HTML elements. The find method finds the first occurrence of a tag:

```
>>> t = soup.find('h1')
>>> t
<h1>Asian Elephant</h1>
```

The find method returns a BeautifulSoup Tag object which has properties such as name, attrs, and contents. The name property contains the name of the tag, the attrs property contains a list of tag attributes, and the contents property contains a list of tag contents:

```
>>> type(t)
<class 'BeautifulSoup.Tag'>
>>> t.name
u'h1'
>>> t.attrs
[]
>>> t.contents
[u'Asian Elephant']
```

This h1 tag has no attributes, so the attrs property is an empty list. The tag contents property is a list of items, even if there is only one content item, as in this example. List indexing can be used to access individual content items.

To access the first item in the list, use index zero, as in the following example:

```
>>> t.contents[0]
u'Asian Elephant'
```

The img tag has two attributes. The Tag object stores the attributes as a list of (attribute name, attribute value) tuples:

```
>>> t2 = soup.find('img')
>>> t2
<img src="lakshmi.jpg" alt="elephant" />
>>> t2.attrs
[(u'src', u'lakshmi.jpg'), (u'alt', u'elephant')]
```

The src attribute specifies the image file and the alt attribute provides a text alternative to be displayed in case there is trouble loading the image. You can access the value of an attribute by using the name, as you use a key in a dictionary:

```
>>> t2['src']
u'lakshmi.jpg'
>>> t2['alt']
u'elephant'
```

You can also loop through the attribute list in a way that's reminiscent to looping through a dictionary getting both key and value for each entry. For the attribute list, you can use two comma separated variables between the for and in keywords to get both the names and the values of the attributes:

```
>>> for name, value in t2.attrs:
...        print 'Name: ' + name + ' Value: ' + value
...
Name: src Value: lakshmi.jpg
Name: alt Value: elephant
```

Another soup object method named findAll returns a list of Tag objects for tags with the specified name. The following code finds all the list item (li) tags in htmlList:

```
>>> htmlList = '\n <ul>\n <li>elev</li>\n <li>landcov</li>\n
<li>soilsid</li>\n <li>getty_cover</li>\n </ul>\n'
>>> soup2 = BeautifulSoup.BeautifulSoup(htmlList)
>>> tags = soup2.findAll('li')
>>> tags
[<li>elev</li>, <li>landcov</li>, <li>soilsid</li>, <li>getty_cover</li>]
```

Loop through this list to retrieve the individual items. Access the first item in the tag contents lists by using a zero index:

```
>>> for t in tags:
...        print t.contents[0]
...
elev
landcov
soilsid
getty_cover
```

So far, the examples have used hard-coded HTML strings to create the soup object. In practice, useful HTML content will usually come from HTML files (instead of hard-coded strings). A soup object can also be created from a text file object open for reading. Example 20.5 opens an HTML file in read mode and creates a `soup` object from `infile`, the file object. Then it parses the HTML content using several of the soup navigation methods. The FOR-loop prints the two hypertext links found in the 'elephant2.html':

```
>>> Link: elephant1.html
Link: https://www.google.com/
```

Example 20.5

```
# getLinks.py
# Purpose: Read and print the links in an html file.
import sys
basedir = 'C:/gispy/'
mPath = 'data/ch20/htmlExamplePages/elephant2.html'
sys.path.append(basedir + 'sample_scripts/ch20')
import BeautifulSoup
# Read the HTML file contents.
with open(basedir + mPath, 'r') as infile:
    # Create a soup object and find all the hyperlinks.
    soup = BeautifulSoup.BeautifulSoup(infile)
    linkTags = soup.findAll('a')
    # Print each hyperlink reference.
    for linkTag in linkTags:
        print 'Link: {0}'.format(linkTag['href'])
```

20.2   Fetching and Uncompressing Data

Python can also enable you to *fetch* web content (retrieve it from a Web site), so that you can automatically harvest online GIS content that is made available through Web site links.

20.2.1   Fetching HTML

You can use the built-in module named `urllib2` to *fetch* (i.e., get the contents of) online html pages, by importing the module and calling the `urlopen` function. This function takes a URL (a web address) as input and returns a `file` object open for reading (like calling the built-in `open` function in `'r'` mode). You can use the standard Python text file reading functionality on the response. The following code fetches HTML from Google's Web site, calls `read` (which returns the entire contents of the HTML file as a string), and prints the number of characters in the HTML (the length of the string):

```
#fetchHTML.py
import urllib2
url = 'http://www.google.com'
response = urllib2.urlopen(url)
contents = response.read()
response.close()
print len(contents)
```

Once you fetch the page and read the contents into a string variable, you can parse the content string using the `BeautifulSoup` module as in the previous examples:

```
soup = BeautifulSoup.BeautifulSoup(contents)
>>> pics = soup.findAll('img')
```

When you run this code on the Google page, you may see different results from day to day, since the content is dynamic. The following output lists one image with seven image attributes defined:

```
>>> pics
[<img alt="Google" height="95" src="/intl/en_ALL/images/srpr/
logo1w.png" width="275" id="hplogo" onload="window.lol&&lol()"
style="padding:28px 0 14px" />]
```

20.2.2 *Fetching Compressed Data*

Online GIS data is often stored in *compressed files*, such as Zip or KMZ files that archive multiple files within a single file. You can also fetch these sorts of files, but they need to be handled differently. These are *binary* (non-text) file formats. The '.shp' extension file from a shapefile dataset is another example of a binary file. Try opening a '.shp' file in Notepad to view the indecipherable contents. Binary files consist mostly of unprintable characters and are not readable in a standard text editor.

The code for fetching and reading compressed files is the same as the code for fetching and reading HTML files. However, for binary compressed files, a binary `file` object is returned from the `urlopen` function, so the output from the `read` command is a binary string. Instead of using the `BeautifulSoup` module to parse the contents, you need to save the contents to your computer by writing them to a file. You can do so by opening a file for output in wb (write binary) mode and calling the write command. In Example 20.6, the binary file object is fetched and the contents of the response are read, as usual. The remainder of the code differs from HTML handling. A new file, outFile, is opened for writing in binary mode and the binary contents are written to the file, a new zip file is written in the output directory. The input for the script was a local Zip file URL, one starting with file:///, followed by the full path file name. The same function can be used for online URLs such as those starting with http:// or ftp://. Additionally, the same approach works for KMZ files.

Example 20.6

```
# fetchZip.py
# Purpose: Fetch a zip file and place it in an output directory.
import os, urllib2

def fetchZip(url, outputDir):
    '''Fetch binary web content located at 'url'
    and write it in the output directory'''
    response = urllib2.urlopen(url)
    binContents = response.read()
    response.close()

    # Save zip file to output dir (write it in 'wb' mode).
    outFileName = outputDir + os.path.basename(url)
    with open(outFileName,'wb') as outf:
        outf.write(binContents)
outputDir = 'C:/gispy/scratch/'
theURL = 'file:///C:/gispy/data/ch20/getty.zip'
fetchZip( theURL, outputDir )
print '{0}{1}created.'.format(outputDir, os.path.basename(theURL))
```

20.2.3 Expanding Compressed Data

Once the Zip or KMZ file archive has been fetched and saved locally, it can be decompressed using Python. The built-in zipfile module can handle extracting the files from a Zip or KMZ archive. You work with a Zipfile object which has namelist and extractall methods. Example 20.7 shows how to do this for a Zip file. The unzipArchive function takes two arguments, the name of a Zip or KMZ file and a destination for the output. It works like a standard file extraction program, extracting files to the given location. It does so by first, creating a Zipfile object zipObj. Then it extracts the contents of each archived file to the destination directory (dest). Next it gets a lists of the files in the archive using the namelist method and loops through the list of archived files to report that they have been extracted. The script's printed output appears as follows:

```
Unzip C:/gispy/data/ch20/park.zip to C:/gispy/scratch/park/
Extract file: park.prj ...
Extract file: park.sbn ...
Extract file: park.sbx ...
(... and so forth)
```

Example 20.7

```
# extractFiles.py
# Purpose: Extract files from an archive;
#     Place the files into an output directory.
# Usage: No script arguments

import os, zipfile

def unzipArchive(archiveName, dest):
    '''Extract files from an archive
    and save them in the destination directory'''
    print 'Unzip {0} to {1}'.format( archiveName, dest )
    # Get a Zipfile object.
    with zipfile.ZipFile(archiveName, 'r') as zipObj:
        zipObj.extractall(dest)
        # Report the list of files extracted from the archive.
        archiveList = zipObj.namelist()
        for fileName in archiveList:
            print ' Extract file: {0} ...'.format(fileName)
    print 'Extraction complete.'

archive = 'park.zip'
baseDir = 'C:/gispy/'
archiveFullName = baseDir + 'data/ch20/' + archive
destination = baseDir + 'scratch/' +     \
                os.path.splitext(archive)[0] + '/'
if not os.path.exists(destination):
    os.makedirs(destination)
unzipArchive(archiveFullName, destination)
```

To collect online data, you can combine fetching and parsing HTML with fetching and unzipping compressed files. Many government and agency Web sites link to compressed data that can be downloaded by a script. For example, the government of North Carolina hosts an online data repository for Wake County (search online for 'wakegov data download'). The Data Download page links to zipfiles containing transportation, census, and other local GIS datasets that are updated periodically. The following workflow would be used to harvest these datasets:

```
FETCH the data download page.
CREATE a soup object.
GET a list of the links in the data download page.
FOR each link
    IF the link references a Zip file THEN
        CALL fetchZip to fetch the Zip file, saving it locally.
        CALL unzipArchive to extract the compressed files.
    ENDIF
ENDFOR
```

The sample script 'wakeStreets.py' provides a partial-Python implementation; A few lines of code are left as an exercise.

20.3 Working with KML

Now that you know something about HTML and writing/reading it with Python, you'll also be able to work with other markup languages. Keyhole Markup Language (KML) files are of particular interest for GIS work, since this is a markup file format for geographic information. Like HTML, KML uses tags wrapped around content to encode data in a text file. KML files contain tags relating to places with geographic positions. They are designed to be viewed in Google Earth (Figure 20.5) and other GIS. ArcGIS can import these datasets into shapefiles for geographic analysis, using the ArcGIS KML to Layer (Conversion) tool, but this tool does not necessarily

Figure 20.5 A KML placemark balloon from 'restaurants.kml' viewed in Google Earth.

import all of the information encoded in the file. Geographic features are imported, but important attributes may not be imported. Since KML is a markup language, Python scripting with BeautifulSoup can be used to solve this problem and parse the data. Then ArcGIS cursors can be used to build tables from the parsed data.

20.3.1   The Structure of KML

To use Python for parsing KML, you'll need some understanding of the structure of a KML file. The most common way to specify a geographic feature is to use a Placemark tag. Each placemark tag contains a Geometry object, such as a point, line, or polygon (specified with Point, LineString, and Polygon tags). The location of an object is further nested within coordinate tags. Placemark properties such as a name and a description can also be specified.

Example 20.8 shows a KML file containing two points. The first line indicates the file type. XML, which stands for Extensible Markup Language, contains tags that define the structure of a document (such as <chapter> and <section> tags). KML is a special type of XML file for encoding geospatial objects. The KML opening and closing tags are wrapped around all of the KML content. The Document tag allows you to put more than one placemark in a single file. The file in Example 20.8 only contains two placemarks, one for each of the two points. The tags inside the placemark specify a name and description for each point. The contents of these tags appear in the balloon attached to the placemark when it is selected in Google Earth (see Figure 20.5). The Point tags are wrapped around a pair of coordinates tags with comma separated longitude, latitude, and altitude values for the placemark.

Example 20.8: KML code from the 'C:\gispy\data\ch20\restaurants.kml' file.

```
<?xml version="1.0" encoding="UTF-8"?>
<kml xmlns="http://www.opengis.net/kml/2.2">
 <Document>
   <Placemark>
     <name>Bubba's Tofu Gumbo</name>
        <description>Tofu Gumbo and Zydeco!<br />
                   Score: 97</description>
     <Point>
       <coordinates>-90,30,0</coordinates>
     </Point>
   </Placemark>
   <Placemark>
     <name>Joe Bob's Good Cookin'</name>
     <description>The best tree top grits n' greens restaurant
                south of the Mason-Dixon line.<br />
                Score: 94</description>
     <Point>
       <coordinates>-78,35,0</coordinates>
```

```
      </Point>
    </Placemark>
  </Document>
</kml>
```

20.3.2 Parsing KML

The ArcGIS KML to Layer (Conversion) tool may not be sufficient for bringing KML data into ArcGIS for analysis. A common problem is that more than one piece of information can be encoded within the description. For example, the descriptions in 'restaurants.html', contain a blurb and an inspection score. The KML to Layer tool does not create a field for each of these pieces of information, so analysis can't be performed on description content. Figure 20.6 shows the field values for one of the restaurants in a file created by converting with this tool. Average scores can't be calculated because the descriptive blurb and the score, which are both within the description tag, are both imported together into the `PopupInfo` field. Whereas, we would prefer to obtain the blurb and score information in separate fields as shown in Figure 20.7. The `BeautifulSoup` module can be used to extract this information.

Field	Value
OID	2
Shape	Point
Name	Bubba's Tofu Gumbo
PopupInfo	Tofu Gumbo and Zydeco! score: 97

Figure 20.6 ArcGIS conversion results. The `PopupInfo` field contains all of the `description` tag contents.

Field	Value
FID	0
Shape	Point
name	Bubba's Tofu Gumbo
blurb	Tofu Gumbo and Zydeco!
score	97

Figure 20.7 Script conversion results. The `blurb` and `score` fields contains the `description` tag contents.

Example 20.9

```
# parseKMLrestaurants.py
# Purpose: Print KML placemark names and descriptions.
import sys

baseDir = 'C:/gispy/'
sys.path.append(baseDir +'sample_scripts/ch20')
import BeautifulSoup

fileName = baseDir + 'data/ch20/restaurants.kml'

# Get the KML soup.
with open(fileName, 'r') as kmlCode:
    soup = BeautifulSoup.BeautifulSoup(kmlCode)

# Print the names and descriptions.
names = soup.findAll('name')
descriptions = soup.findAll('description')
for name, description in zip(names, descriptions):
    print name.contents[0]
    print '\t{0}'.format(description.contents)
```

Parsing KML with the BeautifulSoup module is just like parsing HTML with this module; Only the tags differ. Example 20.9 parses the 'restaurants.kml' file and prints the names and description of each placemark using the name and description tags. This example imports BeautifulSoup, then opens the KML file for reading. KML files are ASCII text, so the kmlCode text file object open for reading KML code, can be used to create the soup object. The Python built-in zip function is then used in this example to loop through both lists simultaneously and print each name and description. The output is as follows:

```
Joe Bob's Good Cookin'
    [u'Tofu Gumbo and Zydeco!', <br />, u'Score: 97']
Bubba's Tofu Gumbo
    [u"The best tree top grits n' greens restaurant south of the
        Mason-Dixon line.", <br />, u'Score: 94']
```

Remember that tag contents are always returned as a list, even when there is only one value. The name tag contents list simply contains a name. Example 20.9 prints the name by indexing the first item in the list. The description contains a list of three items, the descriptive blurb, a line break, and the restaurant's sanitation score:

```
>>> description.contents
[u'Tofu Gumbo and Zydeco!', <br />, u'Score: 97']
```

To enable GIS analysis of the sanitation scores, we can extract their numeric value from the description contents list. The score is the third item in the list, so it can be retrieved with index 2, but the numeric portion of this item needs to be separated from the string portion. The following lines of code use string methods and casting to retrieve the score as a float:

```
>>> scoreString = description.contents[2]
>>> scoreString
u'Score: 97'
>>> scoreList = scoreString.split(':')
>>> scoreList
['Score', ' 97']
>>> score = float(scoreList[1])
>>> score
97.0
```

20.3.3 Converting KML to Shapefile

Once you've determined how to parse the desired information from a KML file, you can use an insert cursor to convert the extracted values into a GIS shapefile format. A shapefile can be created from a KML file using the following steps:

```
CALL the Create Feature Class (Data Management) tool
SET the field names and types
FOR each field name
    CALL the Add Field (Data Management) tool
ENDFOR
CREATE a soup object from the KML file contents
GET tag lists from the soup (findAll)
CREATE an insert cursor
FOR each item in the tag lists
    GET the value for each field
    PUT field values in a list
    INSERT the new row into the shapefile
ENDFOR
```

Example 20.10 shows an excerpt from sample script 'restaurantKML2shp.py' that represents an implementation of the FOR each item in tag lists loop to handle the 'restaurants.kml' file. Other KML files with point features that have multiple pieces of information within the description tag can be created by making modifications to the portions of this script that specify the field and types and the loop contents.

Example 20.10

```
# Excerpt from restaurantKML2shp.py...
coordinates = soup.findAll('coordinates')
names = soup.findAll('name')
descriptions = soup.findAll('description')
# Populate the shapefile.
with arcpy.da.InsertCursor( fc, [ 'SHAPE@XY' ] + fieldNames ) as ic:
    for c,n,d in zip(coordinates, names, descriptions):
        # Get field values.
        [x, y, z] = c.contents[0].split(',')
        myPoint = arcpy.Point(x, y)
        name = n.contents[0]
        blurb = d.contents[0]
        scoreString = d.contents[2]
        scoreList = scoreString.split(':')
        score = float(scoreList[1])
        # Put row values in a list & insert the new row.
        newRow = [myPoint, name, blurb, score]
        ic.insertRow(newRow))
```

20.4 Discussion

This chapter briefly introduced two markup languages, illustrated writing and reading markup files with standard built-in Python file objects and functions, and showed how to use Python to fetch and consume Web content. The BeautifulSoup module for parsing markup languages was used to parse HTML and KML content. Finally, insert cursors were used to import parsed KML into a shapefile. One difficulty in parsing markup files is that there can be errors in the markup code. Also, since the content in markup files varies, writing code to step through with the debugger to watch soup Tag objects and their contents is a pragmatic approach for developing scripts with BeautifulSoup. Once you see the tag contents (such as the KML description tag list contents), you can refine your code to consume the information you need. This code may need to be highly customized to handle your data, which explains the difficulty in creating a general tool that handles any HTML or KML content. Though this chapter only used HTML and KML examples, these approaches can be adapted for other markup language files, such as XML files.

20.5 Key Terms

HTML	Wrapped tags
KML	Beginning and ending tags
HTML lists	Fetch
HTML tables	URL
Tags	Compressed file
Start tag and end tag	Binary file
Tag content	The `BeautifulSoup` module
Tag attributes	

20.6 Exercises

1. **HTMLvocab** For the HTML in the code sample below, identify the following components: (a) the start or single tag names, (b) the names of attributes used in each tag, (c) the values of the attribute, and (d) the tag contents (the data between the opening and closing tags).

```
<img src="tree.gif" alt="tree image" width="80" height="100" />
<a href="http://www.sierraclub.org">Sierra Club</a>
<br />
<ol type='a'>
<li>Sweet Gum</li>
<li>Maple</li>
<li>Oak</li>
</ol>
```

2. **tagMatch** Match the tag HTML or KML tag name with its purpose.

Markup tag name	Purpose
1. placemark	A. bold font setting
2. a	B. line break
3. br	C. hyperlink
4. tr	D. feature
5. ul	E. embedded picture
6. coordinates	F. bulleted list
7. h1	G. table row
8. linestring	H. geometric feature
9. img	I. header
10. b	J. latitude, longitude, and (optionally) altitude

3. Follow the instructions below to practice creating HTML files with Python code. Write one script for each part (a-c). Each script should create an output HTML file in the 'C:/gispy/scratch' directory.

   (a) **writeHTML1.py** Write a Python script to create an HTML page named 'images1.html' by hard-coding the HTML strings and the output directory

path ('C:/gispy/scratch/images1.html'). The HTML page should contain a
header ('Butterfly garden') and two embedded images (Use a relative path
to 'butterfly.jpg' and 'flower.jpg' in the 'C:/gispy/data/ch20/pics' directory). Display the images as 87×65 (w×h) pixel thumbnails.

Example input:
(No arguments required)

Example output: C:/gispy/scratch/images1.html created. When opened in a
Web browser the output looks like this:

Butterfly garden

(b) **writeHTML2.py** Write a Python script to create an HTML page named
'images2.html' with dynamic content. The script should take two arguments, a directory containing images and an output directory where the
HTML will be created. The page should have a header that is the name of
the image directory and the page body should provide a link (with a relative
path) to every JPEG image found in the given directory with each link on a
separate line. Use the os.path.relpath function to find the relative
path from the HTML to the images. (For example,
```
relPath = os.path.relpath(imageDir, outDir))
```

Example input:
C:/gispy/data/ch20/pics/ C:/gispy/scratch/

Example output:
```
>>> C:\gispy\scratch\images2.html created.
```

When opened in a Web browser the output looks like this:

pics

[butterfly.jpg](#)
[flower.jpg](#)
[jack.jpg](#)
[lakshmi.jpg](#)

(c) **writeHTML3.py** Write a Python script to create an HTML page named 'images3.html' with dynamic content. The script should take two arguments, a directory containing images and an output directory where the HTML will be created. The page should have a header which is the name of the image directory and the page body should embed every JPEG image found in the given directory as 87×65 (w×h) pixel thumbnails. Use relative paths to the images and use the `os.path.relpath` function to find the relative path from the HTML to the images. (For example,
`relPath = os.path.relpath(imageDir, outDir)`)

Example input: C:/gispy/data/ch20/pics

Example output:
```
>>> C:/gispy/scratch/images3.html created.
```

When opened in a Web browser the output looks like this:

pics

4. **writeRastersHTML.py** Write a Python script to create an HTML page named 'rasters.html' with dynamic content. The script should take two arguments, a workspace containing rasters and an output directory. The page header should say 'Rasters in' plus the name of the workspace and the page body should show a bulleted list of the rasters found in the given workspace. You can use the `python2htmlList` function from the 'printHTMLList.py' sample script to generate the HTML list.

Example input:
C:/gispy/data/ch20/rastTester.gdb C:/gispy/scratch

Example output:
```
>>> C:/gispy/scratch/rasters.html created.
```

When opened in a Web browser the output looks like this (only the first six bullets are shown):

Rasters in rastTester.gdb

- elev
- landcov
- soilsid
- getty_cover
- landc197
- landuse

5. **wakeStreets.py** The sample script 'wakeStreets.py' is missing code which needs to be filled in. Practice handling HTML and Zip files by replacing the five ### comments with one or more lines of code each. This script takes one argument, the URL for a download page for Wake County, NC data and an output directory. An example is given here; if the example input URL is no longer correct, search online for 'Data download Wake GIS'. When the code is complete, the script will fetch and unzip a 'Wake_Streets' Zip file. This script assumes that 'BeautifulSoup.py' resides in the same directory as this script.

Example input:
http://www.wakegov.com/gis/services/Pages/data.aspx C:/gispy/scratch/

Example output:
```
>>> Unzip C:/gispy/scratch/Wake_Streets_20xx_xx.zip to
C:/gispy/scratch/Wake_Streets_20xx_xx/
Extract file: StreetsMetadata.doc ...
Extract file: Wake_Streets_20xx_xx.dbf ...
Extract file: Wake_Streets_20xx_xx.prj ...
(and so forth)
Extraction complete.
```
xx will be replaced by the current date information in the form YY_MM.

6. **writeFieldsHTML.py** Write a script that writes an HTML file containing a report of the name of an input data file and a list of the fields in the file. The file name should be formatted as a header and the fields should be formatted as a bulleted list using html tags. The script should take two arguments, the full path file name of an input file and an output directory.

Example input:
C:/gispy/data/ch20/park.shp C:/gispy/scratch

Example output:
```
>>> C:/gispy/scratch/park.html created.
```

When opened in a web browser the output file looks like this:

park.shp Fields

- FID
- Shape
- COVER
- RECNO

7. **writeUniqueValuesHTML.py** The sample script named 'writeUniqueValues HTML.py' is missing code which needs to be filled in. Practice handling HTML and lists by replacing the five ### comments with one or more lines of code each. File locations are hard-coded, so no arguments are needed. When the code is complete, the script will create an HTML page like the one shown here:

Unique values in park.shp field COVER:

- woods
- other
- orch

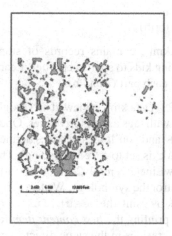

8. **parseHTMLimages.py** Write a script that takes an http URL as input and parses the HTML on the page using the `BeautifulSoup` module in the sample scripts directory. Assume 'BeautifulSoup.py' resides in the same directory as this script. The purpose of the script is to parse the given input for all of the image tags and print the number of images found and the source for each image. Assume that 'BeautifulSoup.py' resides in the same directory as this script.

Example input: http://www.nytimes.com/

Example output (only 3 of the 69 image sources are shown here):
```
>>> 69 images found.
image src: http://ad.doubleclick.net/ad/N5928.276948.NYTIMES/
B3868511.6;sz=184x90;
image src: http://ad.doubleclick.net/ad/N5928.276948.NYTIMES/
B3868511.7;sz=184x90;
pc=nyt227382A353135;
image src: http://i1.nyt.com/images/misc/nytlogo379x64.gif
...
```

9. **extractMany.py** Write a script that takes two arguments, an input directory and an output directory. The script should unzip every Zip and KMZ file in the input directory and write the results to the output directory. Use the `unzipAr-chive` function from sample script 'extractFiles.py' to perform the extraction for each file.

Example input:
C:/gispy/data/Ch20 C:/gispy/scratch

Example output:
```
>>> Unzip C:/gispy/data/Ch20/park.kmz to C:/gispy/scratch
Extract file: doc.kml ...
Unzip C:/gispy/data/Ch20/getty.zip to C:/gispy/scratch
Extract file: COVER63p.prj ...
Extract file: COVER63p.sbn ...
Extract file: COVER63p.sbx ... (and so forth)
```

10. The KML file, 'Sample_7_Day_GPS_Data.kml', contains records of stops (points) and routes (linestrings) of parents taking kids to sports activities. Learn more about KML code by writing a script for each part (a and b).

    (a) **printStyles.py** When 'Sample_7_Day_GPS_Data.kml' is viewed in 'Google Earth', the stops and routes are displayed with tags and colored lines. Open the file in a text editor such as 'Notepad++' and you'll see the styles defined in the KML header. The 'normalPlacemark' is set to a 'red-stars.png'. The names 'YellowLineGreenPoly1', 'YellowLineGreenPoly2', and so forth are set to various colors. These styles control the symbology. Write a script to print the symbology for each placemark by using the `BeautifulSoup` module to find all the 'styleurl' tags and printing the *first content item* for each tag. Assume that 'BeautifulSoup.py' resides in the same directory as this script. The script should take one argument, the full path to a KML file.

    Example input:
    C:/gispy/data/Ch20/Sample_7_Day_GPS_Data.kml

    Example output:
    ```
    >>> #highlightPlacemark
    #yellowLineGreenPoly1
    #normalPlacemark
    #yellowLineGreenPoly2
    #normalPlacemark
    #yellowLineGreenPoly3
    ... (and so forth)
    ```

    (b) **parseKMLDescription.py** In 'Sample_7_Day GPS_Data.kml', the route descriptions contains driving time, distance traveled, maximum speed, and average speed. The description tag contents for the linestring placemark return these statistics in a list, as shown in the following example:

```
[<br />, u'Driving Time: 00h:01m:49s', <br />, u'Distance
Traveled: 0 mile', <br />, u'Maximum Speed: 1.25 mile/hour',
<br />, u'Average Speed: 0.05 mile/hour']
```

The sample script 'parseKMLDescription.py' finds all the placemarks and their description tag contents (descriptionList). Add code to the script that parses the descriptionList to retrieve the distance traveled in each description. Create a function named getDistance, that takes descriptionList as an argument and returns the numeric distance. Print each distance, sum the values, and finally print the total overall distance. Assume that 'BeautifulSoup.py' resides in the same directory as this script. The script should take one argument, the full path to a KML file.

Example input: C:/gispy/data/Ch20/Sample_7_Day_GPS_Data.kml

Example output:
```
>>> Current distance: 0.0
Current distance: 0.01
Current distance: 13.32
Current distance: 5.9
...
Current distance: 0.04
Total distance 183.36 miles.
```

11. **parseGPSLog.py** The sample script 'parseGPSLog.py' is missing code which needs to be filled in. Practice parsing KML code by replacing the nine ### comments with one or more lines of code each. The KML file, 'Sample_7_ Day_GPS_Data.kml', contains records of stops and routes of parents taking kids to sports activities. When the code is complete, this script should create a point shapefile containing the KML points with the GPS stop names, dates and times. The script assumes that 'BeautifulSoup.py' resides in the same directory as this script. This script takes two arguments, the directory where 'Sample_7_ Day_GPS_Data.kml' resides and an output directory.

Example input:
C:/gispy/data/ch20 C:/gispy/scratch

Example output:
```
>>> 35.782101 -78.677338 0
...
35.781952 -78.67717 0
Skipping this Linestring placemark.
--Adding entries to C:/gispy/scratch/Sample_7_Day_GPS_Data.shp--
--Output created: C:/gispy/scratch/Sample_7_Day_GPS_Data.shp--
```

Chapter 21
Classes

Abstract Classes are the central construct for object-oriented programming (OOP). Knowing how to design your own objects will deepen your understanding of Python and help with learning other OOP languages. Object-oriented programming requires a paradigm shift from functional programming, which essentially works by grouping related steps into reusable functions. In functional programming, to make an omelet, you get six eggs and repeat break_egg until all the eggs are broken. In contrast, OOP revolves around the objects involved in the problem. Methods and properties belong to object classes; Making an omelet involves Egg objects. The Egg objects can have a shell property and a break_shell method and you could make each egg object break its shell. It may sound strange at first, but thinking about objects you have already been using will help. When you create a list, you're creating an instance of a list object. Then you apply list methods to it, such as, mylist.sort()). Similarly, you have used geoprocessing objects, such as Describe objects which have properties such as dataType and dataPath. Throughout this book you have used objects, methods, properties, and dot notation. All of these are aspects of OOP. Up to this point, you haven't designed your own objects; In this chapter you'll learn the syntax to do so. The examples in this chapter demonstrate basic OOP syntax and some of the benefits.

Chapter Objectives
After reading this chapter, you'll be able to do the following:

- List some benefits of object-oriented programming.
- Explain the terms 'object', 'class', 'property', and 'method'.
- Define a class and create an object instance.
- Create class properties and methods.
- Modify object properties and call object methods.
- Call a class defined in a user-defined module.

21.1   Why Use OOP?

User-defined functions increase code reusability. However, they have their limitations. Functions don't store information like variables do. Every time a function is run, it starts afresh. Functions and variables are often closely related to each other and need to interact with each other. As an example, suppose you are studying park maintenance and utilization using a dataset of hiking trails. Each trail has an identification number. You could store the trail IDs in a list as follows:

```
>>> trailList = [1, 2, 5, 10, 15]
```

As the database is updated, you may need to modify the trail list. To do so, you could append or remove IDs from the list:

```
>>> trailList.append(16)
>>> trailList.remove(5)
```

However, each land trail has additional information associated with it, such as the vegetation classification for the trail. How can you keep track of the vegetation? You might think of using a dictionary of trail IDs and vegetation classifications, as in the following example:

```
>>> trailVegetation = {1: 'barren', 2: 'some bare ground',
    5: 'stunted vegetation', 10: 'barren', 15: 'over-grown'}
```

Then you might need another dictionary for the trail lengths, and other dictionaries for the trail surfaces, the benches, the trail maintenance organizations, the manmade structures near each trail, the grade, and so forth. Functions need to access various trail data by passing in arguments and using the trail IDs to identify the

value of each attribute for that trail. For example, a function calculating the costs for
maintaining a trail needs to know the trail length and vegetation classification. The
function, `calculateCost`, defined in Example 21.1 passes these as arguments.
An upcoming example will show a different way to do this using OOP.

Example 21.1

```
def calculateCost(trail_ID, vegetation, length):
    '''Calculate trail maintenance based on
       vegetation and length.'''
    rate = 1000
    if vegetation[trail_ID] == 'barren':
        rate = 800
    elif vegetation[trail_ID] == 'some bare ground':
        rate = 900
    cost = length[trail_ID]*rate
    return cost
```

As the code grows, it becomes difficult to keep track of which functions need
which variables and how everything relates. With all of these variables separate but
depending on the same set of properties, maintaining the variables correctly can
grow cumbersome. Suppose you need to remove the trail with ID number 5. Each
variable that references that trail must be modified to reflect the removal. The trail 5
entry for the vegetation dictionary, the length dictionary, the volunteer dictionary,
and so forth must all be deleted. This process could be made easier by grouping
these trail attributes with the trail. Then when a trail needs to be added, updated or
removed, the trail information is all together in one place. Object-oriented programming does this by grouping variables and related functions so the functions can act
on the variables and interact with each other more smoothly.

Classes group closely related functions and variables together. They also provide a convenient way to work with a group of objects that have common attributes, such as a set of trails that each have a length, a vegetation classification, and
a volunteer organization. A class allows you to design a basic trail and each time
you need to create a new trail, you can just specify the values of its attributes. This
concept of grouping functions and variables related to a particular type of item is
the key principle of OOP. Classes are the container for these related functions and
variables.

You design objects by creating an object template, called a *class*. Which acts as
a blueprint for the object. The class specifies functions and variables associated with
the object. In this way, the class structure provides a means for grouping related
functions and variables.

21.2   Defining a Class

To define a Python class, use the `class` keyword followed by the class name and a colon. The contents of the class are indented to indicate that they are a related block of code. Python class names usually use upper camel case capitalization. The format for a Python class is as follows:

```
class ClassName:
    '''Class docstring.
    '''
    code statement(s)
```

Example 21.2 shows a `Trail` class definition. Think of a class definition as a blueprint. It isn't creating an object itself; It simply describes how to make one. You can create multiple instances of objects from the blueprint. The `Trail` class has three properties, `ID`, `length`, and `vegetation` and it defines three functions or *methods*. The term 'method' is used to refer to functions that are defined within a class. The `Trail` class contains the methods, __init__ (a special method), `cal-culateCost`, and `reportInfo`. The argument list for all three of these methods starts with a special variable named `self`. Next we will discuss how to create an object from the class blueprint and the special role of the __init__ method and the `self` variable in creating objects.

Example 21.2: A `Trail` class, created using the `class` keyword and indenting all of the contents.

```
# excerpt from trailExample.py
class Trail:
    '''Pedestrian path.

    Properties:
        ID          A unique identifier
        length:     Length in kilometers
        vegetation: Plant growth on the trail
    '''
    def __init__(self, tid, theLength, theVegetation):
        '''Initialize trail properties.'''
        self.ID = tid
        self.length = theLength
        self.vegetation = theVegetation

    def calculateCost(self):
        '''Calculate maintenance costs based
           on vegetation and length.'''
```

```
        rate = 1000
        if self.vegetation == 'barren':
            rate = 800
        elif self.vegetation == 'some bare ground':
            rate = 900
        cost = self.length*rate
        return cost

    def reportInfo(self):
        '''Print trail properties'''
        print 'ID: {0}'.format(self.ID)
        print 'Length: {0}'.format(self.length)
        print 'Vegetation: {0}'.format(self.vegetation)
```

21.3 Object Initialization and Self

Once the Trail class is defined, you can create a Trail object, a specific instance of a Trail with a particular ID, length, and vegetation classification. The syntax for creating an object instance looks similar to calling a function that returns a value. You call the class, using the class name followed by parentheses containing arguments and store the return value in a variable, as in Example 21.3. If you attempt to run Example 21.3 without running Example 21.2 beforehand, you will receive a NameError, since Trail is not yet defined.

When you call a class, the __init__ function inside the class is automatically invoked. The __init__ method for the Trail class sets the values of self.ID, self.length, and self.vegetation. The terms __init__ and self need some additional explanation.

- __init__ is a special method which is automatically called when an object of that class is created. It is used to create an initial state for the object (similar to a constructor in other languages such as C++ and Java). Use one of these inside your class to initialize object properties. Note that the name __init__ requires two underscores on the front and two on the back; otherwise, it will not be recognized as the initialization method.
- self, as the word suggests, refers to the particular instance of the object itself. This means that when you set self.vegetation with an assignment statement inside of __init__, the vegetation property for the current object is initialized. You should use self as the first parameter in the list of any function defined within a class. This allows you to refer to object properties within other methods without passing them as arguments. For example, calculateCost uses self.vegetation and self.length without passing these into this method. Though you could use a name other than self as a placeholder for the first position in the parameter list, it is best to conform to convention so that your code is consistent with others.

When a class is invoked, the caller must provide arguments for the argument list in the __init__ method. When the Trail class is invoked in Example 21.3, the caller provides three arguments. The argument list should correspond to the __init__ argument list, except for self which is passed implicitly into all class methods. Since the self variable is passed implicitly, only three, not four, arguments are needed to create a Trail object. The three arguments, an identification number, a length, and a vegetation classification, must be specified in this order in the class call. The __init__ method constructs an instance of the object and initializes its properties. The class returns an object instance, which can be saved using an assignment statement. Example 21.3 saves the returned object in a variable named myTrail.

Example 21.3: Creating an object instance. This statement creates a Trail instance called myTrail.

```
myTrail = Trail(1, 2.3, 'barren')
```

21.4   Using Class Objects

A class blueprint can define object properties and methods. An instance of a Trail object has three properties (ID, length, and vegetation), and two methods (calculateCost and reportInfo) in addition to the __init__ method. Notice the differences between the class method, calculateCost, in Example 21.2 and the stand-alone calculateCost function defined in Example 21.1. Example 21.1 uses three parameters (trail_ID, vegetation and length); Whereas, Example 21.2 only uses self as a parameter. The class method does not need to use a trail ID number to access its vegetation classification or the length, because the Trail object stores this information about itself. It can simply refer to self.vegetation and self.length. Variables that are assigned a value within the class __init__ method as self.variableName (such as self.length and self.vegetation) are object *properties* or *attributes*. General Python documentation uses the term 'attribute', but ArcGIS Resources documentation uses the term 'property'. This book generally uses the term 'property', for consistency with the arcpy documentation. This is the same terminology we used earlier to discuss arcpy object properties, such as the dataType property of the arcpy Describe object. Using properties and methods from user-defined classes has the same format as using other objects. Once a Trail object has been created, you can access its properties and methods using dot notation. The following code creates a Trail object, prints the length and vegetation properties and runs the calculateCost method:

```
>>> myTrail2 = Trail(2, 5.0, 'some bare ground')
>>> myTrail2.length
5.0
```

```
>>> myTrail2.vegetation
'some bare ground'
>>> myCost = myTrail2.calculateCost()
>>> myCost
4500.0
```

You don't have to pass trail length or vegetation information into the `calculateCost` function because these properties are stored with the trail itself. This centralization of control on variables is an advantage of OOP over functional programming. You can change the values of object properties and the value stored by the object will be modified. The following code changes the length property of `myTrail2` to 2.1 and then `calculateCost` uses that new length to return a much lower cost for this trail:

```
>>> myTrail2.length = 2.1
>>> myTrail2.calculateCost()
1890.0
```

`myTrail2.length` corresponds to `self.length` inside the class definition. The value of `self.length` for this object changed when `myTrail2.length` changed. When referring to an object property from outside of the class definition, use the variable name before the dot (e.g., `myTrail2.length`); When referring to an object property inside the class definition, use `self` before the dot (e.g., `self.length`).

A class blueprint allows you to create multiple object instances. The examples above each create one `Trail` object, but you can make multiple `Trail` objects and place them in a list or dictionary. The sample dataset, 'trails.txt' contains trail ids, lengths, and vegetation classifications, one trail per line (e.g., `1,2.3,barren` is the first line). Example 21.4 reads this data, creates a set of `Trail` objects, and stores them within a dictionary.

Example 21.4: Read a dataset and create `Trail` objects.

```
# excerpt from trailExample.py
data = 'C:/gispy/data/ch21/trails.txt'
trailDict = {}
with open(data, 'r') as f:
    # Read each line.
    for line in f:
        # Strip '\n' from the end and split the line.
        line = line.strip()
        lineList = line.split(',')
        tID = int(lineList[0])
        tLength = float(lineList[1])
        tVeg = lineList[2]
```

```
# Create a Trail object.
theTrail = Trail(tID, tLength, tVeg)
# Add the Trail object to the dictionary.
trailDict[traID] = theTrail
```

The `trailDict` dictionary uses trail identification numbers as keys:

```
>>> trailDict.keys()
[1, 2, 10, 5, 15]
```

The dictionary values are `Trail` objects:

```
>>> trailDict.values()
[<__main__.Trail instance at 0x18E8C530>,
 <__main__.Trail instance at 0x18E8C5D0>,
 <__main__.Trail instance at 0x18E8C620>,
 <__main__.Trail instance at 0x18E8C5F8>,
 <__main__.Trail instance at 0x18E8C670>]
```

The `reportInfo` method prints the object properties:

```
>>> for t in trailDict.values():
...         t.reportInfo()
ID: 1
Length: 2.3
Vegetation: barren
ID: 2
Length: 5
Vegetation: some bare ground
ID: 10
Length: 1.6
Vegetation: barren
ID: 5
Length: 4.2
Vegetation: stunted vegetation
ID: 15
Length: 20.0
Vegetation: over-grown
```

An additional `Trail` object can easily be added to the dictionary:

```
anotherTrail - Trail(3, 2.56, 'barren')
trailDict[3] = anotherTrail
```

Or an object can be deleted from the dictionary, using a dictionary key to identify the `Trail` object to be removed:

```
>>> del trailDict[5]
```

Contrast this object-oriented means of storing the trail information with the functional approach. The functional approach creates a list for IDs, dictionaries for lengths and vegetation classifications and then each of these would need to be modified to remove a trail. An OOP `Trail` object can easily be deleted, using the `del` keyword. Deleting the object takes care of removing all of its properties too. Suppose you need to remove all trails longer than 10 km. With the OOP approach, we can also easily delete trails based on one of their properties. The following code deletes long trails:

```
# Delete trails whose length exceeds 10 km.
for t in trailDict.values():
    if t.length > 10:
        del trailDict[t.ID]
```

With the functional approach, the programmer is responsible for keeping track of all attributes associated with a particular ID. Deleting trails based on length would require using more than one loop and deleting each attributes one by one, as in the following implementation:

```
# functionalTrailDelete.py
trailList = [ 1, 2, 5, 10, 15]
trailVegetation = {1: 'barren', 2: 'some bare ground',
5: 'stunted vegetation', 10: 'barren', 15: 'over-grown'}
trailLength = {1: 2.3, 2: 5.0, 15: 4.2, 10: 1.6, 5: 20.0}

# Determine which trails to delete.
longT = []
for k, v in trailLength.items():
    if v > 10:
        longT.append(k)
# Delete the long trails and each corresponding properties.
for i in longT:
    trailList.remove(i)
    del trailLength[i]
    del trailVegetation[i]
```

With OOP, it's also easy to modify objects based on their properties. The following code adds 1.2 km to the `Trail` length if the vegetation is 'barren':

```
# Increase barren property lengths by $5000.
for t in trailDict.values():
    if t.vegetation == 'barren':
        t.length = t.length + 1.2
```

Performing these deletions and modifications within the functional approach would again require managing multiple data structures to maintain the accuracy of each property. The OOP approach simplifies these operations. An additional method can be added to the class by simply adding another indented function within the class that uses `self` as the first parameter:

```
def calculateCrowding(self, visits, track):
    '''Calculate number of visitors/100 m
        (count double for narrow trails).'''
    if track == 'single':
        val = 2*visits/(self.length*10)
    else: # default unit is square meters
        val = visits/(self.length*10)
    return round(val,2)
```

Since `self` is passed implicitly, only two arguments are needed to call this method (one less argument than the length of the parameter list):

```
>>> t = Trail(11, 9.5, 'barren')
>>> t.calculateCrowding(20, 'single')
0.42
```

Extensibility is another OOP quality that strengthens reusability. An additional property can be added to a class that's already in use by adding an optional argument to the __init__ method argument list. For example, a `visits` property can be added to the `Trail` class by modifying the __init__ method as follows:

```
class Trail:
    def __init__(self, tid, length, vegetation, visits = 0):
        '''Initialize trail properties.'''
        self.ID = tid
        self.length = length
        self.vegetation = vegetation
        self.visits = visits
```

Then the `Trail` objects can be created in one of two ways:

```
>>> t1 = Trail(1, 14.5, 'over-grown')
>>> t2 = Trail(2, 3.4, 'some bare ground', 3.5)
```

When only three arguments are used, the `visits` property value defaults to zero;

```
>>> t1.visits
0
>>> t2.visits
3.5
```

21.5   Where to Define a Class

Organization

header comments

import modules

define class
    define __init__ method
    define other methods

define additional functions

code statements that create
objects and do other things

Just like user-defined function definitions, class definitions must precede any class calls, else a `NameError` will occur. Generally, a class should be placed at the top of a script after the header comments and module imports. Within the class, the __init__ method should be defined first, followed by any other class methods. Next, stand-alone user-defined functions should be defined. Finally, the main flow of the code should be placed last. This code may contain statements that create objects and perform other tasks. Example 21.5 shows code that contains header comments, followed by a class definition, followed by code that creates objects and uses them.

Example 21.5: Defining a class, creating an object instance, and using the object properties and methods.

```python
# highwayInfo.py
# Purpose: Define a highway class.
class Highway:
    '''
    Major road class.
    Properties:
        name          None-numerical road name.
        length        Length based on a network dataset.
        travelTime    Estimated time it takes to travel the
                      full length of this highway.
    '''
    def __init__(self, name, length, tTime):
        '''Initialize a Highway object.'''
        self.name = name
        self.length = length
        self.travelTime = tTime

    def calculateAvgSpeed(self):
        '''Calculate the average speed.'''
        avgSpeed = self.length/float(self.travelTime)
        return avgSpeed

    def getOrientation( self, number ):
        '''Determine highway orientation based on the hwy number.'''
        if number%2 == 1: # if the number is odd...
            orientation = 'NS'
        else:  # else the number is even.
            orientation = 'EW'
        return orientation

    def report(self):
        '''Print highway properties.'''
        print
        print '''{0} Highway
-----------
{1} km long
{2} hours travel time
        '''.format(self.name, self.length, self.travelTime)

if __name__ == '__main__':
    hwy = highway('Lincoln', 4946, 100)
    hwy.report()

favoriteHighway = highway('Blue Ridge Parkway', 755, 19)
favoriteHighway.report()
```

Also, just like user-defined functions, the class definition may either be placed within the script that calls the class or within a separate user-defined module, so that it can be easily re-used. A separate user-module should place test code inside a conditional expression that checks __name__ against '__main__' as discussed in Chapter 16.

In Example 21.5 sample script, 'highwayInfo.py', defines a class and then creates two objects instances, hwy and favoriteHighway. The first object, hwy, is created within the conditional construct, but the second object is created outside of the conditional construct to demonstrate the difference. The hwy object will only be created when the __name__ is set to '__main__'. The report method, which prints the object properties, is called on both objects. When 'highwayInfo.py' is run as the main script, it creates two objects and prints a report for each:

```
>>> Lincoln Highway
-----------
4946 km long
100 hours travel time

Blue Ridge Parkway Highway
-----------
755 km long
19 hours travel time
```

When the highwayInfo module is imported (instead of being run as the main script), hwy is not created and the report method is not called for this object. But favoriteHighway is still created and the report method is called. The output printed by the import command shows this difference:

```
>>> import highwayInfo
Blue Ridge Parkway Highway
-----------
755 km long
19 hours travel time
```

To improve this script, the code that creates and uses favoriteHighway should be moved inside of the conditional construct so that it is not executed when the module is imported.

21.6 Classes Within a Separate User-Defined Module

Recall that to call a function defined in a separate user-defined module, you import the function and then use dot notation, placing the dot between the module name and function name. The same format applies to classes. To use a class within a separate user-defined module, import the module, and then create an object instance by using dot notation between the module name and class name. For example, the following code imports highwayInfo and creates a `Highway` object named `myHwy`:

```
import highwayInfo
myHwy = highwayInfo.Highway('Pacific Highway', 496, 5)
```

The object can then be called by its name, as usual. The module name is not used when referring to object instances. For example, the following code uses the `myHwy` object:

```
>>> myHwy.report()
Pacific Highway Highway
-----------
496 km long
5 hours travel time
```

Example 21.6 puts these steps together, importing the module, creating an object, and using it.

Example 21.6: Calling the Highway class in highwayInfo, from a separate script.

```
# handleHighways.py
# Purpose: Demonstrate using a class defined in a separate
#          user-defined module.
# Usage:   No arguments needed, but highwayInfo.py must be in
#          the same directory.
import highwayInfo
myHwy = highwayInfo.Highway('Pacific Highway', 496, 5)

spdLimit = myHwy.calculateAvgSpeed()
orient = myHwy.getOrientation(5)
print 'Avg travel speed: {0} km/hr'.format(spdLimit)
print 'Orientation: {0}'.format(orient)
```

21.7   Discussion

This chapter introduced basic syntax for creating user-defined objects. It focused on a few pertinent goals:

- Defining, instantiating, and using objects in Python.
- Exploring some advantages of OOP.
- Looking under the hood to expose OOP from the programmer's side and provide some deeper understanding of built-in objects.

For simple projects, functional programming is appropriate and sufficient. But for larger projects OOP can greatly improve reusability and maintainability of code. Associating properties and methods with an object provides a convenient way to manage the object-related data, particularly when multiple instances of an object are involved.

Designing object-oriented programs takes some adjustment when you're accustomed to functional programming. A good way to hone this skill is to practice thinking in terms of objects, properties, and methods. For example, which objects, properties, and methods can you identify to characterize the archeological study data analysis described in Example 21.7?

Example 21.7: Identify objects, properties, and methods to characterize the following scenario:

Field surveys are being planned for archeological study sites. Scripts will be used to estimate the acreage to be surveyed and the associated costs, and to plan 'shovel tests' and store the field notes. Shovel tests are small holes, 30-50 cm in diameter, dug to search for artifacts. The presence or absence and nature of artifacts is recorded for each shovel test along with the stratum, soil color, and texture. Evidence of past changes such as floods, bulldozing, or plowing is also noted. For these sites, the shovel tests are positioned in a grid for extensive coverage. An appropriate sampling interval (e.g., 30m, 20m, 10m, 5m) is determined for each site. A GIS can assign these positions as X and Y coordinates based on the coordinate system of the study site.

An OOP structure for the domain described in Example 21.7 could be formulated as follows:

Objects
Study sites, shovel tests, positions

Properties
Study site properties: acreage, cost, set of shovel tests, sample interval
Shovel test properties: ID, position, artifacts (present/absent), artifacts description, stratum, soil color, soil texture, past terrain changes, depth
Position properties: shovel test ID, X, Y

Methods
Study site methods: compute acreage, compute cost, calculate sampling interval, determine shovel test grid positions
Shovel test methods: initialize ID and position, set artifacts, set stratum, set soil color, set soil texture, set past terrain changes
Position methods: initialize shovel test ID, X, and Y.

The number of objects, methods, and properties may expand as the project progresses. Fortunately, OOP creates a flexible framework for growing projects.

Some of the OOP syntax can be confusing on first exposure. Here are a few key syntax details worth repeating:

- Refer to object properties with `self` before the dot when inside the class. Refer to object properties with the *property name* before the dot when outside the class.
- Object properties do not need to be passed as arguments into methods other than `__init__`.
- Use `self` as the first parameter when writing a function inside a class. (If a method does not need any input from the user, you still need to list `self` as the only parameter when writing a function inside a class.)
- When calling a class method, pass in one less argument than the length of the parameter list, since `self` is implicitly passed. (If a class method only lists the `self` parameter, don't use any arguments.)

21.8   Key Terms

The `class` keyword   Properties
Object-oriented programming   Methods
Functional programming   The `self` argument
A class versus an object instance   The `__init__` method
Instantiating an object instance

21.9   Exercises

1. **air_classes** Plan the OOP components for a script that would support daily air traffic analysis of airlines and their flights. The script will use Airline, Flight, and Seat objects. Identify properties and methods for each of these objects based on the following description:

    Each airline name is stored along with a set of numbered flights. The origin and destination airports of each flight are stored and can be used to estimate the length of flight paths. Flight departure and arrival times are stored, so that the flight duration can be derived. Each flight has a set of seats organized in rows and columns. Each seat has a fare rate and a status (occupied or vacant) which can be modified when a new reservation is made. The class (first, business, or economy) can be determined based on the row. Flights can be added and removed. The capacity, the load factor, and the total fare for a flight can be calculated. The total number of flights for each airline for a day can also be derived.

    Airline properties (2):
    Flight properties (5):
    Seat properties (4):

Airline methods (3):
Flight methods (5):
Seat methods (2):

2. **seismic_classes** Plan the OOP components for a script that collects data samples from sensors. The script will use Sensor and DataSample objects. Identify properties and methods for each of these objects based on the following description:

   A network of 14 sensors will be lowered via helicopter into volcanic basins to measure seismic activity and weather conditions. Each sensor station will have an ID, an XYZ location, and daily sample count. Each sample will record a seismic signal as well as the start time of the reading, the air quality, temperature, humidity, and wind speed. The system needs to be able to identify invalid readings and identify seismic signals that exceed a significant threshold. The sensor will report significant seismic signals real-time. It will also routinely purge invalid samples, and then compute daily summary statistics for each reading. Finally, each sensor will need to transmit the summary results to a monitoring center.

   Sensor properties (3):
   DataSample properties (6):

   Sensor methods (4):
   DataSample methods (2):

3. **lineSegments.py** Add code to sample script 'lineSegments.py' to take four arguments representing the x and y values of two points, x1, y1, x2, and y2. For example, an input of 1 2 4 3 would represent two points, (1,2) and (4,3). Create an instance of a LineSegment object for the line segment from the first point to the second point (e.g., (1,2) to (4,3)).

   Add a method `printSegment` to print the values of the endpoints as Endpoint 1 and Endpoint 2 as shown below. Call `printSegment`. Then use the existing class methods to calculate the slope and length of the line segment and print these results. Next use the translateY method to shift the line segment down 3 units. Then call `printSegment` again.

   Example input: 1 2 4 3

   Example output:
```
>>> Endpoint 1:( 1.0, 2.0 )
Endpoint 2: ( 4.0, 3.0 )
Slope: 0.3
Length: 3.2
Endpoint 1:( 1.0, -1.0 )
Endpoint 2: ( 4.0, 0.0 )
```

4. **treeAnalysis.py** Add to the code found in sample script, 'treeAnalysis.py', to
build a Tree class for manipulating forestry data like the data in RDUforest.txt:

BLOCK PLOT SPECIES DBH 5 91 SG 14
5 91 LP 23
5 91 LP 17
5 91 LP 18
...

Each row represents a tree found in the specified blocks and plots near RDU
airport. SG (Sour Gum), LG (Loblolly Pine), and so forth are abbreviations for
the tree species. The diameter breast height (DBH), the trunk diameter at 4.5 ft
from the ground, is a measure of the tree's maturity. The Tree class should have
four properties, `block`, `plot`, `species`, and `dbh`, initialized in an `__init__`
method. The class should also have three additional methods. A method named
`report` is given already. Add a `calculateDIB` method that will calculate
the diameter inside bark (DIB), assuming the DIB is DBH times 0.917. Then add
a `calculateHeight` method to calculate height based on species type using
the equations given in the script comments. Once the class is built, create four
tree objects corresponding to the first four rows of 'RDUforest.txt'. Add code as
instructed where you see ### comments. The output should look like this:

```
>>> Tree 1 Species: SG
Tree 1 DIB: 12.838

Report Tree 1
-------------

Block: 5
Plot: 91
Species: SG
DBH: 14
DIB: 12.838
Height: 100.9

Tree 2 DBH: 23
Tree 2 Height: 87.175
Tree 3 block: 5
Tree 3 plot: 91

Report Tree 4
-------------

Block: 5
Plot: 91
Species: LP
DBH: 18
DIB: 16.506
Height: 87.05
```

5. **campusCrimes.py** Add code to sample script 'campusCrimes.py' to create an Incident class to handle crime incidents reported on a university campus. An Incident object should have ID, time, type, location, narrative, and status properties. The class should have __init__, brief, isMorning, and resolve methods, as described in the ### comments. The code should then call these methods as described in the script ### comments. The script takes one argument, the full path file name of a crime report.

Example input: C:/gispy/data/ch21/crimes.csv

Example output:
```
1251: Breaking & Entering, Pending
Report that someone had pried steel door from hinges. Nothing
was found to be missing.

1251: Breaking & Entering, Resolved
Report that someone had pried steel door from hinges. Nothing
was found to be missing.

Did incident 1313 occur in the morning? False
1313: Larceny, Bummer
Non-student reported book bag had been stolen.

1320: Fire Alarm, Resolved
Units responded to alarm caused by cooking.
```

6. **birds.py** Write a script that creates a Bird class. A Bird object should have name, habitat, and weight properties. The class should include __init__, talk, and diet methods. The __init__ method should initialize the properties. The talk method should print 'Squawk!', if the name property has the value of 'Toucan' and should print 'Tweet' otherwise. The diet method should return (not print) 'fruit' if the habitat is 'South America', 'bugs' if the habitat is 'North America', and 'fish' otherwise. Write test code that only runs when you run 'birds.py' as the main script. The test code should create a Bird instance, using a variable, named bigBird; Set its name to 'Condor', its habitat to 'Sesame Street', and its weight to 2000. In the test code, call the talk method three times. Then call the diet method on bigBird and print the return results. When 'birds.py' is run as the main script, it will print:

```
>>> Tweet
Tweet
Tweet
Condors eat fish.
```

Then restart your IDE and place sample script 'famousBirds.py' in the same directory as your script. The 'famousBirds.py' script will import your script.

Run it to check the printout when your script is not the main script. The output should appear as follows:

```
>>> Squawk!
Squawk!
Toucans eat fruit.
Sam gained an ounce. He weighs 1.9625 lbs
Tweet
Parrots eat fish.
```

Chapter 22
User Interfaces for File and Folder Selection

Abstract Running a script via an IDE requires some expertise, particularly when the script takes arguments. When you create scripts to share with others, your target audience may not be familiar with how to run a Python script. To make it easier, you can create a graphical user interface, a more intuitive means for the user to provide information for the script than command line arguments. *Graphical user interfaces* (GUIs) allow the user to interact with the script via windows labeled with explicit instructions. A Python script that uses GUIs to collect arguments can be run by double-clicking on it in Windows Explorer, avoiding an IDE altogether. Developing highly sophisticated custom GUIs is an advanced programming skill. However, you can generate certain GUIs with just a few lines of Python code. This chapter covers some easy to program GUIs for entering text or browsing to files.

Chapter Objectives
After reading this chapter, you'll be able to do the following:

- Get user input with graphical user interfaces.
- Get user input to open and save files.
- Customize interfaces using optional arguments.

22.1 A Simple Interface with `raw_input`

The built-in `raw_input` function launches a rudimentary GUI for text entry. When the script reaches a `raw_input` command, the user is prompted for input with a dialog box like the one shown in Figure 22.1. A *dialog box* is a type of GUI that uses a window to gather information from the user or inform the user of something, so called because it facilitates communications (a dialog) between the computer and the user. The dialog box in Figure 22.1 has two buttons, a label, and a text box. The label says 'Enter a workspace'. The white recessed box is called a *text box*.

This allows the user to type text which will be returned to the script that called the `raw_input` function.

The `raw_input` command in Example 22.1 is the code that generates the GUI in Figure 22.1. The `raw_input` function takes one argument, a string message to be displayed as the text box label (the argument in Example 22.1 is 'Enter a work-

© Springer International Publishing Switzerland 2015   435
L. Tateosian, *Python For ArcGIS*, DOI 10.1007/978-3-319-18398-5_22

Figure 22.1 `raw_input`
dialog box.

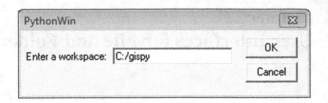

space:' and this same message is displayed in the dialog box label). You can place any string message here, but the label is usually used to prompt the user to enter a value. The response typed by the user in the text box is returned to the script when the user selects the 'OK' button; This response is the return value of the `raw_input` function. In Example 22.1, the return value is being stored in `arcpy.env.work-space`. When the user types 'C:/gispy' in the text box and presses 'OK', the value of `arcpy.env.workspace` is set to 'C:/gispy'. The `raw_input` function return value is always a string, even if the user enters a number. You can cast numerical entries (This is similar to handling numeric `sys.argv` values). The `raw_input` function generates a simple general purpose GUI that can be used in a variety of situations with some limitations. One limitation is that the user is allowed to enter anything in the text box, so your script may need to anticipate common errors. For example, if the script casts a number garnered from the user, it can include code to handle `TypeError` exceptions. As a second example, there may be a typo in the workspace path. You can use error handling for this as well, but we'll discuss a GUI that constrains the input to avert this particular mistake altogether by using a specialized GUI. File names are such a common GIS scripting input need that it's worth learning about a few specialized GUIs for this purpose. The remainder of the GUIs discussed in chapter are designed for handling file and directory interactions.

Example 22.1

```
# askWorkspaceRaw.py
import arcpy
arcpy.env.workspace = raw_input('Enter a workspace:')
print arcpy.env.workspace
```

Printed output:
```
>>> C:/gispy
```

22.2   File Handling with `tkFileDialog`

This built-in `tkFileDialog` module is one of several `tk` (or tool kit) modules that provide functionality for interfaces. The `tkFileDialog` module functions generate *file dialog boxes* for browsing to files. This eliminates the need to type file paths and enforces path accuracy. The appearance of the file dialog boxes is controlled by the operating system. For example, the screen shots shown in this chapter are generated by Windows 7. Windows 8 file dialogs may look different, but the functionality is the same. The file dialogs come in several different flavors for opening files, saving files, and choosing directories. There are several functions that get file or directory names from the user. Others functions return `file` objects open for reading or writing. Optional arguments can be used to control the appearance and behavior of the file dialogs. The upcoming sections discuss functions for getting file names and `file` objects, and the optional arguments for these functions.

22.2.1   Getting File and Directory Names

File paths are a common input for scripts. The `tkFileDialog` method `askopenfilename` creates a file browsing GUI, like the dialog box in Figure 22.2. When execution reaches the `askopenfilename` call in Example 22.2, this dialog is launched. When the user selects a file and clicks the 'Open' button, the full path name of the file is returned to the script. The assignment statement in Example 22.2 stores the return value in the `fc` variable. The script prints the file name, calls the `arcpy Describe` method, and uses the `Describe` object to print the data type of the input file. Optional arguments for `askopenfilename` allow the dialog to be customized. The two options used in Example 22.2 are setting the dialog box title and the type of files displayed by the browser. Here we only want the user to select a shapefile, so we specify this as the file type—more about these options in a moment.

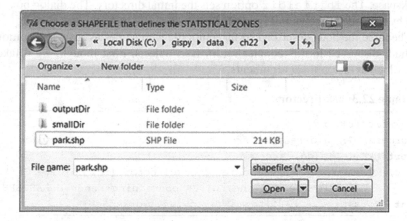

Figure 22.2   A dialog box launched with the Python code in Example 22.1.

Example 22.2

```
# getFileName.py
# Purpose: Get a shapefile name from the user and print the shape type.
import tkFileDialog, arcpy
fc = tkFileDialog.askopenfilename(filetypes=[('shapefiles','*.shp')],
    title='Choose a SHAPEFILE that defines the STATISTICAL ZONES')
print 'fc = {0}'.format(fc)
desc = arcpy.Describe(fc)
print 'Shape type = {0}'.format(desc.shapeType)
```

Printed output:
```
>>> fc = C:/gispy/data/ch22/park.shp
Shape type = Polygon
```

The asksaveasfilename method is similar to the askopenfilename method in that it returns a file name. However, since it's asking the user for a name for saving a file, it warns the user when a chosen name already exists. The user can cancel or overwrite the existing file. Calling this method doesn't actually create the file you name. Rather it returns a string file name that the script can use for some output it is creating. The return value from asksaveasfilename is used as an output file name in the following code which makes a copy of the input file:

```
fname = tkFileDialog.asksaveasfilename(initialfile='output.txt',
title='Save output as...')
arcpy.Copy_management('C:/gispy/data/ch22/precip.txt', fname)
```

Directory paths are another input that scripts often require. Example 22.3 uses the tkFileDialog module method, askdirectory, to set the geoprocessing workspace. The initialdir option sets the initial directory. The dialog box created by the code is shown in Figure 22.3.

The three methods described so far return file path names as strings. Additional methods discussed in this chapter return file objects. First though, we'll take a closer look at the optional arguments.

Example 22.3: askdirectory

```
# askDirectory.py
# Purpose: Get a directory from the user and set the workspace.
import tkFileDialog, arcpy
arcpy.env.workspace = tkFileDialog.askdirectory(initialdir='C:/',
        title='Select the FOLDER containing Landuse RASTERS')
print 'Workspace = {0}'.format(arcpy.env.workspace)
```

Printed output:
```
>>> Workspace = C:\buzzard
```

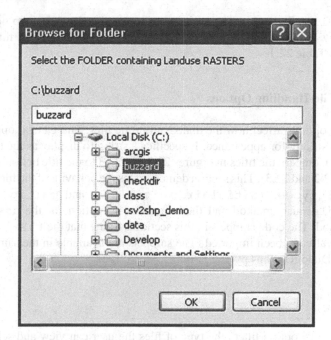

Figure 22.3 askdirectory dialog box.

Table 22.1 `tkFileDialog` method options.

tkFileDialog method	**Options**
askopenfilename	-defaultextension, -filetypes, -initialdir, -initialfile, -multiple, -parent, -title, -typevariable
askdirectory	-initialdir, -mustexist, -parent, or -title
asksaveasfilename	-defaultextension, -filetypes, -initialdir, -initialfile, -parent, -title, -typevariable
askopenfile	Same options as askopenfilename
asksaveasfile	Same options as asksaveasfilename

22.2.2   Options

`tkFileDialog` method options allow you to customize the appearance and behavior of the dialog boxes. These are specified as *keyword arguments*. All, none, or a subset of the options may be used. Throughout this book, we've been working with *positional arguments*. Positional arguments are ones which must be provided in the order specified in the function signature. Unlike positional arguments, the ordering of keyword arguments does not matter. Instead of order, assignment statements are used to specify which options are being set; The option name goes on the left of the equals sign and the value on the right. Arguments must still be separated by commas (e.g., the `askdirectory` call in Example 22.3 uses two options, initialdir and title; The assignment statements for these two options are separated by a comma). The options for the methods discussed in this chapter are listed in Table 22.1.

22.2.2.1 File-Handling Options

Most of the options affect how the dialogs filter the files. The title option is the only option strictly for appearance; It specifies a string to display as the title of the dialog box (Compare the titles in Figures 22.2 and 22.3 to the title option settings in Examples 22.2 and 22.3). This section demonstrates the behavior of the file-handling options, filetypes, initialdir, initialfile, and multiple for the askopenfilename method and the mustexist option for the askdirectory method. The code snippets in this section assume that the tkFileDialog module has already been imported (The snippets are available in the sample script named 'fileDialogOptions.py').

File Types

The filetypes option filters the type of files the user can view and select in the browser. You can set this option to a single file type or multiple file types. If this option is not set, then all file types are available. To specify file types, set this option to a list of tuples. To restrict browsing to a single file type, use a list containing only one tuple. Each tuple should contain two items: the file type name and the file extension. The following code creates such a tuple with elements "shapefile" and "*.shp" and uses the tuple in an askopenfilename call:

```
>>> t = ('shapefiles','*.shp')
>>> fname1 = tkFileDialog.askopenfilename(filetypes=[t])
```

The filetypes option needs to be set to a Python list, even if only one tuple is specified. The square braces around the t embed the tuple in a list. The file type options appear in the lower right corner of the dialog. For example, the dialog box in Figure 22.2 was also generated with filetypes=[('shapefiles','*. shp')].

To specify more than one file type, use a list of tuples. The following code uses two tuples ('csv (Comma delimited)','*.csv') and ('Text Files','*. txt') to specify two file types:

```
>>> fname2 = tkFileDialog.askopenfilename(
                filetypes=[('csv (Comma delimited)','*.csv'),
                    ('Text Files','*.txt')])
```

When multiple types are specified, you can click on the file type in the dialog and select alternative file types. The files shown in the browser are filtered by the selected type.

Initial Directory

The `initialdir` option specifies the initial browsing directory. Using `'./'` sets the initial directory to the Python current working directory. The following code makes 'C:/gispy' the default directory:

```
>>> fname3 = tkFileDialog.askopenfilename(initialdir = 'C:/gispy')
```

Initial File

The `initialfile` option specifies an initial filename. If this file does not exist in the specified directory, a warning is displayed and the user is forced to select another file, as in the following code, which points to a nonexistent file:

```
>>> fname4 = tkFileDialog.askopenfilename(initialfile = 'bogus.shp')
```

Multiple Files

The `multiple` option allows the user to select multiple files. This is a Boolean variable set to `False` by default, only allowing the user to select one file. When this option is set to `True`, the return value is a string of space separated file names. Example 22.4 lets the user select multiple files and loops to print the individual file names. The individual names are extracted by splitting the string on spaces. This returns a list of one or more file names (you'll need to take another approach if the file names contain spaces).

Example 22.4

```
# excerpt from fileDialogOptions.py
# Purpose: Vary file dialog options to get file and directory
#          names from user and print the results.

fnames = tkFileDialog.askopenfilename(
         title='Test multiple selections allowed', multiple=True)
files = fnames.split()
print 'Name list:'
for fname in files: # for each file selected by the user
    print '    {0}'.format(fname)
```

Existing Directories

The mustexist option specifies if the user must select an existing directory. This boolean variable is set to False by default. If the user types a non-existent directory name when this is set to True, a warning appears and the user is forced to cancel or select an existing directory. This option is not available for the askopenfile- name, since this method is specifically designed to get existing file names (the asksaveasfilename method can be used instead to get a new file name). The following code restricts the user's directory selection to an existing directory:

```
>>> inputDir = tkFileDialog.askdirectory(mustexist = True)
```

Modify the code samples in the 'fileDialogOptions.py' to gain a deeper understanding of how these options work. One additional option, the parent option, is not related to file handling and requires some explanation. This is discussed next.

22.2.2.2 Close the Tk Window

The parent option enables you to force the Tk window to close. When you run the tkFileDialog examples in this chapter, you will notice that a small window entitled 'Tk' is launched in the background (see Figure 22.4). When you run the script from an IDE such as PythonWin or PyScripter, the 'Tk' window persists after you exit the script and won't close until you exit the IDE. This is the default window for the tk-based applications. If you were building a custom GUI, you would add buttons and text boxes to this window and these buttons could provide a means to close this window. But calling the tkFileDialog method directly does not use the 'Tk' window, so we need to add a few lines of code to force it to close.

Figure 22.4 The tk window is opened when a tk file dialog command is called, but does not close when the dialog box is exited.

Forcing the 'Tk' window to close involves setting the `parent` option and using another module called `Tkinter` (for interface tool kit). The `Tkinter` module has a `Tk` method that returns a `Tk` object. Setting the `tkFileDialog` parent option to that `Tk` object allows you to destroy it after the file dialog has been closed. The following code imports both `tkFileDialog` and `Tkinter`, creates a tk object named `tkObj`, sets the parent option of the `askdirectory` method, and then destroys the `tk` object, which closes the default window:

```
import tkFileDialog, Tkinter
tkObj = Tkinter.Tk()

inputDir = tkFileDialog.askdirectory(parent=tkObj)
tkObj.destroy()
```

In fact, by adding one more line of code, you can prevent the `Tk` window from being displayed at all. The `Tk` method named `withdraw` hides the window. You must still destroy the `Tk` window, else it will be running hidden in the background. Example 22.5 withdraws the `Tk` window and destroys it. There is nothing special about the name `tkObj`; Example 22.5 names the `Tk` object `fatherWilliam` (after the Lewis Carroll poem). The 'withdraw — set-parent — destroy' approach can be used for any of the `tkFileDialog` methods.

```
# askAndDestroy.py
# Purpose: Get a directory from the user
#              and suppress the default Tk window.

import tkFileDialog, Tkinter
# Get a tk object
fatherWilliam = Tkinter.Tk()

# Hide the tk window
fatherWilliam.withdraw()

inputDir = tkFileDialog.askopenfilename(parent=fatherWilliam)

# Destroy the tk window
fatherWilliam.destroy()
```

22.2.3   Opening Files for Reading and Writing

The `askopenfilename`, `asksaveasfilename`, and `askdirectory` methods return the names of the files. Other `tkFileDialog` methods, `askopenfile` and `asksaveasfile`, return a file object open for reading or writing.

askopenfile and askopenfilename both force the user to browse to an existing file; However, while askopenfilename returns a string filename, the askopenfile opens the file and returns a file object, which is opened in read mode. This file object can be treated just like a file object that is created using the built-in open command with mode set to 'r'. In other words, you can use methods such as read, readline, and readlines. For example, the following code opens a file specified by the user, reads a line in the file, and closes the file.

```
fobject = tkFileDialog.askopenfile(
          filetypes=[('shapefiles','*.shp')],
          initialfile='data.txt', title='Open a data file...')
firstLine = fobject.readline()
fobject.close()
```

Analogously, the asksaveasfile returns a file object in write mode. Since these methods return file objects, they can be used within a WITH statement. The following code prompts the user for an output file name and writes a message in the file:

```
myTitle = 'Select an output file name'
with tkFileDialog.asksaveasfile(title=myTitle) as ofile:
    ofile.write('I like tkFileDialog')
```

22.3 Discussion

This chapter presented some simple GUI techniques, including the raw_input function and several tkFileDialog methods. The raw_input function returns a string version of the characters entered by the user. The tkFileDialog method provide several types of functionality. Several distinctions between the file dialog methods determine which method to select for use in a script.

One distinction is between those that return the names of files and those that return file objects. The tkFileDialog method names that end with 'filename' return the names of the files; These should be used when the script does not need to open the file for reading or writing. The askdirectory method also falls into this category (it returns a directory name). The tkFileDialog method names that end with 'file' return file objects open for reading or writing. When you call these methods, you need to use the file object close command or a WITH statement to avoid locking.

Another distinction is between those that are designed for input and output files. The methods names that start with 'askopen' are for getting an input file to read or use the file contents in some way. The method names that start with 'asksaveas' are for allowing the user to set output file names or create an empty file for writing.

Table 22.2 Selected tkFileDialog methods.

Method	Action	Return
askopenfilename	Ask for the *name* of an existing file	String file name or names
askdirectory	Ask for the *name* of a directory	String directory name
asksaveasfilename	Ask for a *name* to save a file	String file name
askopenfile	Ask for a file to *open for reading*	file object
asksaveasfile	Ask for a file to *open for writing*	file object

Table 22.2 summarizes the distinctions between the tkFileDialog methods discussed in this chapter.

The GUI implementations in this chapter do have some limitations. Only one function is called at a time. To collect both an input directory and an output file name, you need to call two methods sequentially and the GUIs will appear in succession. The raw_input method can handle any text input, but all values are returned as strings. When you collect a number with raw_input you need to cast it. Also, raw_input provides no validation. For example, when you ask for a number, the user can enter something other than a number, so you would have to check that in your script. When writing code for GUIs, there is a tradeoff between simplicity and customizability. There are a number of popular Python packages for building Python GUIs, such as Tk, wxWidgets, and pyQT, but these involve a steep learning curve. ArcGIS provides an alternative middle ground with a relatively easy to learn tool for building GUIs that are more complex than those presented in this chapter, though less flexible than stand-alone custom Python GUIs. Chapter 23 presents this ESRI alternative using ArcGIS toolboxes to build the interfaces.

22.4   Key Terms

Graphical user interface (GUI)
Dialog box
File dialog box
Text box
The built-in raw_input function

filename vs. file methods
askopen vs. asksaveas methods
positional arguments
keyword arguments

22.5   Exercises

1. **radiusRaw.py** Write a script which uses a `raw_input` GUI to get the radius of a circle and then calculates and prints the area of the circle. Use the `math` module to get the value of pi. Catch a named exception and set the radius to 1 if the user enters a non-numeric value. Print the results as in the following examples:

   Example1 input: User enters HELLO

   Example1 output:
   ```
   >>> Radius must be numeric; 'HELLO' is not numeric.
   Default radius of 1 used.
   Radius = 1   Area = 3.14159265359
   ```

   Example2 input: User enters 5

   Example2 output:
   ```
   >>> Radius = 5.0    Area = 78.5398163397
   ```

2. **boogieWoogie.py** Use sample script 'boogieWoogie.py' with the following instructions:

   (a) Run the script, take two screen shots of the file dialog that demonstrates each of the options and annotate the image, labeling the area in the file dialog affected by each option name: Label the first four options in the first screen shot (`filetypes`, `title`, `initialdir`, and `initialfile`) and use a second screen shot to label the area affected by the `multiple` option.
   (b) Add code to the script to hide and destroy the default 'Tk' window.
   (c) Add code to the script to print the file names returned by the file dialog, one per line.

3. **bufferDialog.py** Write a script that uses two `tkFileDialog` GUIs and one `raw_input` GUI to get an input shapefile name, an output shapefile name, and a buffer distance. Set the initial input file to 'park.shp', set the initial input directory to 'C:/gispy/data/ch22', set the initial output directory to 'C:/gispy/scratch', and set the initial output file to 'out.shp'. Finally, call the Buffer (Analysis) tool using the three values returned by the GUIs. The script should destroy the default 'Tk' window.

4. Match the task with the most appropriate GUI function.

Scripting task	GUI function
1. Count the number of lines in a file	A. `askdirectory`
2. Get a user's age	B. `askopenfilename`
3. Get the name of a KML file to import	C. `asksaveasfile`
4. Get the name of an output directory	D. `raw_input`
5. Get a name to use for a raster the script will create	E. `asksaveasfilename`
6. Write a log of `arcpy` messages in a file	F. `askopenfile`

5. **superman.py** The sample script 'superman.py' has some errors, but when they are repaired, it will use the `raw_input` function to prompt the user, giving the user three chances to guess who Clark Kent is.

If you guess incorrectly three times it should print:

```
>>> Haha! You don't know Superman!
```

If you guess correctly (Superman), it should print:

```
>>> You are right!
```

Watch the video 'superman.swf' (found in the sample scripts directory) to see a demonstration of the correctly working script. Use the IDE to identify the syntax errors and the debugger to identify the logic errors (There are eight errors in total). Once you fix the syntax errors, turn on the debugging toolbar. Set a breakpoint on line 10, run to the breakpoint, and use the Watch window to watch `answer` and `count` as you step through the rest of the script.

6. **numChars.py** Correct the five errors in sample script 'numChars.py'. When corrected, the script should get an existing file from the user, read the contents, and print the number of characters contained in the file. Find a solution without using the built-in `open` function. Also, be sure to withdraw and destroy the default 'Tk' window.

Example input: User selects 'precip.txt'

Example output:
```
>>> 'C:/gispy/data/ch22/precip.txt' has 43283 characters.
```

7. **dirGUI.py** Write a script that uses a `tkFileDialog` GUI to get an input directory from a user and prints a list of the raster files in the directory. Set the title for the GUI to 'Select a raster directory' and the initial directory to 'C:/gispy/data/ch22/smallDir'. The script should withdraw and destroy the default 'Tk' window.

Example input: User selects smallDir

Example output:
```
>>> Directory = C:/gispy/data/ch22/smallDir
[u'dog.JPG', u'pines.JPG', u'tree.gif']
```

8. **fieldFileGUI.py** Write a script which gets a single shapefile or dBASE file from the user with a GUI and prints the field names of the selected file. Set the title to 'Pick a file'. Set the initial directory to 'C:/gispy/data/ch22/'. Set the initial file to 'park.shp'. Accept files with '.shp' or '.dbf' extensions. The script should destroy the default 'Tk' window.

Example input: User selects 'park.dbf'

Example output:
```
>>> C:/gispy/data/ch22/park.dbf has the following fields:
FID
Shape
COVER
RECNO
```

9. **batchAscii2Raster.py** Write a script which converts batches of ESRI ASCII format files to raster with the ASCII to Raster (Conversion) tool. Use a file dialog to let the user select one or more ASCII input files (with '.asc' or '.txt' file extensions). Name the output files 'out0', 'out1', and so forth. Set the GUI initial directory to 'C:/gispy/data/ch22/'. Also use a file dialog to get an output directory from the user. Set the default directory to 'C:/gispy/scratch'. The script should withdraw and destroy the default 'Tk' window.

Example input1: User browses to 'precip.txt' and 'ASCIIout.txt'

Example printed output1:
```
>>> User selections: [u'C:/gispy/data/ch22/precip.txt',
u'C:/gispy/data/ch22/ASCIIout.txt']
Created output raster: C:/gispy/scratch/out0
Created output raster: C:/gispy/scratch/out1
```

Example input2: User browses to 'smallPrec.asc'

Example print output2:
```
>>> User selections: [u'C:/gispy/data/ch22/smallPrec.asc']
Created output raster: C:/gispy/scratch/out0
```

Chapter 23
ArcGIS Python GUIs

Abstract The graphical user interfaces (GUIs) discussed in the previous chapter create file browsing interfaces with simple Python calls to the `tkFileDialog` module. Some applications may require additional input (other than files/directories). Or you may want to create one GUI dialog that accepts multiple input parameters. If the application is meant to be run in an environment where ArcGIS is installed, *Script Tools* and *Python toolboxes* provide a solution for building GUIs with these characteristics. 'Script Tool' is an ESRI term for an ArcGIS construct that resides within a custom toolbox and points to an underlying Python script. By using a Script Tool, you can create a custom GUI that looks similar to the built-in ArcGIS tool GUIs. You can also add a button to one of the ArcGIS menus, so that users can launch the tool with one button click. Python toolboxes are another way to create GUIs that look just like Script Tool GUIs. A Python toolbox is a text file containing Python classes to define the toolbox and tools. Script Tools are a good way to learn about the GUI options available; Python toolboxes are an efficient way to develop tools, once you understand these options. This chapter introduces Script Tools and then steps through the various customization techniques. Last, Python toolboxes are discussed.

Chapter Objectives
After reading this chapter, you'll be able to do the following:

- Create an ArcGIS graphical user interface for a Python script.
- Specify GIS data type interface parameters.
- Set types, default values, direction, multiplicity, and other properties of parameters.
- Enable a user to input hand-digitized points, lines, or polygons features.
- Create a toolbar button to launch the GUI.
- Add output to a map.
- Set the symbology of an output parameter.
- Display progress as processes run.
- Implement dynamic behavior in a GUI.
- Build a Python toolbox GUI.
- Explain the difference between a Script Tool and a Python toolbox tool.

© Springer International Publishing Switzerland 2015   449
L. Tateosian, *Python For ArcGIS*, DOI 10.1007/978-3-319-18398-5_23

23.1   Creating a Script Tool

A Script Tool can be thought of as a wrapper for running a Python script; It points to a Python script, passes user input into the script, runs the script, and receives output from the script. The workflow for building a Script Tool involves creating a Python script and a toolbox first, then using the Script Tool wizard to set up the Script Tool. Script Tools reside within custom ArcGIS toolboxes. A custom toolbox is one created by a user, not one of the built-in ArcGIS toolboxes (In Figure 23.1, 'practice.tbx' is a custom toolbox). You create a toolbox, then add a Script Tool to it. This launches a Script Tool wizard. The wizard steps you through setting up the Script Tool properties, including a list of parameters which are used to pass arguments to the Python script. Once created, the Script Tool appears in ArcCatalog in the table of contents under its toolbox. The icon for a Script Tool looks like a scroll (In Figure 23.2, 'printTextFiles' is a Script Tool).

Note To create a Script Tool, you need three things:

1. A Python script.
2. A custom toolbox.
3. A list of desired script parameters and their types.

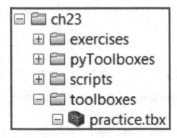

Figure 23.1 A custom toolbox, 'practice.tbx' viewed in ArcCatalog.

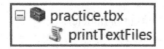

Figure 23.2 A ScriptTool, 'printTextFiles' viewed in ArcCatalog.

Let's step through an example. Since a Script Tool is merely a pointer to a script, we first need an existing script. We'll use a sample script which lists the text files with '.txt' extensions within a directory (see 'textLister.py' in Example 23.1) and we'll create a custom toolbox to house a new Script Tool. This sample script doesn't take any parameters. Upcoming examples will show how to use parameters. This example simply focuses on creating and running a Script Tool. Get some hands-on experience by following these steps:

1. Open and run 'C:\gispy\sample_scripts\ch23\scripts\textLister.py' in an IDE such as PythonWin or PyScripter. It should print text file names:

```
>>> cfactors.txt
crop_yield.txt
poem.txt
RDUforest.txt
report.txt
wheatYield.txt
xyData2.txt
```

2. Create a toolbox by browsing to 'C:\gispy\sample_scripts\ch23\toolboxes' in ArcCatalog.
3. Add a Script Tool to the new toolbox by right-clicking on 'practice.tbx' in ArcCatalog and choosing Add > Script... This launches a Script Tool wizard to customize the Script Tool step by step.
4. In the first pane of the wizard, set the script 'Name' and 'Label' to 'printText-Files'. The label appears in the ArcCatalog table of contents; It can have spaces, but the name can't. For simplicity, you can use the same value for both. Check 'Store relative path names'. This is an important choice, as we'll discuss shortly. Click 'Next >' to go to the next wizard pane.
5. We want this Script Tool to point to 'textLister.py'. To set the 'Script File', browse to 'textLister.py' in 'C:\gispy\sample_scripts\ch23\scripts'. Leave the default values for the check boxes. Running a Script Tool in-process improves efficiency, so this is usually the preferred approach. Click 'Next >' to proceed.
6. The final pane pertains to script parameters. Since 'textLister.py' needs no arguments, we'll leave this page as-is. Finally, click 'Finish' to exit the wizard. You can always come back and change any of the choices made during the initial setup by right-clicking on the Script Tool and selecting 'Properties...'.
7. A Script Tool named 'printTextFiles' appears in ArcCatalog as in Figure 23.2.

Run the tool by double-clicking on it. A GUI with several buttons and a help panel appears as in Figure 23.3. The left panel states 'This tool has no parameters'. This is what we expect to see, since we didn't set up any parameters. Click 'OK' to run the Script Tool. When the Script Tool run has completed, it launches a window to report success or failure. We refer to this as the 'Geoprocessing Window'.

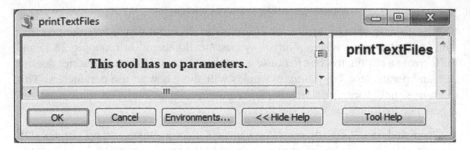

Figure 23.3 A Script Tool GUI with no parameters.

Example 23.1: Simple script for illustrating Script Tools.

```
# textLister.py
# Purpose: Print the text file (.txt) names in the directory.
import arcpy, os
myDir = r'C:\gispy\data\ch23\smallDir'
fileList = os.listdir(myDir)
for f in fileList:
    if f.endswith(".txt"):
        print f

arcpy.AddMessage('And I like pie!')
```

This example showed the basic steps for creating and running a Script Tool. You can check your work against the 'printTextFiles' tool in 'practiceExamples.tbx'. Right-click on the Script Tool and select 'Properties...' to view the properties of an existing Script Tool. Meanwhile, did you notice that the Geoprocessing Window reported that the script ran successfully and mentioned something about pie, but did not print any text file names? This is because standard Python print statements do not appear in the Geoprocessing Window. The next section discusses two techniques for communicating with the user within Script Tools. Also, this GUI didn't take any input from the user. We'll be getting to that shortly.

Note The ArcToolbox 'Add toolbox' commands puts a toolbox in the default toolbox location. To control the location of the toolbox, right-click on the target directory in ArcCatalog and select New > Toolbox.

23.1.1   *Printing from a Script Tool*

Until now, we've used the Python `print` keyword to communicate with the user. The print statement takes a string and displays it in the Interactive Window of an IDE. But printed messages don't appear in the Geoprocessing Window. To get

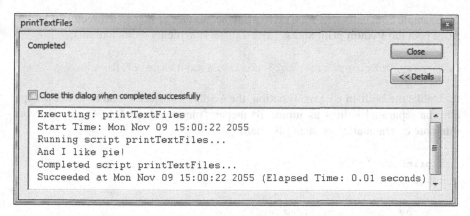

Figure 23.4 The 'Geoprocessing Window' which appears after a Script Tool has completed.

around this, you can use the `arcpy AddMessage` method. The `AddMessage` method takes one argument, a string message, and prints it in the Geoprocessing Window. In the Geoprocessing Window in Figure 23.4, as you may have already guessed, the following line generates the message (`'And I like pie!'`) wedged between the standard messages 'Running script...' and 'Completed script...':

```
arcpy.AddMessage('And I like pie!')
```

Python's print statement doesn't display in the Geoprocessing Window and `AddMessage` doesn't print in the Interactive Window:

```
>>> import arcpy
>>> message = 'And I like pie!'
>>> print message
And I like pie!
>>> arcpy.AddMessage(message)
>>>
```

This leads to a dichotomy: If you only use print statements, they will not appear in the Geoprocessing Window when you run a Script Tool. However, if you only use `AddMessage`, the message will not appear in the Interactive Window when you run the underlying script in an IDE. Since neither will print in both environments, the best solution is to use both types of statements to print output messages. For convenience, you can define a function to take a message and report it both ways:

```
def printArc(message):
    '''Print message for Script Tool and standard output.'''
    print message
    arcpy.AddMessage(message)
```

Example 23.2 calls `printArc` to print the number of files in a directory. Output from both the Python print and `AddMessage` statements looks like this:

```
Directory C:/gispy/data/ch23/smallDir contains 27 files.
```

Unlike the built-in `print` function, the `AddMessage` method does not accept comma separated values as input. To prepare messages for `AddMessage`, you must use concatenation or string formatting.

```
>>> print 5, 'miles'
5 miles
```

```
>>> arcpy.AddMessage(5,'miles')
TypeError: AddMessage() takes exactly 1 argument (2 given)
```

Example 23.2: Print and AddMessage.

```
# print4ScriptTools.py
# Purpose: Prints a directory's file count using
#          both 'print' and 'AddMessage'

import arcpy, os

def printArc(message):
    '''Print message for Script Tool and standard output.'''
    print message
    arcpy.AddMessage(message)

myDir = r'C:\gispy\data\ch23\smallDir'
# Lists all the files in the given directory.
fileList = os.listdir(myDir)
myMessage = 'Directory {0} contains {1} files.'.format(myDir,
                                            len(fileList))

printArc(myMessage)
```

23.1.2   Making a Script Tool Button

So far, we've been running Script Tools by double-clicking on them in ArcToolbox. You can also set up a button shortcut for a Script Tool. You can customize ArcMap (or ArcCatalog) to include this button on a toolbar for quick access to the tool. The button icon and text can be customized to suit the application.

To create a button on an ArcMap toolbar to launch a Script Tool, use the following steps:

1. Select 'Customize' > 'Customize mode...'
2. Select the 'Commands' tab
3. Under 'Categories', scroll down to select '[Geoprocessing tool]'
4. Select 'Add Tools...'
5. Browse to the custom toolbox > Select the Script Tool in toolbox > Select 'Add'. The Script Tool appears under 'Commands'.
6. Click on the Script Tool under 'Commands' and while holding the left mouse key down, drag the Script Tool between any existing buttons on any ArcMap toolbar. A black vertical bar appears when you're between buttons in a position where the new button can be placed. Release the mouse button to drop it into position.
7. Before closing the 'Customize' dialog, right-click on the tool in the toolbar to modify the button text and image, as desired.
8. Click 'Close' on the 'Customize' dialog to save changes.

Changes to the location, icon image, and text can be made by reopening the 'Customize' dialog ('Customize' > 'Customize mode...'). When not in customize mode, the button is locked. To remove the button, open the 'Customize' dialog and drag it back into the commands window.

Clicking the button launches the same GUI as double-clicking on the tool in ArcToolbox. This can be useful for sharing custom applications with novice ArcGIS users. All irrelevant toolbars can be hidden to reduce confusion.

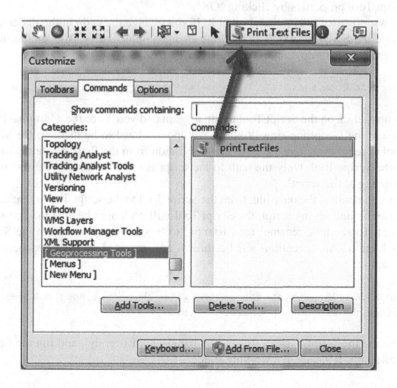

23.1.3   *Pointing to a Script*

Before we create GUIs with Script Tools, it's important to understand a few things about Script Tools and their Python scripts. Figure 23.2 shows the Script Tool in ArcCatalog, but if you browse to the same directory ('C:\gispy\sample_scripts\ch23\toolboxes') in Windows Explorer, you see toolbox files, but no Script Tool file. Script Tools do not appear in Windows Explorer; They are stored as part of a '.tbx' file. To see evidence of this, check the current file size of the 'practice.tbx' file (~6 KB). Add a dummy Script Tool to the toolbox (right-click on 'practice.tbx' in ArcCatalog and choose Add>Script...). Accept all the defaults (Just click 'Next', 'Next', 'Finish' without browsing to a script, etc.). Now check the size of the 'practice.tbx' file again (~7 KB). It's larger because it is now storing the additional Script Tool.

A Script Tool is part of the toolbox. To share a Script Tool with someone else, you need to give them the toolbox; the Script Tool will be visible to them when they view the toolbox in ArcCatalog. Since the Script Tool is just a pointer to a script, you must also give them the script. You may be wondering, what it means to say the Script Tool is a 'pointer to a script'? And will the Script Tool still point to the script when the toolbox is moved? To explore the relationship between Script Tools and scripts try some experiments, as described in the following steps:

1. Right click on the 'printTextFiles' Script Tool, select 'Properties...', and then select the 'Source' tab. Note the Python script specified by the 'Script file' path on this tab ('C:\gispy\sample_scripts\ch23\scripts\textLister.py'). Close the Script Tool properties by clicking 'OK'.
2. Browse to 'textLister.py' and open it. If you change the code in the underlying script, the Script Tool will adopt these changes immediately. Add another message to 'textLister.py' and save the script:

```
arcpy.AddMessage('***I like kale***!')
```

3. Double-click on the Script Tool to run it again and you'll see the additional message in the Geoprocessing Window. No updates need to be made to the Script Tool to see this change, because no information from the Python script is stored in the Script Tool. Only the path to the script is stored. This is what meant by 'pointing at the script'.
4. Since the path is the only link from the Script Tool to the script, if you rename or move the underlying script, the Script Tool will no longer be able to access the script. To see this, rename 'textLister.py' to 'textLister2.py' and run the Script Tool again. An exception will be thrown and reported in the Geoprocessing Window:

```
ERROR 000576: Script associated with this tool does not exist.
Failed to execute (printTextFiles).
```

5. Rename the script back to its original name, 'textLister.py', and run the Script Tool again. It's back to a working state.

6. Now move 'textLister.py' up one directory to 'C:\gispy\sample_scripts\ch23\'. Run the Script Tool again. You get the same error as when it was renamed. Though the script does exist, the Script Tool can't find it. Move the script back into the 'C:\gispy\sample_scripts\ch23\scripts' directory.

Take care when you share a Script Tool, so that the user doesn't encounter Error 000576. Maintain a relative path between the Script Tool and the script and set the Script Tool to use relative paths. Open the 'printTextFiles' Script Tool properties and select the 'General' tab. The 'Store relative path names' checkbox should be checked. This means we can move the Script Tool as long as we move the script to the same relative location. Relative to 'example.tbx', 'textPrinter.py' is one directory up out of 'toolboxes' and then one step down into a 'scripts' directory. Remember that moving the Script Tool is accomplished by moving the toolbox. To see what happens when you don't maintain a relative path with the script try the following:

1. Make a copy of the 'practice' toolbox and place it under 'C:\gispy\sample_scripts\ch23\sandbox\toolboxes\'.
2. Run the new copy of the Script Tool inside; It reports Error 000576.
3. To correct the error, you need to recreate the original relative path to the script. How can you achieve this?
4. Make a copy of the 'scripts' directory under 'C:\gispy\sample_scripts\ch23\sandbox'.
5. Test the new Script Tool again to see that it works. For this example, a relative path is maintained by copying both the 'toolboxes' directory and the 'scripts' directory into the same parent directory.

Script Tool Concepts to Remember

- Script Tools do not appear in Windows Explorer. They are part of a '.tbx' file.
- Since a Script Tool points to a script, you can change the script and see the updates immediately when you run the tool.
- If you rename, move, or remove the underlying script, the Script Tool will be broken.
- For portability, check 'Store relative path names'. Then copy the toolbox and the script maintaining the relative path.
- Script Tools don't have a debugger, so it's important to test Python scripts thoroughly outside the Script Tool.
- A simple way to maintain relative paths is to set up the Script Tools and scripts in the same directory and share that directory.

23.2   Creating a GUI

Now that you know how to create Script Tools and how the tools relate to underlying scripts, you're ready to use Script Tools to create GUIs. When you set up a Script Tool, the last pane of the Script Tool wizard contains a table for parameters. For the 'printTextFiles' example, we left this table empty. This is why this Script Tool says 'This tool has no parameters' when it is launched. In this section, we'll use the Script Tool wizard parameter pane to set up some parameters. This pane contains two boxes, one for the parameters and one for the parameter properties (Figure 23.5). A parameter is added by specifying a display name and data type in the parameter table. Parameter properties can be adjusted for each parameter. This section steps through an example that adds some simple parameters. Sections 23.2.1 and 23.2.2 contain details on parameter data types and properties.

When items are added to the parameter table, the Script Tool generates a widget for the parameters in the list. User interface elements (e.g., text boxes, buttons, check boxes, combo boxes, and list boxes) are commonly referred to as *widgets*. Widgets help the user make input choices by constraining the way input is accepted. For example, they may help the user browse to a file or select amongst several mutually exclusive choices.

Figure 23.5   The Script Tool wizard parameters page.

The Script Tool automatically adds a widget to the GUI for each input parameter. To see this, we'll create a Script Tool with three parameters. The sample script named 'deleter.py' (Example 23.2) deletes files in a workspace based on file type and name. For input, the script needs a workspace, a file type, and a wild card. If the data type is 'raster' or 'feature class', it deletes files of that type within the workspace that contain the wild card string in their names. If any other data type is given, it deletes files which have a file extension matching the wild card. Create a GUI for this script by building a Script Tool that lists three parameters with the following steps:

1. View 'C:/gispy/data/ch23/rastTester.gdb' in ArcCatalog and observe that there are over a dozen rasters in this workspace labeled '_out'. The file geodatabase 'rastTester_1.gdb' and 'rastTester_2.gdb' are identical copies of this workspace. Open and run 'deleter.py' in PythonWin with the following arguments to see how it works:

   C:/gispy/data/ch23/rastTester_1.gdb raster _out

   When the script has stopped running, view 'rastTester_1.gdb' in ArcCatalog to confirm that all of the rasters with names containing '_out' have been deleted. No debugging is available within Script Tools; Hence, it is essential to test Python scripts before building Script Tools and then test modifications to the script outside of the Script Tool.

2. Add a second Script Tool to the 'practice.tbx' toolbox (Right-click on the toolbox in ArcCatalog and choose 'Add > Script...').

3. Name and label it 'deleteFiles'. Add a description of the Script Tool's purpose ('Delete files from a workspace based on their type and name.'). The description appears in the tool help box when the tool is launched (Figure 23.7). Also, check 'Store relative path names'. Click 'Next >' to proceed.

4. For the 'Script File', browse to 'deleter.py' in 'C:\gispy\sample_scripts\ch23\scripts'. Click 'Next >' to proceed.

5. Next, we'll add three script parameters, as needed by 'deleter.py'. Click in the left column in the parameter table and add display names as shown in Figure 23.6. Click in the right column and select the data types, Workspace, String, and String again (see Figure 23.6). You can use the mouse to scroll through the data types,

Display Name	Data Type
Select a workspace	Workspace
Specify a data type	String
Enter a wildcard	String
@	

Figure 23.6   Parameter table for 'deleteFiles' Script Tool.

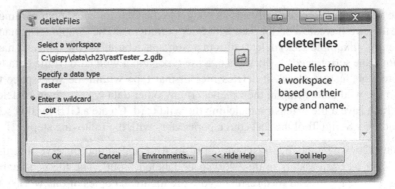

Figure 23.7 A Script Tool GUI with three parameters.

but for greater efficiently, type the first letter of the data type (e.g., 'w' for 'Workspace') and then use the 'down arrow' key to scroll to the desired data type.

6. For now, we'll use the default parameter properties, so click 'Finish' to exit the wizard.

7. Run the new Script Tool, 'deleteFiles', by double-clicking on it. The GUI has three widgets (a file browsing text box and two basic text boxes, in this case), as in Figure 23.7, one for each parameter. The three display names appear above the boxes to prompt the user.

8. Specify the input as shown in Figure 23.7 (Browse to 'rastTester_2.gdb' in the first box and specify 'raster' and '_out' in the second and third boxes). Click 'OK' to run the tool. Then check the results in 'rastTester_2.gdb' by viewing it in ArcCatalog.

Adding three parameters in the 'deleteFiles' example created three boxes in its GUI. The first box has a browsing button, because the data type for this parameter was set to 'Workspace'. This filters the workspace selection to directory, file geodatabase, and any other formats accepted by ArcGIS as a workspace. The browsing mechanism helps the user select a valid existing workspace. The other two widgets are simply text boxes that accept text without restrictions. The type of widget created for each parameter depends on the parameter data type and properties, as discussed next.

Note Script Tools can't be stepped through with a debugger. Test Python scripts outside of the Script Tool in a Python IDE.

Example 23.3

```
# deleter.py
# Purpose: Delete files from a workspace based on
#          their type and name.
```

```
# Usage: workspace datatype (raster, feature class,
#           or other) wildcard
# Sample input: C:/gispy/data/ch23/rastTester.gdb raster _out

import arcpy, os, sys

def printArc(message):
    '''Print message for Script Tool and standard output.'''
    print message
    arcpy.AddMessage(message)

arcpy.env.workspace = sys.argv[1]
fType = sys.argv[2]
wildcard = sys.argv[3]
substring = '*{0}*'.format(wildcard)

if fType == 'raster':
    data = arcpy.ListRasters(substring)
elif fType == 'feature class':
    data = arcpy.ListFeatureClasses(substring)
else:
    entireDir = os.listdir(arcpy.env.workspace)
    data = []
    for d in entireDir:
        if d.endswith(wildcard):
            data.append(d)
for d in data:
    try:
        arcpy.Delete_management(d)
        printArc( '{0}/{1} deleted.'.format(arcpy.env.workspace, d))
    except arcpy.ExecuteError:
        printArc( arcpy.GetMessages())
```

23.2.1   Using Parameter Data Types

Script Tool parameter data types are Esri data types, not built-in Python data types; They are specialized structures related to ArcGIS data, like the data types that are listed on the geoprocessing tool help parameter tables. A few of the Esri data types have Python built-in data type equivalents as shown in Table 23.1.

Table 23.1 Esri and Python equivalent data types.

Esri data type	Built-in Python data type
Boolean	bool
Double	float
Long	int
String	unicode (string)

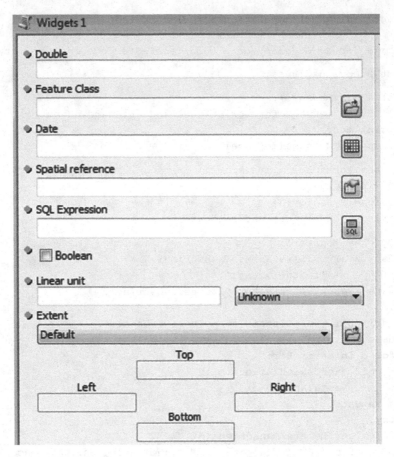

Figure 23.8 The parameter in the 'widgets1' Script Tool GUI are labeled to match their parameter type to demonstrate how the widgets appear.

Script Tool widgets generated by Esri data types are the same widgets used in standard ArcGIS tools to collect these types of data. GUI Figure 23.8 shows a Script Tool GUI example, 'widgets1' (found in 'C:\gispy\sample_scripts\ch23\toolboxes\widgetExamples.tbx'). To demonstrate the widgets for various data types, the display name and data type are set to the same value for each parameter in 'widgets1'. The appearance of the widget created for each parameter depends partially on the parameter data type (and partially on the parameter properties, which we'll get to in a moment). Simple data types like Double, Float, or String generate a box where text or numbers can be entered. Input file (or path) data types, such as Feature Class, Raster Dataset, and Workspace, generate a text box with a file browsing button on the right end. Other types, such as 'Date', 'Spatial Reference', and 'SQL Expression', generate a text box with a distinct button that launches a specialized interface, such as a calendar, a spatial reference browser, or a SQL expression generator. Still others, such as 'Boolean', 'Linear unit', and 'Extent', tailor GUIs to guide user interaction. The 'Boolean' check box constrains the user to only two choices (checked or unchecked). The 'Linear unit' generates a text box on the left, plus a drop-down menu, also know as a 'combo box', on the right. The text box allows the user to specify a numerical value and the combo box constrains the unit to valid linear distance measures. The 'Extent' GUI allows the user to select the 'Default' value or to specify four values in the boxes below the combo box. The 'widgets1' Script Tool pictured in Figure 23.8 is pointing to the 'reportSTargs.py' from Example 23.4. The 'reportSTargs.py' script prints the parameter values received from a Script Tool. When 'widgets1' is run with a set of (arbitrarily chosen) input values, the output in the Geoprocessing Window looks like this:

Number of arguments = 9
Argument 0: C:\gispy\sample_scripts\ch23\scripts\reportSTargs.py
Argument 1: 0.55
Argument 2: C:\gispy\data\ch23\smallDir\trails.shp
Argument 3: 3/12/2014 3:14:41 PM
Argument 4:   GEOGCS['GCS_Chatham_Islands_1979',DATUM['D_Chatham_
    Islands_1979',SPHEROID['International_1924',6378388.0,297.0]],PRIMEM
    ['Greenwich',0.0],UNIT['Degree',0.0174532925199433]];-400 -400 1000000000;
    -100000   10000;-100000   10000;8.98279933943848E-09;0.001;0.001;IsHigh
    Precision
Argument 5: value = 5
Argument 6: true
Argument 7: 3 Kilometers
Argument 8: 0 0 10 10

In the early stages of constructing GUIs, importing `reportSTargs` and calling `printArgs` may be helpful for trouble shooting and understanding the format of incoming values. To do so, place the `reportSTargs` module in the same directory as the script and add the following code to the script:

```
import reportSTargs
reportSTargs.printArgs()
```

The 'widgets1' Script Tool in Figure 23.8 shows the widgets for a few of the data types discussed here. The Script Tool named 'widgets2', also in 'widgetExamples. tbx', points to 'reportSTargs.py' and has a longer parameter list with matching names and types. This Script Tool can be used to preview the widgets that various data types generate. Of course, you can append additional parameters in this tool or build your own Script Tools to see how the widgets work. A Script Tool doesn't even need to be pointed to a Python script to merely experiment with parameter widget appearance.

23.2.2   Using Parameter Properties

Parameter properties provide additional tailoring for the input widgets. The box at the bottom of the parameters input pane (see Figure 23.5) controls these properties. Take a look at the parameter list for '01_optionalParam' in the 'propertyExamples' toolbox (Right-click on '01_optionalParam' in ArcCatalog, select 'Properties…' and then select the parameters tab). Double click on one of the parameter names in the top box. An @ symbol appears to the left of its name to show that it is selected:

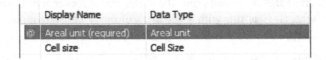

When you select a parameter in the top box, the parameter properties box updates to display the properties for that particular parameter. You may not have noticed this, since many data types start out with the same property values. A few data types do have unique default values auto-populated. To see this, click on 'Cell Size' and then 'Compression' in the '01_optionalParam' parameter list and watch the parameter 'Default' value change from 'MAXOF' to 'LZ77' (these are names of the default algorithms for these data types). Table 23.2 lists the parameter properties with descriptions. This section discusses the parameter properties examples in the 'propertyExamples.tbx' toolbox.

23.2.2.1   Type

The 'Type' property designates a parameter as 'Required', 'Optional', or 'Derived'. By default, parameters are 'Required'. A dot appears next to required parameters on the GUI until they are filled in. The tool will not run unless required parameters are specified. Optional parameters can be left blank. Script Tool '01_optionalParam' takes one required argument (an areal unit) and four optional arguments ('Cell

Table 23.2

Property	Description
Type	Required, optional, or derived
Direction	Input or output
Multivalue	Accept a list of values or just one value
Default or schema	A schema for feature set or record set data types; a default value for all other data types
Environment	Set the default value based on an environment setting
Filter	Restrict the input values
Obtained from	Set a values on information from another parameter
Symbology	Set symbology to display output

Size', 'Compression', 'Double', 'Feature Class' types). It points to 'reportSTargs. py'. When the tool is run with the areal unit set to '8 SquareKilometers' without specifying any of the optional arguments, it reports the following:

Argument 0: C:\gispy\sample_scripts\ch23\scripts\reportSTargs.py
Argument 1: 8 SquareKilometers
Argument 2: MAXOF
Argument 3: LZ77
Argument 4: #
Argument 5: #

When a tool is run with an optional parameter left blank, the default value for that data type is used. Some data types have a special default value (e.g., MAXOF and LZ77 for cell size and compression data types). Other data types, such as Double, Feature class, and String, have a generic default value, a hash sign (#). The user doesn't see the hash sign, but this is what the underlying script receives. The Python script needs to check for the hash sign when handling these types of optional parameters. Script Tool '02_optionalParam' takes one required argument (a base) and one optional argument (a power). This tool points to 'exponentiator.py' which calculates the base number raised to a power (e.g., if the input is 5 and 2, the tool prints '5.0 raised to the 2.0 is 25.0'). If the optional argument is omitted, the script uses a power of 1. To handle the optional argument, the underlying script checks for a hash sign:

```
if sys.argv[2] == '#':
    power = 1
    reportSTargs.printArc('No exponent provided. Using \
                          default power of 1.')
else:
    power = float(sys.argv[2])
```

With this approach, the user must provide hash signs for the 'optional arguments' when running the script outside of the Script Tool, else the script will raise an IndexError exception. Required arguments should be placed at the beginning of the parameter list (with optional parameters at the end).

An example of the third parameter type, 'Derived', will be discussed in the next section, as this is closely related to the 'Direction' property.

23.2.2.2 Direction

The 'Direction' property designates a parameter as either 'Input' or 'Output'. By default, parameters are 'Input'. Input values are information the script needs to perform its tasks, such as a dataset or workspace to use. The Script Tool examples so far in this chapter have only used input parameters. Input parameters can be required or optional, but not derived.

The output direction is used for Script Tool parameters that represent output generated by the tool. This may be one or more datasets created by the tool. It may be a modified preexisting dataset (e.g., a dataset with a new field added, as shown in an upcoming example, '14_derivedObtainedFrom'). It may be a Boolean. It may be numerical values resulting from Script Tool calculations. Any types of results generated by standard ArcToolbox tools could be output from a custom Script Tool. Like standard tools, custom Script Tools can be used as tools in ModelBuilder models. The output (Boolean, numerical results, dataset, and so forth) would then be passed along to the output ovals in the models.

Output parameters can be required, optional, or derived. Output parameters with a 'Required' or 'Optional' type allow the user to set the name of new output datasets that will be created by the Script Tool. The '03_requiredOutput' Script Tool has two parameters, one required input feature class and one required output feature class. It points to a script named 'copier.py' that makes a copy of the first argument and names it as specified by the second argument, as in the following code:

```
arcpy.Copy_management(sys.argv[1], sys.argv[2])
```

When a tool with output data parameters is run in ArcMap, the geoprocessing output can be automatically added to the table of contents. Go to Geoprocessing menu > Geoprocessing options and check 'Add results of geoprocessing output to the display'. Run the '03_requiredOutput' on any input feature class. When the tool run is completed, the output copy of the input feature class should be automatically displayed on the map.

The combination of 'Required' type with 'output' direction can only be used for output that will be created by the script, not for modifications to existing data (if you try to select an existing dataset, the GUI will raise an error or warning). Also, it should only be used when you want to allow the user to select the location and name of the output. If the output is a modification of existing data or if the script itself determines the output name and location, output should be a 'Derived' type parameter. Input can't be derived, when you select 'Derived' in the 'Type' property, the 'Direction' property is automatically set to 'Output'. To display derived output results, you need to use an `arcpy` method named `SetParameterAsText`. Script Tool '04_derivedOutput1' has only one parameter, a derived output. This tool points to script 'buffer1.py' (Example 23.4) which buffers a hard-coded shapefile and passes the output buffer file name back to the Script Tool using the following statement:

```
arcpy.SetParameterAsText(0, outputFile)
```

The `SetParameterAsText` method takes two arguments, a number and a string. The number specifies the Script Tool parameter index. The second argument is a string representing the name of the output. In this case, it's the output file name. The `SetParameterAsText` method does not count the script path name, rather it only counts the parameters in the Script Tool list (using zero-based indexing). In this example, the Script Tool has only one parameter, so the index for this derived output parameter is zero.

Example 23.4

```
# buffer1.py
# Purpose:    Buffer a hard-coded file and send the result
#             to a Script Tool.

import arcpy
arcpy.env.overwriteOutput = True

fileToBuffer = 'C:/gispy/data/ch23/smallDir/randpts.shp'
distance = '500 meters'
outputFile = 'C:/gispy/scratch/randptsBuffer.shp'

arcpy.Buffer_analysis(fileToBuffer, outputFile, distance)

arcpy.SetParameterAsText(0, outputFile)
```

When the Script Tool has multiple parameters, care must be taken to index the output correctly. Script Tool '05_derivedOutput2' points to 'buffer2.py' (Example 23.5) and demonstrates how this works. Example 23.5, like Example 23.4, buffers a file, but in this case, the file to be buffered and the buffer distance are garnered from user input. The Script Tool, '05_derivedOutput1', points to this script and lists three parameters, a file to buffer, a buffer distance, and an output derived feature class, the buffer output. Example 23.5 calls SetParameterAsText with an index of 2 because the output file is the third one in the Script Tool parameter list. The first entry in the sys.argv list is the script name, so indexing for sys.argv user arguments starts with 1; Whereas, indexing for the SetParameterAsText method starts with zero. As long as the derived output parameters are listed last, you can simply start with the same index number as the last sys.argv index.

Example 23.5

```
# buffer2.py
# Purpose:    Buffer an input file by an input distance
#             and send the result to a Script Tool.

import arcpy, os, sys
arcpy.env.overwriteOutput = True

fileToBuffer = sys.argv[1]
distance = sys.argv[2]
arcpy.env.workspace = os.path.dirname(fileToBuffer)
outputFile ='C:/gispy/scratch/Buff'

arcpy.Buffer_analysis(fileToBuffer, outputFile, distance)

arcpy.SetParameterAsText(2, outputFile)
```

23.2.2.3   Multivalue

The multivalue property can be 'Yes' or 'No'; the default is 'No'. When this property is set to 'Yes', the parameter will accept a list of values. Not all data types can have multiple values (e.g., Boolean parameters can not handle multiple values). A multivalue input parameter allows the user to select multiple input files and a multivalue derived output parameter allows a Script Tool to output multiple files and add them to a map automatically. When multiple values are passed into one parameter, the script receives these values as a semi-colon delimited string. Script Tool '06_multValueIn' has a multivalue raster input parameter. Run this tool and select several rasters. The output will look something like this:

```
Input string:
C:\gispy\data\ch23\rastTester.gdb\aspect;C:\gispy\data\ch23\
rastTester.gdb\elev;C:\gispy\data\ch23\rastTester.gdb\landcov
Input file: C:\gispy\data\ch23\rastTester.gdb\aspect
Input file: C:\gispy\data\ch23\rastTester.gdb\elev
Input file: C:\gispy\data\ch23\rastTester.gdb\landcov
```

To consume the individual input rasters, the script, 'multiIn.py' (Example 23.6), splits the multivalue input string on the semicolon. This returns a list of items that can be processed within a loop.

Example 23.6

```
# multiIn.py
# Purpose: Parse a semicolon delimited input string.
# Usage: semicolon_delimited_string
import reportSTargs, sys

inputString = sys.argv[1]

reportSTargs.printArc('Input string: {0}'.format(inputString))

inputList = inputString.split(';')

for i in inputList:
    reportSTargs.printArc ('Input file: {0}'.format(i))
```

For multivalue derived output, the set of output names needs to be compiled into a semi-colon separated string to be passed back to the script. This can then be accomplished using the string join method. Script Tool '07_ multiValOut' has three parameters, an input folder, an input linear unit, and a derived multivalue output with a shapefile data type. It points to the 'bufferAll.py' script shown in Example 23.7. The script buffers every shapefile in the input folder, creating one output for each input shapefile. The names of the output files that are successfully created are collected in a list. Append the list only inside the try block to avoid adding files that failed to be created. After all files are created, the list is joined with a semicolon and the Script Tool output is set to the resulting string with the `SetParameterAsText` method.

Example 23.7

```
# bufferAll.py
# Purpose: Buffer all the feature classes in an input folder by
#           the input distance and send the output file names to
#           the Script Tool.
# Usage: working_directory linear_unit
# Sample input: C:/gispy/data/ch23/smallDir "0.2 miles"

import arcpy, reportSTargs, sys

arcpy.env.overwriteOutput = True
arcpy.env.workspace = sys.argv[1]
distance = sys.argv[2]

fcs = arcpy.ListFeatureClasses()
outList =[]
for fc in fcs:
    reportSTargs.printArc('Processing: {0}'.format(fc))
    outputFile = fc[:-4] + 'Out.shp'
    try:
        arcpy.Buffer_analysis(fc, outputFile, distance)
        reportSTargs.printArc('Created {0}'.format(outputFile))
        outList.append(outputFile)
    except arcpy.ExecuteError:
        reportSTargs.printArc(arcpy.GetMessages())

results = ";".join(outList)
reportSTargs.printArc(results)

arcpy.SetParameterAsText(2, results)
```

Both the input and output from a script can have multivalue set to true; Just use the string `split` and `join` methods as needed. A script can also have a combinations of non-multivalue and multi-value input or output. The script just needs to treat the multi-value input or output as one entity when receiving or returning it.

23.2.2.4   Default or Schema

For 'Feature Set' or 'Record Set' data types, this property supplies a training set, a 'Schema'. For all other data types this provides a default value for the parameter. The default value is intuitive, so we'll first look at just one example with some tips on how to use the default. Sample script '08_defaultValues' is identical to sample script '01_optionalParam', except the 'Default' property has been used to specify a value for each parameter. If you run the tool without modifying any of the arguments, you'll get the following arguments:

```
Argument 0: C:\gispy\sample_scripts\ch23\scripts\reportSTargs.py
Argument 1: 5 SquareKilometers
Argument 2: MINOF
Argument 3: 'JPEG' 75
Argument 4: 1.5
Argument 5: C:\gispy\data\ch23\smallDir\trails.shp
```

The defaults must be specified in a way that the tool recognizes. This is simple for well known numerical and string data types, but for many attributes, you may need to run the tool to see the choices for default values and how they are specified. For example, try to modify the default for the first parameter, the areal unit, to 200 m^2. To find out how to specify this, run the tool with this value selected. The output prints 'Argument 1: 200 SquareMeters', that is, 200 followed by a space, followed by Square and Meters (which are not space separated). Similarly, you can also run the tool and select various values for cell size or compression to see the sets of valid values.

When default data values are specified, relative paths should be used whenever possible. Paths should be specified relative to the position of the toolbox. The default for the last parameter in '08_defaultValues' was set using a relative path. The toolbox examples for this chapter are in 'C:\gispy\sample_scripts\ch23\toolboxes'. The relative path to the 'trails.shp' dataset is up three directories and down two; Hence, the relative path is specified as '..\..\..\data\ch23\smallDir\trails.shp'. When this default value was set, it looked like this:

```
Default          ..\..\..\data\ch23\smallDir\trails.shp
```

But when the tool is opened again it shows the derived full path file name ('C:\gispy\data\ch23\smallDir\trails.shp'). This means that when you open the tool, it will not appear to be a relative path. To determine if a relative path has been used, move the toolbox to another directory, and the default path will be changed if a relative path was used.

When a parameter's data type is set to 'Feature Set' (or 'Record Set'), the fourth property listed in the property box says 'Schema' instead of 'Default'. For these parameter data types, this property supplies a template or training set (a 'Schema'). A feature set Script Tool parameter can be used to digitize new features or to select existing features. The feature can then be used as input for a Script Tool. The feature type (polygon, polyline, etc.) depends on the 'Schema'. The 'Schema' can be a template that you create with symbology tailor-made for a specific application or you can use a more generic feature class or feature layer.

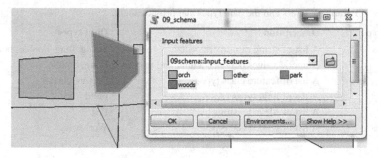

Figure 23.9 With the '09_shema' toolbox, the user can select a land cover type for digitizing polygons of those types.

Script Tool '09_schema' lists two parameters, a 'Required', 'Input' Feature Set and a 'Derived', 'Output' Shapefile. The first parameter has the 'Schema' set to a polygon layer file, 'C:\gispy\data\ch23\training\COVER4.lyr' which uses color categories for land cover types. When the tool is launched in ArcMap, the feature set parameter shows the symbology options from this layer file (Depending on the computer's speed, it may take a moment for the symbology display to populate). The user can select a land cover type and digitize a polygon with this type. In Figure 23.9 three polygons were digitized, one with 'orch' cover, one with 'other' cover and one with 'woods' cover. The Script Tool points to 'getFeature.py' (Example 23.8) which copies the features to a shapefile and sets the derived output to the result. The output is added to the map.

Example 23.8

```
# getFeatures.py
# Purpose: Copy the digitized feature set input into a shapefile.
#          and send this to the Script Tool as output.
import arcpy, sys
arcpy.env.overwriteOutput = True
fs = sys.argv[1]
outputFeat = 'C:/gispy/scratch/getFeaturesOutput.shp'
arcpy.CopyFeatures_management(fs, outputFeat)
arcpy.SetParameterAsText(1, outputFeat)
```

To use the tool to select existing features, click on the pull-down menu and select a layer map, then select one feature or hold down the shift key to select more than one feature from the selected map layer. You can try this by running the tool in 'featureSetExample.mxd' and selecting 'workzones' features in this way.

Table 23.3 Parameter filters and the data types that provide them.

Filter type	Description	Data types
Value list	A list of string or numeric values	String, long, double, boolean
Range	Numeric values between a min. and max	Long, double
Feature class	A list of feature class types	Feature class
Field	A list of field types	Field
File	A list of file extensions	File
Workspace	A list of workspace types	Workspace

23.2.2.5   Environment

This property can be used to set the default value of a parameter to an environment setting. For example, you can use the current workspace, the scratch workspace, the output coordinate system, and so forth, of the map document to set the default value for a parameter. To use this, leave the default value blank and instead select an environment setting from the drop-down list in the 'environment' property. The '10_environ' Script Tool sets the environment property of its only parameter to the scratch workspace. If this tool is run from ArcCatalog inside of ArcMap within the 'featureSetExample' map document, the default value for the 'Workspace' parameter is set to 'C:\gispy\data\ch23', because this is the value of the scratch workspace environment setting for this map.

23.2.2.6   Filter

The filter property restricts the values that can be entered. There are six types of filters and the type of filter that can be used depends on the parameter's data type as listed in Table 23.3. Other data types only give 'None' as a choice for the 'Filter' property. To use this property, click on the parameter name, then click in the filter box in the parameter property list. When you select the filter type, a GUI is launched to guide you in specifying the constraints. Script Tools '11_filterValueList' and '12_filterRangeList' use value list and range filters.

Script Tool '11_filterValueList' takes a string argument, a United States region. To help the user select a valid table, the regions are given in a value list. Double-click on the parameter and then on its filter property to see the 'List of Values' (Figure 23.10). These values are simply typed into the list; Add another item to the list and you'll see it when you run the tool. The value list creates a combo box (Figure 23.11). This tool points to 'regional.py' (Example 23.9) which prints the states in the input region. If 'New England' is selected, the output is:

```
--States in New England--
Maine
Vermont
```

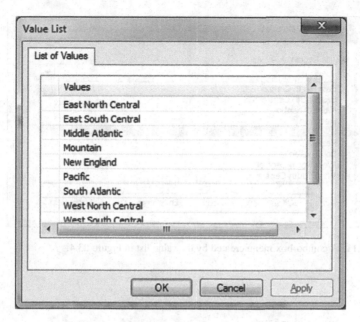

Figure 23.10 A list of values in the value list filter.

```
New Hampshire
Massachusetts
Connecticut
Rhode Island
```

Example 23.9

```python
# regional.py
# Purpose: Print the names of states in the input region.
import arcpy, reportSTargs, sys
region = sys.argv[1]
inf = 'C:/gispy/data/ch23/USA/USA_States_Generalized.shp'
fields = ['SUB_REGION', 'STATE_NAME']
sc = arcpy.da.SearchCursor(inf, fields)
reportSTargs.printArc('\n--States in {0}--'.format(region))
for row in sc:
    if row[0] == region:
        reportSTargs.printArc(row[1])

reportSTargs.printArc('\n')
del sc
```

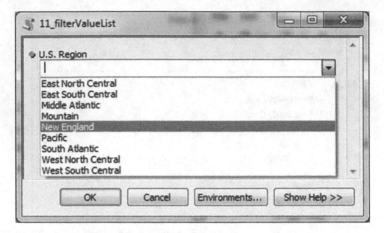

Figure 23.11 A combo-box menu created by the value list in Figure 23.4.

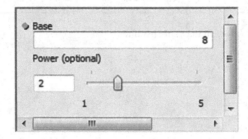

Figure 23.12 The second parameter, 'Power', is a long filtered to have a range of 1 to 5.

'12_filterRangeList' points to the 'exponentiator.py' script used in a previous example, but in this case, the exponent is restricted to the integers from 1 through 5 by using the range filter with the minimum set to 1 and the maximum set to 5. If the range filter is used for a long (integer) data type, the widget is a slider bar like the one for the parameter labeled 'Power' in Figure 23.12. For Double data types, it displays a text box, but no slider.

The remaining filters are only available for the data types by the same name. In other words, 'Feature Class' filters can only be used on 'Feature Class' type parameters, 'Field' filters can only be used on 'Field' type parameters, and so forth. The 'Feature Class' filter can be used to restrict a feature class input to point, multipoint, polygon, polyline, annotation, or dimension feature classes. '15_symbology' described in an upcoming section uses the feature class filter to limit the input feature class to polygon type files, so that the script can perform a feature to point conversion on the user input. Likewise, the 'Field' type filter can be used to restrict a field input based on a field data type (Short, Long, Float, Double, Text, and so

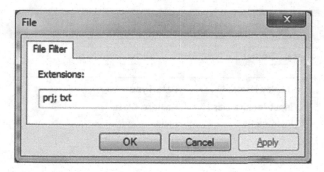

Figure 23.13   Setting a parameter file filter to only accept projection (prj) and text (txt) files.

forth). The 'File' filter serves the same purpose as the 'filetypes' option in the `tkFileDialog` module; For 'File' type parameters, it filters the type of files the user can view and select in the file browser. When you select the 'File' filter, a dialog (like the one in Figure 23.13) opens to prompt you for extensions. Enter semicolon separated extensions. Similarly, for 'Workspace' type parameters, the 'Workspace' filters the selection to file systems, local databases, or remote databases.

23.2.2.7   Obtained from

The essential idea of the 'Obtained from' property is to use information from one parameter to generate or constrain the value of another parameter. It can be used for input or derived type parameters. In Script Tool '13_inputObtainedFrom', the 'Field' values in the second parameter are obtained from the input feature class (Figure 23.14). When you run the tool the 'Field name' combo box choices change based on the 'Input feature class' selection (Figure 23.15). This property can only be set for certain parameter data types, 'Field', 'SQL expression', 'Linear Units', 'Coordinate System', and a few others.

You can also use this 'Obtained from' property to handle output data that is derived from modifying an input file. For example, input data might be modified by an update cursor or a field may be added, as mentioned in the discussion on the parameter 'Type' property. Script Tool '14_derivedObtainedFrom' points to 'combineFields.py' (Example 23.10) which adds a new field to the input dataset by combining two fields. The tool lists five parameters: an input file, two fields to combine, and a new field name (Figure 23.16). The fifth parameter is a derived output parameter obtained from the input file; It corresponds to the updated version of the input dataset. If this tool is run in ArcMap, the updated dataset is added to the map (or if that dataset is already a map layer, the layer is updated).

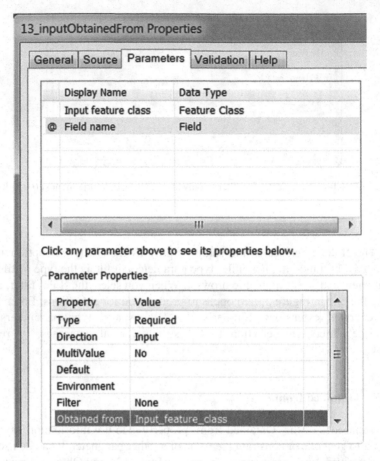

Figure 23.14 Using an input feature class to automatically populate a list of field names.

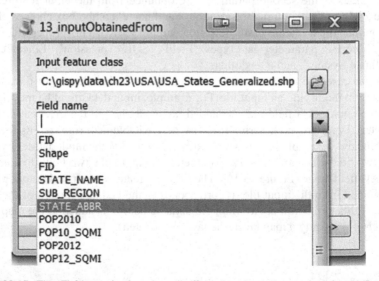

Figure 23.15 The 'Field name' values dynamically update depending on the selected 'Input feature class'.

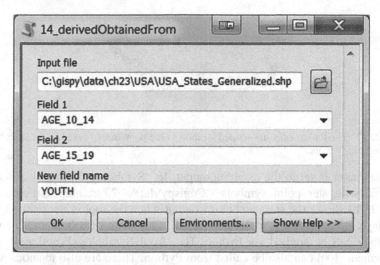

Figure 23.16 Four input parameters appear in the GUI, but the fifth one does not, because it is a derived output parameter.

Example 23.10

```
# combineFields.py
# Purpose: Create a new field that is the sum of two existing fields.
import arcpy, sys
dataset = sys.argv[1]
field1 = sys.argv[2]
field2 = sys.argv[3]
newfield = sys.argv[4]

arcpy.AddField_management(dataset, newfield)
expression = '!{0}!+!{1}!'.format(field1, field2)
arcpy.CalculateField_management(dataset, newfield,
                                expression, 'PYTHON')

arcpy.SetParameterAsText(4,dataset)
```

23.2.2.8 Symbology

The symbology property can be used to set the visual representation of output parameters. To do so, you need to use ArcMap to predefine the desired symbology and export the layer as a layer file (with an '.lyr' extension). Then you can set the symbology property of the output parameter to the location of the layer file. When the output is created, it will be added to the map and displayed with the same symbology as the layer file. Script Tool '15_symbology' points to a script called 'feature-2point.py' (Example 23.11) which takes an input polygon layer and calls the 'FeatureToPoint' tool to find the centroid of each layer. The derived output feature

Figure 23.17 The tool assigns star-shaped symbols to the point output based on the layer file in the symbology property.

class parameter is set to the resulting output. Its 'Symbology' property is set to a layer file with star point symbols ('C:\gispy\data\ch23\training\starPoints.lyr'). Figure 23.17 shows the star symbology output when the tool is run on the United States layer in 'symbologyPropExample.mxd'. This is one of several ways in which symbology can be applied via Python. The 'ApplySymbologyFromLayer' (Data Management) tool can also be called from Python. There are also methods within the `arcpy mapping` module, discussed in the upcoming chapter. All of these methods require a training dataset that has the desired symbology pre-configured.

Example 23.11

```
# feature2point.py
# Purpose: Find the centroids of the input polygons.

import arcpy, sys

arcpy.env.overwriteOutput = True
inputFile = sys.argv[1]
outputFile = 'C:/gispy/scratch/Points.shp'

# Find points based on the input.
arcpy.FeatureToPoint_management(inputFile, outputFile)

# Return the results to the tool.
arcpy.SetParameterAsText(1, outputFile)
```

23.3   Showing Progress

While a Script Tool performs lengthy processes, the user may be uncertain about the progress. To avoid this, the tool can communicate updates as progress occurs. There are several `arcpy` progress bar methods designed to provide custom feedback as the tool works. When a Script Tool is run, the geoprocessing window shows an oscillating progress bar (a trail of boxes moving to and fro) to let the user know that some progress is being made (Figure 23.18). The geoprocessing dialog title shows the name of the Script Tool (e.g., 'deleteFiles') and below that a label tells the user what the tool is doing. The default label states that the Script Tool is being executed (e.g., 'Executing deleteFiles'). With the `arcpy SetProgressor`

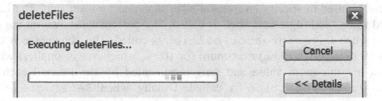

Figure 23.18 The standard progress bar and label.

Figure 23.19 The 'default' progress bar with a custom label.

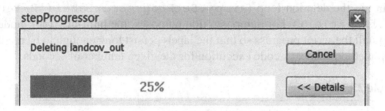

Figure 23.20 The 'step' progress bar with a custom label.

method, you can modify the behavior of this progress indicator. There are two progress modes:

1. The 'default' mode shows the same oscillating bar, but you can change the label. The following code initializes the default progress bar with a `'Hello'` message and then updates the message to `'Deleting elev_out'`:

```
arcpy.SetProgressor( 'default', 'Hello' )
arcpy.SetProgressorLabel('Deleting elev_out')
```

   Figure 23.19 shows the geoprocessing window after the second line of code is run.

2. The 'step' mode displays a percentage bar. The following code initializes the default progress bar with a `'Hello'` message and then updates the message to 'Deleting Int_rand1':

```
arcpy.SetProgressor('step', 'Hello', 0, 4)
arcpy.SetProgressorLabel('Deleting landcov_out')
arcpy.SetProgressorPosition()
```

   Figure 23.20 shows the geoprocessing window after these three lines of code are run. The third and fourth argument in the `SetProgressor` method specify a minimum and maximum value for the progressor position. An optional fifth argument can be used to specify an interval other than one for each step. The label is updated when 'SetProgressorLabel' is called. The percentage shown on the bar is updated when SetProgressorPosition is called. The percentage is calculated as $(s*interval)*100\%)/maximum$ (or 100%, whichever is smaller) where s starts at the minimum value and gets incremented by the interval each time SetProgressorPosition is called. Usually when SetProgressor is called, the minimum is set to zero and the maximum is set to the total number of steps being tracked. In this example, there are four files being deleted, so maximum is set to four. In Figure 23.20, the bar shows 25% progress because the SetProgressorPosition has been called once and the maximum is four $(1/4=25\%)$.

The 'defaultProgressor' Script Tool in the 'progressorExamples' toolbox points to 'defaultProgressor.py' (Example 23.12), which uses the default progress bar and updates the labels. This tool is based on the 'deleteFiles' tool (discussed earlier in the chapter) that deletes files as specified by the user. A few lines of code have been added for the progress commands. The script initializes the default progressor label (arcpy.SetProgressor('default', message)) and then updates the label inside the deletion loop (arcpy.SetProgressorLabel('Deleting {0}'.format(d))). For demonstration purposes, the built-in time module is used to stall the script progress so that the labels persist long enough to be read. The sleep method suspends code execution for the given number of seconds.

Example 23.12

```
# Excerpt from defaultProgressor.py
ws = arcpy.env.workspace
message = """Delete '{0}' files from {1}""".format(wildcard,ws)
arcpy.SetProgressor('default', message)
time.sleep(3)
printArc(message)
for d in data:
    try:
        arcpy.SetProgressorLabel('Deleting {0}'.format(d))
        arcpy.Delete_management(d)
        printArc('{0}/{1} deleted'.format(ws, d))
        time.sleep(3)
    except arcpy.ExecuteError:
        printArc(arcpy.GetMessages())
```

If the number of steps are known at the outset, as in our 'deleteFiles' example, which gathers a list of files to delete, each accomplishment can be reported as a percentage of the overall progress. The number of files to be processed (e.g., deleted, buffered, copied, etc.) can be used as the total number of steps. The 'stepProgressor' Script Tool points to 'stepProgressor.py' (Example 23.13), which is identical to 'defaultProgressor.py', except it demonstrates using the 'step' progress mode. The script initializes the step progressor using four arguments, the mode ('step'), the label message, the minimum step value (0), and the maximum step value (`len(data)`). The script deletes each file in a list, so the total number of files, the length of this list, is used as the maximum step value. Inside the deletion loop, the progress label is updated (`arcpy.SetProgressorLabel('Deleting {0}'.format(d))`) and the bar position is updated (`arcpy.SetProgressorPosition()`). Again, calls to the sleep method are only for demonstration purposes and can be removed for practical applications.

Example 23.13

```
# Excerpt from stepProgressor.py

# Initialize the progressor.
message = """Preparing to delete '{0}' files \
            from {1}""".format(wildcard, arcpy.env.workspace)
arcpy.SetProgressor('step', message, 0, len(data))
time.sleep(3)
printArc(message)
for d in data:
    try:
        # Update progress label
        arcpy.SetProgressorLabel('Deleting {0}'.format(d))
        arcpy.Delete_management(d)
        printArc('{0}/{1} deleted'.format(arcpy.env.workspace, d))
        time.sleep(3)
    except arcpy.ExecuteError:
        printArc(arcpy.GetMessages())
    # Update progress bar percent.
    arcpy.SetProgressorPosition()
```

23.4   Validating Input

Script Tool parameters properties and tool progress methods allow you to control tool behavior in many ways. An important advantage of providing a Script Tool graphical user interface with a script is that it constrains the values users can specify; The GUI verifies and confirms that the values are valid or it posts error messages when they are not. This is referred to as *validation*. The validation process occurs while the user is interacting with the GUI. The Script Tool automatically runs internal validation based on the parameter specifications. *Internal validation* is the set of basic checks the tool performs on parameters (E.g., Are all required inputs filled in? Does the input dataset exist? Is the input a raster when it's supposed to be raster? And so forth). This protects the user from problems that can occur due to invalid input.

Some applications may require input verification that is not included in the internal validation. A 'ToolValidator' Python class embedded within the Script Tool allows you to exact even more control over the tool. Suppose, for example, you want to update a user's choices for one parameter based on another parameter in a way that the 'Obtained From' property doesn't support, or that you want to restrict the range of a non-numeric parameter, such as a date. This kind of control can be programmed by modifying the `ToolValidator` class.

The ToolValidator has to be edited using a prescribed procedure that launches an editor from within ArcDesktop. To view the `ToolValidator` class for a Script Tool, you must open the tool properties and select the 'Validation' tab. This tab displays the `ToolValidator` class. The code can not be edited in place in the 'Validation' tab; instead, you must click the 'Edit' button on this tab to launch an editor. The geoprocessing options specify which editor is used. To modify the editor, close the Script Tool properties and select the ArcCatalog (or ArcMap) 'Geoprocessing' menu > 'Geoprocessing Options…' > 'Script Tool Editor/Debugger'. For the 'Editor', select the full path file name of the executable for your preferred IDE (e.g., pythonwin.exe or pyscripter.exe). You may need to search Windows Explorer for the executable location.

When you edit the 'ToolValidator' script, you need to follow a particular workflow for updates to take effect. We'll use the '01_favorites_um' tool to demonstrate the steps for editing 'ToolValidator' code. This tool prompts the user for a favorite positive number. Run the tool with a positive number and then a negative number to see that it accepts either input and merely prints a message either way. 'ToolValidator' code can be added to throw an error message when input is negative. Open the '01_favorites_um' tool properties and use the following steps to edit the tool:

1. Open the tool properties (Right-click on the tool > 'Properties')
2. Select the 'Validation' tab.
3. Click 'Edit…'. The script will open in your IDE.
4. Modify the script by adding three lines of code to the `updateMessages` method as shown in Example 23.14. (We'll return to the details of this code shortly.)
5. Save the script in the IDE and close it.
6. Click OK on Validation tab.
7. Run the tool to test the results. Run the tool with a negative number to see the error message the new code created.

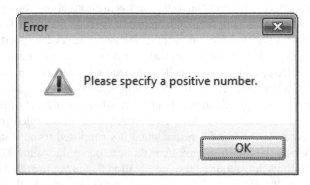

Example 23.14: ToolValidator code from Script Tool, '01_favorites_um', in validatorExamples.tbx.

```python
# Setting an error message
def updateMessages(self):
    """Modify the messages created by internal validation for
    each tool parameter.    This method is called after
    internal validation."""
    if self.params[0].altered:
        if self.params[0].value <= 0:
            self.params[0].setErrorMessage("Please specify \
                                            a positive number.")
    return
```

This example demonstrated how to access and modify the `ToolValidator` class. Next we'll discuss its methods and attributes.

23.4.1 The ToolValidator Class

Example 23.15

```python
class ToolValidator(object):
    """Class for validating a tool's parameter values and
    controlling the behavior of the tool's dialog."""

    def __init__(self):
        """Setup arcpy and the list of tool parameters."""
        self.params = arcpy.GetParameterInfo()

    def initializeParameters(self):
        """Refine the properties of a tool's parameters. This method is
        called when the tool is opened."""
        return

    def updateParameters(self):
        """Modify the values and properties of parameters before
        internal validation is performed. This method is called
        whenever a parameter has been changed."""
        return

    def updateMessages(self):
        """Modify the messages created by internal validation
        for each tool parameter. This method is called after
        internal validation."""
        return
```

The `ToolValidator` class has four methods (`__init__` , `initialize Parameters`, `updateParameters`, and `updateMessages`) and a class property named `params`. By default, three of the methods are empty except for docstrings and the `return` keyword. These methods can be modified but should not be deleted or renamed and the `return` keywords should not be deleted. The structure of the 'ToolValidator' class (shown in Example 23.15) should look familiar from the chapter on user-defined classes. The `__init__` method, is activated when an instance of the `ToolValidator` class is created. The Script Tool performs this instantiation behind the scenes when the tool is launched. The other three methods are also triggered automatically by the Script Tool, based on various events. Most code that you would want to add involves using the `self.params` list, which contains a list of `Parameter` objects. To understand the modifications that can be made to the `ToolValidator` methods, you need to be familiar with the `Parameter` object properties and methods.

23.4.1.1   Initializing Parameter Objects

The modifications that you can make usually involve the class attribute named `params` which is defined in the `__init__` method. The `getParameterInfo` `arcpy` method returns a list of `Parameter` objects and this is stored in `self. params`:

```
def __init__ (self):
    self.params = arcpy.GetParameterInfo()
```

This list of `Parameter` objects corresponds to the parameters in the list on the Script Tool properties 'Parameters' tab, with zero-based indexing starting at the top of the list. For example, in the '01_favorites_um' Script Tool, self.params[0] refers to the 'Favorite positive number' parameter and self.params[1] refers to the 'Favorite color' parameter (Figure 23.21).

`Parameter` objects have some of the properties that are specified on the 'Parameters' tab, such as, `name` (Display name with underscores substituted for spaces), `direction`, `datatype`, and `symbology`. These are read-only properties. Other properties, such as, `category`, `value`, `enabled`, `altered`, and `filter.list` can be modified. Search for 'ToolValidator Parameter object' on the

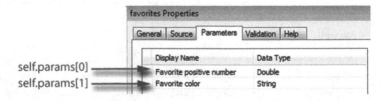

Figure 23.21 The `self.params` Python list corresponds to the tool parameter list.

ArcGIS Resources site for documentation of the `Parameter` object properties and methods. The upcoming examples demonstrate ways to use these properties and methods.

23.4.1.2 An `updateMessages` Example

Let's return to Example 23.14. This code forces the user to enter a positive favorite number into the '01_favorites_um' Script Tool, using the `updateMessages` method. As the docstring states, the `updateMessages` method is called just after internal validation (the set of checks automatically performed by the tool). If an invalid input is used, internal validation detects this problem and throws an error when you update the parameter or try to run the tool. For some applications, you may want to add your own checks for conditions that can't be constrained by other means. For example, though you can use a filter to force the user to select a number between 1 and 5, you can't use the filter to specify an open ended range, but `updateMessages` can enforce this. The purpose of `updateMessages` is to enable custom checks, errors, and warning messages. Example 23.14 uses `altered` and `value` Parameter properties and the `setErrorMessage` Parameter method. The `params` variable is a `ToolValidator` class property, so when you refer to it inside of class methods, you must use `self.params`. The first line of code inside `updateMessages` in Example 23.14 checks if the first parameter in the list has been `altered` by the user:

```
if self.params[0].altered:
```

The `altered` Parameter property is Boolean (`True` or `False`). When the tool is first launched, the `altered` property for each parameter is `False`. This property is set to `True` when the tool detects a change (and back to `False` when appropriate). The next line of code compares the first parameter's `value` to zero:

```
if self.params[0].value <= 0:
```

The Parameter property named `value` has the value passed in by the user. In the Python `sys.argv` variable, all arguments are received as strings. By contrast, the data type of the `value` property is a `Value` object and the behavior depends on the data type of the input. For parameters with an Esri data type in Table 23.1, the Python equivalent data types are used. Many data types are received as strings. However, the `value` of the user's favorite number does not need to be cast to 'float' before it is compared to zero because `self.params[0].value` is derived from a 'Double' type parameter. If the value of this number is not positive, `setErrorMessage` is called:

```
self.params[0].setErrorMessage('Please specify a positive number.')
```

The `setErrorMessage` method is one of several Parameter methods related to messaging (Other methods include `setWarningMessage`, `clearMessage`, `hasError`, and so forth). The `setErrorMessage` method opens a window containing the message and the tool can not be run until the input is changed to a positive value. Putting it all together, the code checks if the parameter has been altered, if so, it checks if its value is positive. If not, it displays a message associated with that parameter:

```
if self.params[0].altered:
    if self.params[0].value <= 0:
        self.params[0].setErrorMessage('Please specify a positive \
                                number.')
```

Generally, the code in `updateMessages` should check a parameter's `altered` condition before checking the parameter's `value`. If no default is set for a parameter and the script doesn't check if it has been altered, an error may appear before the user even has an opportunity to specify a value. To see this occur, remove the statement `if self.params[0].altered:` from the '01_favorites_um' ToolValidator code (dedent the code below it as needed) and then run the tool.

23.4.1.3   An `initializeParameters` Example

The `initializeParameters` method is called just one time. The call occurs when the tool is first launched (after the `ToolValidator` object has been created). This method is only ever called once during a single tool run, it can be used as an alternative means to set default values. Default values can either be set on the 'Parameters' tab using the 'Default or schema' property or they can be set in the `ToolValidator` class using the `Parameter` object `value` attribute. If the default is set in both ways, the value used in the `ToolValidator` code will override whatever value was set on the 'Parameters' tab.

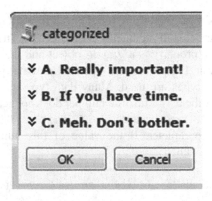

Figure 23.22 The 'categorized' Script Tool has three parameter categories, shown here collapsed.

The `initializeParameters` method can also be used to set up parameter categories. Figure 23.22 shows a tool with three categories. Parameters grouped in categories can be collapsed and expanded. The user can click on the arrows next to the category name to expand or collapse a category. This can help to keep the size of the GUI manageable and organize the parameters logically. Example 23.16 sets up three categories. The code loops through the parameters and sets the `category` property of each `Parameter` object. By looping through the indices and checking the parameter index, the `category` for the first third of the parameters is set to `'A. Really important!'`. The `category` for the middle third is set to `'B. If you have time.'`, and for the last third, it's set to `'C. Meh. Don't bother.'` for the last third. Tools with numerous parameters sometimes group parameters into 'Required' and 'Optional' categories by checking their `parameterType`. This is posed as an exercise at the end of the chapter.

Example 23.16: Code in the 'validatorExamples.tbx/02_categories_ip' ToolValidator.

```
def initializeParameters(self):
    """Assign parameter categories."""
    numParams = len(self.params)
    for index in range(numParams):
        if index < numParams/3.0:
            self.params[index].category = 'A. Really important!'
        elif index < (2*numParams)/3.0:
            self.params[index].category = 'B. If you have time.'
        else:
            self.params[index].category = "C. Meh. Don't bother."
    return
```

23.4.1.4 An `updateParameters` Example

The `updateParameters` method is called each time a parameter is changed. This method can be used to dynamically update one parameter based on another parameter. The '03_rasters_up' Script Tool is designed to allow the user to select a workspace and then select rasters within the workspace. The tool has two parameters, a workspace parameter and a multivalue string parameter. The string data type is used for the second parameter because the string data type allows a value list filter. The code shown in Example 23.17 uses the `filter.list` Parameter property to set the list of items in the filter. The code in `updateParameters` updates the value list of the second parameter based on the first. The code gets a list of rasters in the workspace specified by the user and sets the `filter.list` property to the list of raster names. The `updateMessages` method posts an error if the list is empty (in other words, the selected workspace contains no rasters).

Example 23.17: ToolValidator code from Script Tool, 'validatorExamples.tbx/ 03_rasters_up'.

```python
def updateParameters(self):
    '''Initialize raster list.'''
    if self.params[0].altered:
        arcpy.env.workspace = self.params[0].value
        rasts = arcpy.ListRasters()
        if rasts:
            self.params[1].filter.list = rasts
        else:
            self.params[1].filter.list = []

    return

def updateMessages(self):
    '''Check for rasters.'''
    if self.params[0].altered:
        if not self.params[1].filter.list:
            self.params[0].setErrorMessage('This directory \
                              does not contain any rasters.')
    return
```

When you save the ToolValidator script and select 'Apply', the script is checked internally. If there is a problem with the modified code, the process may hang until you cancel the update (by clicking the 'Cancel' button). Some common mistakes, such as omitting `self` or `value` when needed obstruct the script from being applied. For example, the following code exhibits these two mistakes:

```python
# Incorrect...
self.params[0] = 5
params[1].filter.list = [1,2,3]

# Correct...
self.params[0].value = 5
self.params[1].filter.list = [1,2,3]
```

The `ToolValidator` class can be debugged by creating a stand-alone script that imports the toolbox, specifies the tool, sets the input parameters, instantiates a `ToolValidator` object, and invokes the validation methods. An alternative to the Script Tool referred to as a 'Python toolbox' makes debugging easier by holding all the information within a text file.

23.5   Python Toolboxes

A *Python toolbox* is an ASCII text file with a '.pyt' extension that contains Python code to define a toolbox and one or more tools. Create one in a directory in ArcCatalog (10.1 or higher) with right-click > New > Python toolbox. In ArcCatalog, the Python toolbox (named 'Toolbox.pyt' by default) is displayed with a toolbox icon and expands to show the tools it contains (by default, it contains one named 'Tool'). The Python toolbox icon shows a scroll by the toolbox and the tool uses the same icon as Script Tools:

A tool in a Python toolbox can do everything that a Script Tool can do, but it streamlines the workflow of creating a GUI. With the Script Tool, each parameter must be specified by clicking and selecting the properties on the 'Parameters' tab of the Script Tool wizard. The Python code associated with a Script Tool is contained in a separate file; Whereas, the code for a Python toolbox and one or more tools can all be contained in a single '.pyt' file (It can also import user-defined modules to invoke code in other Python scripts). Script Tools are a convenient way to learn about the GUIs that can be made with ArcGIS, but Python programmers may prefer Python toolboxes, since they are entirely Python based.

Initially, the code in a Python toolbox defines two classes, `Toolbox` and `Tool`. Example 23.18 shows the default Python code generated by creating a Python toolbox. To view or edit this code, right click on the toolbox in ArcCatalog and select 'Edit...'. (Or to open it directly from an IDE, select Open > change the file type to 'All files' > browse to the '.pyt' file). The `Toolbox` class sets the toolbox label and alias. It also sets `self.tools`, to a list containing one tool class. You can add more tools by making a copy of the `Tool` class, pasting it at the bottom of the file, renaming it (e.g., `Tools2`), and then adding the new tool class to the tools list (e.g., `self.tools = [Tool, Tool2]`). Each tool class must contain six methods, `__init__`, `getParameterInfo`, `isLicenced`, `updateParameters`, `updateMessages`, and `execute`. The `__init__` method sets the tool label and description. The label can differ from the name of the class, but the tool list in the Toolbox class must use the class name, not the label. The `getParameterInfo` method initializes the parameters. The `isLicensed` method can be used to keep a tool from being run if the required geoprocessing tools are not available. The `updateParameters` and `updateMessages` method are for validation (like their Script Tool `ToolValidator` equivalents). Lastly, the code that you want to run when the tool is executed can be placed in the `execute` method.

Example 23.18: Python toolbox template.

```python
import arcpy

class Toolbox(object):
    def __init__(self):
        '''Define the toolbox (name of the toolbox is the
        name of the '.pyt' file).'''
        self.label = 'Toolbox'
        self.alias = ''

        # List of tool classes associated with this toolbox
        self.tools = [Tool]
class Tool(object):
    def __init__(self):
        '''Define the tool (tool name is the name of the class).'''
        self.label = 'Tool'
        self.description = ''
        self.canRunInBackground = False

    def getParameterInfo(self):
        '''Define parameter definitions'''
        params = None
        return params

    def isLicensed(self):
        '''Set whether tool is licensed to execute.'''
        return True

    def updateParameters(self, parameters):
        '''Modify parameters before internal validation. Called
        whenever a parameter has been changed.'''
        return

    def updateMessages(self, parameters):
        '''Modify messages created by internal validation. Called
        after internal validation.'''
        return

    def execute(self, parameters, messages):
        '''The source code of the tool.'''
        return
```

The Python toolbox, 'rasterToolbox.pyt' in the 'pyToolboxes' directory contains a single tool class, RastersExample. This tool calculates the Sine of the input rasters. The first parameter gets the user workspace. When this is selected, the rasters within this workspace are listed (using code similar to Example 23.17). Then when the tool is run, it calculates the Sin of each of the rasters selected in the second parameter. We'll use the RastersExample class to compare Script Tools to the tools in Python toolboxes.

> **Note** To edit a Python toolbox, right-click on the toolbox, not the tool(s) it contains.

23.5.1 Setting Up Parameters (`getParameterInfo`)

The biggest difference between the Script Tool and the Python toolbox is that you set up the parameters in the code inside the `Tool` class in the `getParameter-Info` method. This replaces the 'Parameters' tab on the Script Tool wizard. For each parameter, you create a `Parameter` object and set its properties. For example, the following code creates `myParam`, a `Parameter` object:

```
myParam = arcpy.Parameter()
```

Then the following code sets its `name` property:

```
myParam.name = 'My_precious'
```

The parameter name can not contain spaces or special characters. The parameter name is the only required property; However, the `displayName`, `parameter-Type`, `direction`, and `datatype` are conventionally set as well (by default, the display name is set to `None`, the parameter type is set to `'Required'`, the direction is set to `'Input'`, and the data type is set to `'String'`). There is an equivalent Python toolbox Parameter property for each one displayed in the Script Tool property list. The parameters are created and initialized inside the `getParame-terInfo` method and then this function returns a list of the `Parameter` objects. Examples 23.19 shows the `getParameterInfo` method from the 'RastersExample' tool. Both parameters are required input. The first one is a workspace, the second is a string. A workspace filter list is used to restrict the workspace selection to local databases (file or personal geodatabases). In the Script Tool, this would be done by clicking on filter>workspace>Local Database. The second parameter allows multiple values by setting the `multiValue` property to `True`. The last line of the method returns the two `Parameter` objects in a Python list as follows:

```
return [param1, param2]
```

This list is passed in as an argument in the `updateParameters`, `updateMes-sages`, and `execute` methods as discussed shortly.

Example 23.19

```
def getParameterInfo(self):
    '''Set up the parameters and return
    the list of Parameter objects.'''
    # 1_Select_a_workspace_containing_rasters
    param1 = arcpy.Parameter()
    param1.name = '1_Select_a_workspace_containing_rasters'
    param1.displayName = '1. Select a workspace \
                          containing rasters:'
    param1.parameterType = 'Required'
    param1.direction = 'Input'
    param1.datatype = 'Workspace'
    param1.filter.list = ["Local Database"]

    # 2_Select_rasters_within_the_workspace
    param2 = arcpy.Parameter()
    param2.name = '2_Select_rasters_within_the_workspace'
    param2.displayName = '2. Select rasters within \
                          the workspace:'
    param2.parameterType = 'Required'
    param2.direction = 'Input'
    param2.datatype = 'String'
    param2.multiValue = True
    param2.filter.list = []

    return [param1, param2]
```

23.5.2   Checking for Licenses (`isLicensed`)

The `isLicensed` method can be used to check if an extension license is available and prevent the tool from running. The 'RastersExample' tool uses the Spatial Analyst extension to calculate the Sine. This is not available in all ArcGIS desktop products. The code in Example 23.20 prevents the user from running the tool if the Spatial Analyst extension is not licensed.

Example 23.20

```
def isLicensed(self):
    """Prevent the tool from running if the Spatial Analyst extension
    is not available."""
    if arcpy.CheckExtension('Spatial') == 'Available':
        return True # tool can be executed
    else:
        return False # tool can not be executed
```

23.5.3 Validation (*updateParameters* and *updateMessages*)

Like Script Tools, Python toolbox tools perform internal validation and the code can implement custom evaluation. The `updateParameters` and `updateMessages` method have the same functionality as their Script Tool ToolValidator equivalents. Example 23.21 shows the validation included in the `RastersExample` tool class. Comparing this to the code in Example 23.17, they are almost identical. The only difference is how the parameters are referenced in the code. In Example 23.17 the first parameter is `self.params[0]`. In Example 23.21, the first parameter is `parameters[0]`. The Script Tool `ToolValidator` class stores the list of `Parameter` objects as a class property. The `self.params` list is set in the `__init__` method as follows:

```
self.params = arcpy.GetParameterInfo()
```

The Python toolbox `Tool` class handles the parameter list differently. Instead of using a property, the `Tool` class passes the parameter list into the methods that use it. For example, it is the second item in the argument lists for both `updateParameters` and `updateMessages`:

```
def updateParameters(self, parameters):
def updateMessages(self, parameters):
```

This means, you simply index into this 'parameters' list without using the self-dot notation. The `Parameter` objects are the same data type as those used in the Script Tool ToolValidator class, so all of the same methods and properties apply (e.g., `value`, `filter.list`, `altered`, `setErrorMessage`, `enabled`, and so forth...) To show some additional examples comparing Script Tool validation to Python toolbox validation, Python toolbox 'validatorExConverted.pyt' (in the 'pyToolboxes' directory) is equivalent to the 'validatorExamples.tbx' (in the 'toolboxes' directory). For the most part, the validation code is the same, other than the `Parameter` object handling. Unlike the ToolValidator, the Python toolbox has no `initializeParameters` method. This functionality can usually be placed inside the `getParameterInfo` method for a tool in a Python toolbox. Example 23.16 used `initializeParameters` to categorize the parameters. The 'Categories_ip' tool sets these categories in its `getParameterInfo` method before returning the parameter list.

Example 23.21

```
def updateParameters(self, parameters):
    '''Initialize raster list.'''
    if parameters[0].altered:
        arcpy.env.workspace = parameters[0].value
        rasts = arcpy.ListRasters()
```

```
        if rasts:
            parameters[1].filter.list = rasts
        else:
            parameters[1].filter.list = []
    return

def updateMessages(self, parameters):
    '''Check for rasters.'''
    if parameters[0].altered:
        if not parameters[1].filter.list:
            parameters[0].setErrorMessage('This directory does not \
                                    contain any rasters.')
    return
```

23.5.4   Running the Code (*execute*)

A Python toolbox tool is launched by double-clicking on the tool in ArcCatalog. Then values for the parameters are specified in the GUI. Once valid values are selected, the tool can be run by selecting the 'OK' button. When the 'OK' button is selected, the tool class `execute` method is called. This means that when you create a tool for Python toolbox, the code that you want it to run should be placed in the `execute` method. Script Tools point to a stand-alone scripts; By contrast, in a Python toolbox, the main script code can be placed directly in this text file. As in the `updateParameters` and `updateMessages` methods, the list of `Parameter` objects is passed into this method as the variable named `parameters`. Example 23.22 shows the `execute` method for the 'RastersExample' tool. This code gets the list of user-selected rasters and computes the trigonometric Sine of each raster. The code checks out the Spatial Analyst extension, turns on output overwriting, and sets the workspace based on the value of the first parameter. Next, it loops through the list of raster names selected by the user. Instead of using the `value` property, this line of code uses the `values` property:

```
for rast in parameters[1].values:
```

The `values` property of a multivalue string parameter returns a list of strings. The code inside the loop computes and saves an output raster for each name in the list.

Example 23.22

```
def execute(self, parameters, messages):
    '''Calculate the Sine of each input raster.'''
    arcpy.CheckOutExtension('Spatial')
    arcpy.env.overwriteOutput = True
    arcpy.env.workspace = parameters[0].value # Set the workspace
```

```
for rast in parameters[1].values:
    try:
        outSin = arcpy.sa.Sin(rast)
        outSin.save(rast + '_Sin')
        message = '{0}_Sin created in {1}.'.format(rast,
                                        arcpy.env.workspace)
        arcpy.AddMessage(message)
    except:
        message = '{0}_Sin could not be created.'.format(rast)
        arcpy.AddMessage(message)
```

The main code for the execute method can also be placed in a stand-alone user-defined module within one or more functions. Then the module can be imported and its functions called from within the execute method. For example, 'RastersToolboxVersion2.pyt' imports rastModule and calls batchSine. The following concise code is the resulting execute module:

```
def execute(self, parameters, messages):
    '''Calculate the Sine of each input raster.'''
    import rastModule
    wkspace = parameters[0].value
    rasters = parameters[1].values
    rastModule.batchSine(wkspace, rasters)
```

The user-defined module 'rastModule' is shown in Example 23.23. This approach of modularizing execute code improves organization and maintainability. If the GUI is added last, this approach simplifies integration with existing code.

Example 23.23

```
# rastModule.py
import arcpy
def batchSine(workspace, rastList):
    '''Calculate the Sine of each raster in the list.'''
    arcpy.CheckOutExtension('Spatial')
    arcpy.env.overwriteOutput = True
    arcpy.env.workspace = workspace # Set the workspace
    for rast in rastList:
        try:
            outSin = arcpy.sa.Sin(rast)
            outSin.save(rast+'_Sin')
            message = '{0}_Sin created in {1}.'.format(rast,
                                            arcpy.env.workspace)
            arcpy.AddMessage(message)
        except:
            message = '{0}_Sin could not be created.'.format(rast)
            arcpy.AddMessage(message)
```

23.5.5   *Comparing Tools*

To see a few additional differences between Script Tools and Python toolbox tools, compare standard toolbox, 'propertyExamples.tbx' (in the 'toolboxes' directory) with Python toolbox 'propertyExConverted.pyt' (in the 'pyToolboxes' directory). The tools in the '.pyt' files are equivalent to those in the 'tbx' files. The Python toolbox contains one Python class for each tool. The tool class of 'propertyExConverted.pyt' sets a long list of class names (one for each of the 15 tools):

```
self.tools = [ToolOptionalParam1, ToolOptionalParam2, ToolRequiredOutput,
            ToolDerivedOutput1, ToolDerivedOutput2, ToolmultiValueIn,
            ToolMultiValueOut, ToolDefaultVals, ToolSchema,
            ToolEnvironments, ToolFilterRange, ToolFilterValueList,
            ToolInputObtainedFrom, ToolDerivedObtainedFrom,
            ToolSymbology]
```

Tool names and labels differ slightly in this example since Python class names can not start with a number. For example, the `ToolDerivedOutput2` class creates the '05_derivedOutput2' tool. The code in the `execute` method for this tool (line 249) illustrates one of the things that you should be aware of when writing code for Python toolboxes. The value property of a parameter is a `Value` object. Functions that take strings as input, such as, `os.path.dirname` can not use the `Value` object directly. Instead, you need to use its `value` property to get a string. The following code from the `execute` method for 'ToolDerivedOutput2' sets the 'fileToBuffer' variable in this way:

```
fileToBuffer = parameters[0].value.value
arcpy.env.workspace = os.path.dirname(fileToBuffer)
```

Using `value.value` returns the text string that names the object (instead of the `Value` object). In this case, it is the full path file name of the feature class selected by the user. The `os.path.dirname` function can derive a directory name from this string; whereas, it can not derive a directory name from a `Value` object. The following code throws an error:

```
fileToBuffer = parameters[0].value
arcpy.env.workspace = os.path.dirname(fileToBuffer)
```

To work correctly, some ArcGIS tools require the text string for input, instead of the object. For example, the Calculate Field (Data Management) tool is used in the `ToolDerivedObtainedFrom` class (Example 23.24). If the text name is not used, the tool throws an error saying that the field that was just created does not exist. Not all `Value` objects have a `value` property, so it can't be used all the time, but you can keep in mind that unexpected errors may be resolved by using the text name (as `value.value`) instead of the `Value` object.

Example 23.24

```
def execute(self, parameters, messages):
    # From combineFields.py
    # Purpose: Create a new field that is the sum
    #          of two existing fields.
    dataset = parameters[0].value.value
    field1 = parameters[1].value
    field2 = parameters[2].value
    newfield = parameters[3].value
    arcpy.AddField_management(dataset, newfield)
    expression = '!{0}!+!{1}!'.format(field1, field2)
    arcpy.CalculateField_management(dataset, newfield,
                                    expression, 'PYTHON')

    arcpy.SetParameterAsText(4,dataset)
```

The `ToolDerivedObtainedFrom` class demonstrates another difference between Script Tools and Python toolbox tools. The Python toolbox equivalent of the 'Obtained from' property is `parameterDependencies`. Example 23.25 shows an excerpt from the `getParameterInfo` method. The first parameter is an input 'Feature Class' and the second one is a 'Field'. The tool allows the user to select an input feature class and then a field from that dataset. The fields in the second parameter are obtained from the first parameter by setting the `parameter-Dependencies`. This property takes a Python list of `Parameter` object names. The last line of code in Example 23.25 sets the dependencies of the second parameter to a list containing the first `Parameter` object name (`[param1.name]`). When the user selects the input dataset, the second parameter populates a combo box with the dataset's field names. The fifth parameters in this tool, the derived output dataset, is set up the same way:

```
param5.parameterDependencies = [param1.name]
```

Example 23.25: The `parametersDependencies Parameter` property is equivalent to the 'Obtained from' Script Tool property.

```
def getParameterInfo(self):
    '''Create parameters and set their properties'''
    # Input_file
    param1 = arcpy.Parameter()
    param1.name = 'Input_file'
    param1.displayName = 'Input file'
    param1.parameterType = 'Required'
```

```
param1.direction = 'Input'
param1.datatype = 'Feature Class'

# Field1
param2 = arcpy.Parameter()
param2.name = 'Field1'
param2.displayName = 'Field 1'
param2.parameterType = 'Required'
param2.direction = 'Input'
param2.datatype = 'Field'
param2.parameterDependencies = [param1.name]
```

In Python toolboxes, filter lists are still available and they are accessed with the same notation as in the Script Tool `ToolValidator`, with the `filter.list` `Parameter` property. The 'ToolFilterValueList' tool hard codes the region names with a Python list of strings as in the following code:

```
param1.filter.list = ['East North Central', 'East South Central',
'Middle Atlantic', 'Mountain',
'New England', 'Pacific', 'South Atlantic',
'West North Central', 'West South Central']
```

The range filter for a 'Long' type parameter can be set by specifying the filter type as `'Range'` and setting the filter list as a Python list of two values. For example, the 'ToolFilterRange' tool sets the values of the second parameter to range between one and five with the following code:

```
param2.filter.type = 'Range'
param2.filter.list = [1,5]
```

Python toolboxes can still use relative paths for default datasets, workspaces, schema files, and so forth, but to use a relative path, the code must calculate the absolute path dynamically. For example, to set a default shapefile within the 'ToolDefaultVals' tool, the code uses a custom function named `getAbsPath`. As shown in Example 23.26, `getAbsPath` derives an absolute path from a relative path by using the path of the Python toolbox. Custom functions can be added to a '.pyt' file, just as they can to any other Python script. To make sure they are defined before they are used, functions should be placed just after the import statements, before the class definitions. The following code, from the `ToolDefaultVals` class `getParameterInfo` method, sets a default value for an input dataset using a relative path and dynamically deriving its absolute path by calling the `getAbsPath` function:

```
relativePath = '../../../data/ch23/smallDir/trails.shp'
param5.value = getAbsPath(relativePath)
```

Example 23.26

```
def getAbsPath(relativePath):
    '''Return the absolute path given a relative path to this file'''
    tbxPath = os.path.abspath(file
    tbxDir = os.path.dirname(tbxPath)
    fullPath = os.path.join(tbxDir, relativePath)
    return os.path.abspath(fullPath)
```

23.6 Discussion

Script Tools and Python toolboxes provide a convenient means for creating graphical user interfaces for ArcGIS scripts. Though they have some limitations, they are well-suited for making GUIs to collect ArcGIS data types. Script Tools are a good starting point for learning to use parameter data types and their properties. Python toolboxes facilitate a more streamlined workflow for creating tools by using Python code, instead of a tool wizard, to specify parameter properties. Setting up validation is also more efficient in a Python toolbox tool. When you modify the code, you can save it, refresh the toolbox, and see the results. You can leave the saved '.pyt' file open while testing the tool and you can simply edit it again as needed. Each time you edit Script Tool validation code, you need to open the tool properties, select the 'Edit' button on the validation tab, edit and save the code, apply the changes, close the tool properties, and then run the tool. When you make changes to the code in Python toolboxes, you do need to refresh the file in ArcCatalog to see the changed behavior.

A red X on a tool icon in ArcCatalog means the tool won't run. When you see an X on a Script Tool, it often means that the path to a script was not set as relative or the script has been moved to a different relative position. If this is the case, you can update the 'Script' path to fix the tool (and check 'Use relative pathnames, if desired). Python toolboxes help to avoid this common mistake since they define the toolbox and tools in one file. If the code in a Python toolbox contains an error, a red X will appear as well. For certain kinds of errors, a 'Why' tip will appear on the tool. You can select this tip to learn more about the error (see 'zzz.pyt' for an example).

ArcGIS hosts a downloadable tool for exporting Script Tools to Python toolboxes, but similar to exporting ModelBuilder models to Python scripts, the user needs to have a good understanding of Python to refine the resulting '.pyt' file for full functionality. The template along with sample Python toolboxes provide a good starting point for creating new Python toolboxes.

Graphical user interfaces should usually be added near the end of a GIS project, once the programming functionality has been developed and tested. GUIs are convenient for users, not for developers.

23.7   Key Terms

Script Tool   Validation
Geoprocessing Window   Internal validation
Widget   Python toolbox
Combo box   `Parameter` object
`ToolValidator` class   `Value` object

23.8   Exercises

General instructions: Exercises 1–11 involve creating new Script Tools or copying and modifying existing Script Tools. The toolbox, 'exercises.tbx' in the 'C:\gispy\sample_scripts\ch23\exercises' directory contains sample Script Tools used in the exercise ('exerciseCopy.tbx' is a copy of this one, in case you need to revert to the original state). Place Python scripts in the same directory as the toolboxes and set the Script Tool to use relative path names for scripts. Exercises 12 and 13 involve creating Python toolboxes. Either write all code for these inside the '.pyt' file or import files in the same directory. Display all printed output for exercises 1–13 in both the Interactive Window and the Geoprocessing Window.

1. **triArea** and **triangleArea.py** Write a script, 'triangleArea.py', to calculate the area of the triangle based on the user input of a triangle base and width. Create a Script Tool, 'triArea', that has two float parameters, the base and width of a triangle. Point the tool to the 'triangleArea.py' script.

    Example input: 5.6 9

    Example output:
    ```
    >>> The area of the triangle is 25.2.
    ```

2. **bridgeOfDeath** and **answers.py** Create a script, 'answers.py', to print the user's name, quest, and favorite color, as shown in the example. Create a Script Tool, 'bridgeOfDeath', that poses three questions: 'What is your name?', 'What is your quest?', and 'What is your favorite color?' The tool should have three string parameters to collect the answers. Point the tool to the 'answers.py' script.

    Example input: "Sir Lancelot" "to seek the Holy Grail" blue

    Example output:
    ```
    >>> Your name is Sir Lancelot.
    Your quest is to seek the Holy Grail.
    Your favorite color is blue.
    ```

3. **filterPractice** and **filterFun.py** Write a script, 'filterFun.py' that prints the index and value of the arguments, as shown in the example. Then create a Script Tool, 'filterPractice', that points to the script and use filters to allow the user to:

(a) Choose 'hehe', 'hoho', or 'woot' from a combo box.

(b) Choose an integer in the range of −5 to 5.

(c) Choose one or more Point feature classes.

(d) Choose a text file with a '.csv' or '.txt' extension.

(e) Choose a workspace that is either a file geodatabase or a personal geodatabase.

Example input:
woot -3 C:/gispy/data/ch23/tester.gdb/c1;C:/gispy/data/ch23/tester.gdb/c2
C:/gispy/data/ch23/smallDir/poem.txt C:/gispy/data/ch23/rastTester.gdb/

Example output:
```
>>> Argument 0: C:\gispy\sample_scripts\ch23\exercises\filterFun.py
Argument 1: woot
Argument 2: -3
Argument 3:
C:/gispy/data/ch23/tester.gdb/c1;C:/gispy/data/ch23/tester.gdb/c2
Argument 4: C:/gispy/data/ch23/smallDir/poem.txt
Argument 5: C:/gispy/data/ch23/rastTester.gdb/
```

4. **check4Value** Sample script 'C:\gispy\sample_scripts\ch23\exercises\freqCount.py' counts the number of occurrences of a particular value of a given text field in a file. It takes three argument: a file name, a field name, and a field value. Modify the script so that it prints the count in both the Interactive Window and the Geoprocessing Window. Next, create a Script Tool, 'check4Value' that points to 'freqCount.py' and has three parameters:

(a) The first parameter (File name) should be a required table. Set the default value to C:/gispy/data/ch23/tester.gdb/park.

(b) The second parameter (Field name) should be required input with choices obtained from the first parameter. Set the filter so that only 'text' type fields are displayed.

(c) The third parameter (Field value) should be a required string. Set the default value to 'woods'.

Example input:
C:\gispy\data\ch23\tester.gdb\park COVER woods

Example output:
```
>>> Arg 0 = C:\gispy\sample_scripts\ch23\exercises\freqCount.py
Arg 1 = C:\gispy\data\ch23\tester.gdb\park
Arg 2 = COVER
Arg 3 = woods
The COVER field in C:\gispy\data\ch23\tester.gdb\park contains
213 occurrences of woods.
```

5. **batchCount** and **countEach.py** Write a script, 'countEach.py' that uses the Get Count (Data Management) tool and reports the number of records in each input table, as shown in the example. Then create a Script Tool, 'batchCount', that points to the script and has a multivalue input table parameter.

Example input:
C:\gispy\data\ch23\smallDir\trails.shp;
C:\gispy\data\ch23\smallDir\wheatYield.txt;
C:\gis py\data\ch23\training\xyData2.shp

Example output:
```
>>> C:\gispy\data\ch23\smallDir\trails.shp contains 4 records.
C:\gispy\data\ch23\smallDir\wheatYield.txt contains 124 records.
C:\gispy\data\ch23\training\xyData2.shp contains 8 records.
```

6. **smoother** and **smoothLine.py** Write a script 'smoothLine.py' to smooth the input linear feature class using the 'PAEK' smoothing algorithm and the smoothing tolerance given by the user. Then create a Script Tool, 'smoother', that points to the script and has three parameters:

   - The first parameter (Input feature class) should be a required Polyline feature class. Set the default value to C:/gispy/data/ch23/smallDir/trails.shp.
   - The second parameter (Smoothing tolerance) should be a required input linear unit. Set the default to 2000 meters.
   - The third parameter (Output feature class) should be a derived output feature. Set the symbology to 'C:\gispy\data\ch23\training\wideLines.lyr'. Modify the Python script so that the output feature class is named 'smooth-Line' and is added automatically to the map. Test the tool in ArcMap to confirm that the yellow highway line symbology is applied when the output is added to the map.

7. Script Tool, 'bCopy' in 'C:\gispy\sample_scripts\ch23\exercises\exercises.tbx', copies a batch of raster or feature class datasets from one workspace to another. The tool takes three parameters, source and destination workspaces, and a dataset type. The tool points to script 'bCopy.py' in the 'C:\gispy\sample_scripts\ ch23\exercises' directory. Script Tools 'bCopyProgress1', 'bCopyProgress2', and 'bCopyToMap' are identical to Script Tool 'bCopy'. Also, 'bCopy-Progress1.py', 'bCopyProgress2.py', and 'bCopyToMap.py' are identical to script 'bCopy.py'. For each part below, modify the Script Tool to point to the script with the same base name, then modify the scripts as specified.

   (a) **bCopyProgress1** Modify 'bCopyProgress1.py' to report progress using the default progressor. Update the label before each file is copied, to report which file is being copied next.
   (b) **bCopyProgress2** Modify 'bCopyProgress2.py' to report progress using the step progressor. Update the label before each file is copied, to report which file is being copied next. Update the progress bar percent (the progressor position) after each file is processed.
   (c) **bCopyToMap** Modify 'bCopyToMap.py' to collect a list of the output files created. Use the `SetParameterAsText` method to return a semicolon delimited string of output file names. Then add a fourth parameter to the Script Tool. Make this parameter a multivalue Dataset derived output type,

so that the output will be added to the map. Test the tool in ArcMap to confirm that each copied file is automatically added to the map.

8. **slopeComputer** and **slope.py** Write a script, 'slope.py', to report the slope of the line that passes through points, (x1,y1) and (x2,y2). Create a Script Tool, 'slopeCompute', that has four float input parameters, x1, y1, x2, and y2. Use the ToolValidator `updateMessages` method to throw an error, if the user enters a vertical line.

Example input1: 1 2 1 4

Example output1: Script Tool posts an error message…
```
The slope of this line is undefined (change x1 or x2).
```

Example input2: 1 2 3 4

Example output2:
```
>>> The slope of the line passing through (1.0, 2.0) and (3.0, 4.0)
is 1.0.
```

9. **getDate** and **printDate.py** Write a script, 'printDate.py', to report the month, day, and year of a date passed as a string in the format MM/DD/YYYY. Create a Script Tool, 'getDate', that has one input parameter, a date. Label the parameter 'Select a date in 2013'. Use the ToolValidator `updateMessages` method to throw an error, if the user selects a date that is not in the year 2013. The date is passed in as a `datetime` object so you can create two `datetime` objects, one for the beginning of 2013 and one for the end:

```
dstart = datetime.datetime(2013, 1, 1, 0, 0, 0, 0)
dend = datetime.datetime(2013, 12, 31, 23, 59, 59)
```

Then use comparison operators to check the data input by the user.

Example input1: 11/30/2013 4:00:46 PM

Example output1:
```
Month: 11 Day: 30 Year: 2013
```

Example input2: 5/17/2014 12:08:51 PM

Example output1: Script Tool posts an error message. . .
```
No data available for 2014. Select a date in year 2013.
```

10. **twoCat** Copy '02_categories_ip' from 'validatorExamples.tbx' into 'exercises.tbx' and rename it as 'twoCat'. Modify the `initializeParameters` method to organize the parameters into two categories instead of three. Name the categories "I. Required" and "II. Optional". Set the category of each parameter based on its `parameterType` (if the type is 'Required' set it to the first category; Otherwise, set it to the second category).

11. **pointORpoly** Script 'pointORpoly.py' calls one of four geoprocessing tools based on a tool name input by the user. The 'pointORpoly' Script Tool in

'C:\gispy\sample_scripts\ch23\exercises\exercises.tbx' points to this script. Run the 'pointORpoly' Script Tool five times on a polygon shapefile to test the five tool name options. The first Script Tool parameter is an input point or polygon feature class. The second parameter is a value list of names to specify which tool to call. Currently, any tool can be called for any input. For example, the 'Points to line' tool can be selected for point or polygon input features, even though polygons are not appropriate input. Modify the `updateParameters` method in the Script Tool `ToolValidator` class to change this behavior. Specifically, if the shape type of the first parameter is 'Polygon', change the filter list for the second parameter to `['Feature to point', 'Minimum bounding box']` and if the shape type is 'Point', change the filter list to `['Thiessen polygon', 'Points to line']`.

12. **convertExer.pyt** Create a Python toolbox named 'convertExer.pyt'. For each part of this exercise, copy and paste the `Tool` class template within the '.pyt' file. Rename it as shown in bold. Modify the class, as specified. Add the tool name to the `tools` list in the `Toolbox` class and check the tools.

    (a) **computeTriArea** This Python toolbox tool should have the same parameters and functionality as 'triArea' in #1 above.

    (b) **montyPyBridge** This Python toolbox tool should have the same parameters and functionality as 'bridgeOfDeath' in #2 above.

    (c) **filtersGalore** This Python toolbox tool should have the same parameters and functionality as 'filterPractice' in #3 above.

    (d) **countValues** This Python toolbox tool should have the same parameters and functionality as 'check4value' in #4 above. In the `execute` method, the search cursor will need the field name, not the `Value` object. Set the field name using the `value` property of the `Value` object, as in the following example:

    ```
    fieldname = parameters[1].value.value
    ```

    (e) **bCount** This Python toolbox tool should have the same parameters and functionality as 'batchCount' in #5 above.

13. **enableEx.pyt** The Script Tool 'enableExample' in 'exercises.tbx' has three parameters, an input shapefile, a tool name ('buffer' or 'get count'), and a buffer distance. The third parameter is only applicable when the user chooses to run the buffer analysis. The `updateParameter` in the `ToolValidator` is used to enable/disable the third parameter as needed. Run the tool to see how it works. Then create a Python toolbox, 'enableEx.pyt' that contains a tool class named `enableExample` that has the same functionality as the 'enableExample' Script Tool.

Chapter 24
Mapping Module

Abstract

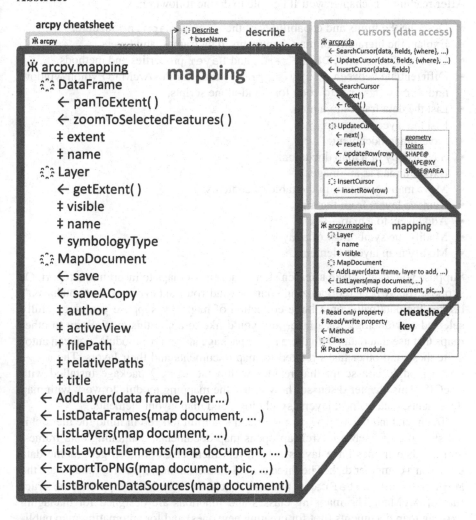

Suppose you want to take a screen shot of dozens of maps to insert into a report. Or suppose you've been reorganizing your data and you want to ensure that you haven't

broken the data paths on a large collection of maps. Or suppose you've carefully
selected a symbology for a layer and you'd like to apply this symbology to other
maps that use that data. With the arcpy package mapping module, you can automate
these and other tasks related to map documents and their layers. The mapping module is a script that resides within the arcpy package installed with ArcGIS. This
chapter discusses how to use the mapping module to work with map documents,
data frames, layers, symbology, and map layout elements.

Chapter Objectives

After reading this chapter, you'll be able to do the following:

- Describe the limits and capabilities of the `arcpy mapping` module.
- Explain the hierarchy of map documents, data frames, and layers.
- Use `MapDocument`, `DataFrame`, and `Layer` properties and methods.
- Differentiate between `arcpy mapping` code for the ArcMap Python window
  and `arcpy mapping` code for stand-alone scripts.
- List the data frames in a map.
- List the layers in a map.
- Save a map document.
- Save a copy of a map document.
- Export a map as an image.
- Move map layers within the table of contents.
- Remove layers from the map.
- Add layers to a map.
- Modify the symbology of a layer.
- Modify map layout elements.

Suppose you want to take a screen shot of dozens of maps to insert into a report. Or
suppose you've been reorganizing your data and you want to ensure that you haven't
broken the data paths on a large collection of maps. Or suppose you've carefully
selected a symbology for a layer and you'd like to apply this symbology to other
maps that use that data. With the `arcpy` package `mapping` module, you can automate these and other tasks related to map documents and their layers. The `map-
ping` module is a script that resides within the `arcpy` package installed with
ArcGIS. This chapter discusses how to use the mapping module to work with map
documents, data frames, layers, symbology, and map layout elements.

   To understand the `mapping` module, it's helpful to keep in mind the hierarchical structure of ArcMap. ArcMap opens map documents, containing data frames,
and the date frames have layers. The layers themselves point to geographic data
stored on a computer disk. The three central classes in the `mapping` module are the
`MapDocument`, `DataFrame`, and `Layer` classes, which correspond to the structure of ArcMap. The mapping classes and functions are designed for managing
existing map documents (not for creating new ones) and for automating map publi-
cation. The module provides access to the `DataFrame` and `Layer` objects with
`ListDataFrames` and `ListLayers` methods that return lists of objects. The
properties of these objects can then be modified. There are also a set of methods for
exporting maps in image formats or PDF (portable document format).

Some examples in this chapter use the ArcMap Python Window. This is a departure from other chapters which work strictly with stand-alone IDEs. The ultimate goal of this chapter still aims for writing stand-alone scripts, ones that you can run outside of ArcMap. However, because the `mapping` module manipulates map document properties, it's useful to test `mapping` module code statements within an open map document. We'll start in the PythonWin Interactive Window and then move to the ArcMap Python Window as we explore the `MapDocument`, `DataFrame`, and `Layer` classes.

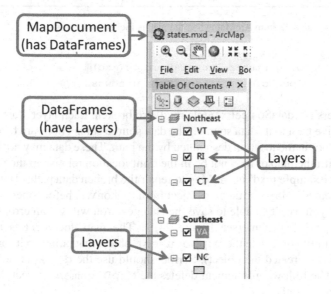

24.1 Map Documents

Manipulating map documents is the main focus of the `mapping` module. For these operations, you first need to create a `MapDocument` object specifying a map. Try the following code in an IDE (such as the PythonWin Interactive Window):

```
>>> myMap = 'C:/gispy/data/ch24/maps/dataSourceExample.mxd'
>>> mxd = arcpy.mapping.MapDocument(myMap)
>>> mxd
<MapDocument object at 0x161618b0[0x16161aa0]>
```

To call functions and classes in user-defined modules, you use dot notation (`moduleName.functionName` and `moduleName.ClassName`). Since the `mapping` is within the `arcpy` package, two dots are needed (`arcpy.mapping.ClassName` or `arcpy.mapping.FunctionName`). This double dot notation is already familiar from ArcGIS environment settings (e.g., `arcpy.env.workspace`). The following code calls a `mapping` function to list the broken data sources in the map document and uses the `MapDocument` object as an argument:

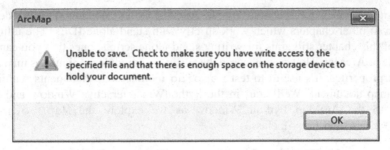

Figure 24.1 A message from ArcMap when the map is locked by Python.

```
>>> arcpy.mapping.ListBrokenDataSources(mxd)
[<map layer u'potHoles'>, <map layer u'township'>]
```

Two layers of 'dataSourceExample.mxd' are broken. Each layer stores an absolute or relative path to its data source. The data source connections are broken when
the data are not in the location designated by the path. These data may not have been
provided with the map or may not be in the right location relative to the map. Open
'dataSourceExample.mxd' in ArcMap to repair the broken data paths. If you the ran
code that created a MapDocument object from PythonWin before opening the map
in ArcMap, you won't be able to save the changes. You will get an error when you
try to save the document (see Figure 24.1). The map document is locked by
PythonWin until you delete the MapDocument object that points to it (or close the
IDE where you created the object). Scripts should use the del keyword to delete
the object. The following statement deletes the MapDocument object, but not the
map document itself:

```
>>> del mxd
```

If the map document object has not been deleted when the map is opened,
changes can not be saved, even after you run the deletion command. So you would
need to close the map document and reopen it after executing the deletion command
to successfully save changes to the map document in ArcMap.

Example 24.1 uses the mapping module to export an image of the map. The
script imports arcpy, sets the workspace, creates a MapDocument object, and
calls the ExportToPNG method. This method captures an image of the 'Layout
view' of a map (as in Figure 24.2). The first argument is the MapDocument
object; The second is the name of the output image file. After this, the script
deletes the MapDocument object so that the map will remain editable. Try the
code with the sample input to see that 'getty_map.png' is created. The exporting
methods do not heed the arcpy geoprocessing workspace environment setting. If
the full path output image file name is not used in the export statement, the image
is placed in the map directory, even if the arcpy workspace variable is set to a
different directory.

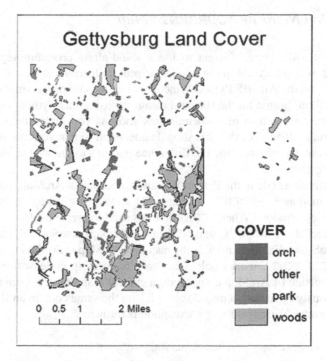

Figure 24.2 Output from Example 24.1, getty_map.png (reduced in size).

Example 24.1: Export the layout view of a map document to a PNG image (Figure 24.2).

```
# mapToPhoto.py
# Purpose: Export the 'Layout view' of a map as a PNG image.
# Usage: fullpath_mxd_filename fullpath_output_png_filename
# Sample input: C:/gispy/data/ch24/maps/landCover.mxd
#               C:/gispy/scratch/getty_map.png
# Note: Portable Network Graphic (PNG) is an image format used for
#       Internet content.
# Many other map export formats are available.

import arcpy, sys

# Full path names of an existing map and an image to create.
mapName = sys.argv[1]
imageName = sys.argv[2]

# Create a MapDocument object.
mxd = arcpy.mapping.MapDocument(mapName)

# Create an image of the map in 'Layout view'
arcpy.mapping.ExportToPNG(mxd, imageName)
print '{0} created.'.format(imageName)

# Delete the MapDocument object.
del mxd
```

24.1.1   Map Name or 'CURRENT' Map

Though it's usually more efficient to use a stand-alone programming IDE, the
Python Window inside ArcMap can be quite helpful for learning about the `mapping`
module. To open the ArcGIS Python Window, select the Geoprocessing>Python or
click the 'Python' button on the standard menu. By using this Python window, you
can see modifications to a map document as soon as you make them. However,
there's one major difference that you should understand so that you can write scripts
that will run outside of ArcMap. The difference pertains to the way you can initialize a `MapDocument` object.

If you're running code in the Python Window embedded in ArcMap, you can refer
to this open map as `'CURRENT'`. In Figure 24.3, code is executed in the ArcMap
Python scripting window. When `'CURRENT'` is used to create the `MapDocument`
object, the filePath is set to 'C:\\gispy\\data\\ch24\\maps\dataSourceExample.mxd'
because 'dataSourceExample.mxd' is the map that ArcMap is displaying.

Using `'CURRENT'` works only when you're running code within ArcMap;
Scripts run outside of ArcMap do not have a current map defined, even if ArcMap
is simultaneously open with a map displayed. Run the same code in an IDE outside
of ArcMap, and a `RuntimeError` exception is thrown:

```
>>> mxd = arcpy.mapping.MapDocument('CURRENT')
RuntimeError: Object: CreateObject cannot open map document
```

Instead of referring to the map as `'CURRENT'` in stand-alone scripts, you must
use the name of the map. In fact, you must specify the path of the map document
(even if it's in your workspace). Otherwise, an error is thrown:

```
>>> arcpy.env.workspace = 'C:/gispy/data/ch24/maps/'
>>> mapName = 'landCover.mxd'
>>> mxd = arcpy.mapping.MapDocument(mapName)
AssertionError: Invalid MXD filename.
```

Figure 24.3 In the ArcMap Python scripting window, `'CURRENT'` can be used instead of a file
name for the open map document.

The `MapDocument` class does not use the `arcpy` workspace environment to search for its input, so you need to add the map's path to its name:

```
>>> # File name of an existing map
>>> mapName = arcpy.env.workspace + '/landCover.mxd'
>>> # Create a MapDocument object.
>>> mxd = arcpy.mapping.MapDocument(mapName)
```

In stand-alone scripts designed to be run outside of ArcMap, you must specify the full file name of the map.

`'CURRENT'` only works for code running inside ArcMap.

One other way to run code within ArcMap, is to use an ArcGIS Script Tool. You can use `'CURRENT'` within an ArcGIS Script Tool, with a couple of caveats. First, by default, ArcGIS Script Tools run in the background. Running a process in the background is just like running a stand-alone script. In order to use `'CURRENT'` in a Script Tool, you have to set the checkbox in the Script Tool properties to run the tool in the foreground. This connects the running code with ArcMap in the same way the ArcGIS Python Window does. Second, when you are operating on an open map (the `'CURRENT'` map), you need to look out for data control conflicts. Locking issues can arise if you're working with data that is open in ArcMap, so there may be certain operations you can't do.

Examples in the next section use the ArcMap Script Window. The keyboard shortcut in PythonWin for efficiently retrieving previous commands is ctrl + UpArrow. In the ArcGIS Python window, just use the UpArrow key to scroll through command history.

As you experiment with code in this environment, you may have to call the `RefreshActiveView` and `RefreshTOC` functions to see the updates appear within your map. Some changes appear automatically, but others require a refresh command to update the display. These are `arcpy` methods, not `mapping` module methods, meaning only one dot is needed. Refreshing the view is only needed when working within the `arcpy` Python window. These statements are usually not needed in stand-alone scripts. Refreshing takes time, so remove 'Refresh' statements from stand-alone scripts, where possible.

24.1.2   *MapDocument Properties*

You may already be familiar with the map documents properties you see when you click on the File menu in ArcMap and select 'Map Document Properties...'. Figure 24.3 uses the `mxd.filePath` to check the file path of the map document.

Python can access additional map document properties, such as `author`, `activeView`, and `title`. To see this in action, open 'states.mxd' in ArcMap, open the ArcMap Python window, and type Python commands in this window.

```
>>> mxd = arcpy.mapping.MapDocument('CURRENT')
```

When you type mxd followed by the dot, you see a menu of `MapDocument` properties and methods. You can select these from the menu with the up and down arrows and tab completion. Some `MapDocument` properties control metadata, such as the map's author:

```
>>> mxd.author
u'Me'
>>> mxd.author = 'Wonderful me!'
>>> mxd.author
u'Wonderful me!'
```

Other properties pertain to viewing the map. The buttons in the bottom left corner of the map modify the active view ('Data' or 'Layout'). The 'Layout' view displays the map surrounds (title, legend, scale, etc.) along with the data. ArcMap will switch between these views if you modify the `activeView`:

```
>>> mxd.activeView
u'Southeast'
>>> mxd.activeView = 'PAGE_LAYOUT'
```

The value of the `activeView` property is either `'PAGE_LAYOUT'` or the name of the active data frame in the data view. The map title can also be changed:

```
>>> mxd.title
u'Hey hey'
>>> mxd.title = 'Eastern US'
>>> arcpy.RefreshActiveView()
```

The `RefreshActiveView` method updates the title on the map. Again, this is an `arcpy` method, not a mapping method, so only one dot is needed.

Another `MapDocument` property, `relativePaths`, is important for portability. Maps can use a relative path or an absolute path to the data sources. When `relativePaths` is set to `True`, the map and data can be moved to a new location without breaking the links to the data, as long as the data stay in the same location, relative to the map. The `relativePaths` property can be set to `True` or `False`:

```
>>> mxd.relativePaths
True
>>> mxd.relativePaths = False
```

The `author`, `activeView`, `title`, and `relativePaths` can be changed within Python because these are 'read and write' properties. Some `MapDocument` properties are 'read only'. For example, the `filePath` property can only be read. Trying to change its value throws a `NameError` exception:

```
>>> mxd.filePath
u'C:\\gispy\\data\\ch24\\maps\\states.mxd'
>>> mxd.filePath = 'C:\\gispy\\data\\ch24\\maps\\US.mxd'
NameError: The attribute 'filePath' is not supported on this
instance of MapDocument.
```

You can check the value of a 'read only' property, but you can't change the value. You can't change the file path in the 'File > Map Document Properties…' window either. Search the ArcGIS Resources site for 'MapDocument (arcpy.mapping)' for a list of the `MapDocument` properties and their read/write capabilities.

24.1.3 Saving Map Documents

There are two `MapDocument` methods for saving changes to map documents, `saveACopy` and `save`. The `saveACopy` method saves a copy. It takes one argument, the new map name. If a full path is not specified, the copy will be saved in the current workspace:

```
>>> mxd.saveACopy('modifiedMap.mxd')
```

To automatically launch ArcMap and view the saved copy, you can use the `os` module `startfile` command, which opens (or tries to open) the document passed in as an argument with the default program associated with that file type. Since this is an `os` command, not an `arcpy` command, the `arcpy` workspace is not used; therefore, you may need to provide the full path file name. The following example launches ArcMap and loads 'landCover.mxd':

```
>>> import os
>>> os.startfile('C:/gispy/data/ch24/maps/landCover.mxd')
```

The `save` method does not require any arguments, but use it with care, as it will overwrite the existing map, even if the `overwriteOutput` environment variable is set to `False`:

```
>>> mxd.save()
```

Notice that there is no method for creating a new map document. The `mapping` module does not allow you to create new maps from scratch. You have to start with an existing map. If you desire to create maps entirely programmatically, you can create an empty map to use as a template, save a copy of the map (instead of over-writing the template), and then add content.

24.2   Working with Data Frames

Since data frames belong to the map document, a map document must be specified to access them. Continuing the ArcMap Python window session associated with the 'states.mxd' map, we'll use our existing `MapDocument` object, `mxd`, to get a list of data frames in the map. The `ListDataFrames` method requires one argument, a `MapDocument` object and it returns a Python list of the `DataFrame` objects in the map document:

```
>>> dfs = arcpy.mapping.ListDataFrames(mxd)
>>> dfs
[<DataFrame object at 0x27b7b950[0x216e2a98]>, <DataFrame object at
0x27b7b8f0[0x216e2a48]>]
```

Each `DataFrame` object has properties. You can use zero-based indexing to access individual `DataFrame` objects. Try the following code to get the first `DataFrame` object in the list and use some `DataFrame` properties:

```
>>> # Get the first data frame.
>>> df = dfs[0]
>>> df.name
u'Northeast'
>>> # Get the second data frame.
>>> df2 = dfs[1]
>>> df2.name
u'Southeast'
>>> df.description
u''
>>> df2.description = "My very cool data frame"
>>> df2.description
u'My very cool data frame'
>>> df2.displayUnits
u'Feet'
```

The `extent` property is an `Extent` object, which stores the corners of the data frame's boundaries as `XMin`, `YMin`, `XMax`, and `YMax` properties:

```
>>> ext = df2.extent
>>> ext.YMin
29.99437527717719
```

You can also zoom to selected features. The following code selects one of the layers, using a Data Management tool and zooms to the selected layer:

```
>>> arcpy.SelectLayerByAttribute_management('VA')
>>> df2.zoomToSelectedFeatures()
```

Since `dfs` is a Python list, you can also navigate it with a FOR-loop:

```
for df in dfs:
    # Print the name of the current data frame.
    print df.name
```

`DataFrame` objects can be used to access data layers. We'll next look at `Layer` object properties and some examples that use `DataFrame` objects and `Layer` objects together.

24.3   Working with Layers

The `mapping` module has a `ListLayers` method for listing the layers in a map. It requires one argument, a `MapDocument` object and it returns a list of `Layer` objects. `Layer` objects have properties that Python can access. Continue typing in the ArcMap Python window of the 'states.mxd' map to try the following code statements:

```
>>> layers = arcpy.mapping.ListLayers(mxd)
>>> myLayer = layers[0]
>>> myLayer.dataSource
u'C:\\gispy\\data\\ch24\\USstates\\VT.shp'
>>> myLayer.isFeatureLayer
True
>>> myLayer.isRasterLayer
False
```

The `getExtent` method returns an `Extent` object. This object provides extent boundaries for the `Layer` object, which may differ from the data frame's boundaries:

```
>>> e = myLayer.getExtent()
>>> e
```

```
<Extent object at 0x146d41d0[0x1472a9e0]>
>>> e.YMin
1219145.8956193451
```

You can zoom to a layer's extent by setting the data frame's extent to the layer's extent:

```
>>> df.extent = myLayer.getExtent()
```

You can also select a set of features and set the data frame extent to the value returned by getSelectedExtent(), to zoom to those features.

The ListLayers method can return all of the layers in the map or a subset of the layers. A MapDocument object is the only required argument. If this is the only argument used, all of the layers are listed:

```
# List ALL of the Layer objects
>>> layers = arcpy.mapping.ListLayers(mxd)
>>> for myLayer in layers:
...         print myLayer.name
...
VT
RI
CT
VA
NC
```

This method also has two optional arguments, a wild card and a DataFrame object. The wild card argument can restrict the results based on a substring (just like the arcpy methods for listing datasets) and the DataFrame object can restrict the list to a particular data frame. To specify a data frame, but not utilize the wild card, you need to use a placeholder ('*') for the wild card.

```
>>> # Get a list of DataFrame objects.
>>> dfs = arcpy.mapping.ListDataFrames(mxd)

>>> # Get the first DataFrame object.
>>> df = dfs[0]

>>> # Get a list of Layer objects in this data frame.
>>> layers = arcpy.mapping.ListLayers(mxd, '*', df)
>>> for myLayer in layers:
...         print myLayer.name
...
VT
RI
CT
```

The mapping module methods that use Layer objects to manipulate layers (MoveLayer, RemoveLayer, AddLayer, and InsertLayer) are discussed next.

24.3.1 Moving, Removing, and Adding Layers

The mapping module can move, remove, and add layers. Because layers are nested within data frames (which are within a map document), to use any of these layer methods, you must first instantiate MapDocument and DataFrame objects. These are then used to specify the Layer object. To see how it works, open 'layer-ManipExample1.mxd'. In this map, a layer containing points is hidden by the polygon layers. To make the points visible, we'll use the MoveLayer method to move this layer to the top of the table of contents. The MoveLayer method allows a layer to be moved before or after a reference layer within the same data frame. To try this, begin by initializing a MapDocument object, a DataFrame object, and a list of Layer objects in the ArcMap Python Window:

```
>>> mxd = arcpy.mapping.MapDocument('CURRENT')
>>> dfs = arcpy.mapping.ListDataFrames(mxd)
>>> df = dfs[0]
>>> lyrs = arcpy.mapping.ListLayers(mxd)
```

Now that we have a list, we need to select the specific Layer objects to use for calling MoveLayer. Moving the 'centers' layer before the 'cover' layer brings it to the top of the table of contents, so we'll use the 'cover' layer as the reference layer. We need to use numbers (not names) to index into the list. We could find the correct indices by looping through the list and checking against the names, but here we'll just hard-code the numbers. The ListLayers function returns the layers in the order they appear in the table of contents from top to bottom. The layer to move is the third layer (index 2) and the reference layer is the first layer (index 0):

```
>>> layerToMove = lyrs[2]
>>> layerToMove.name
u'centers'
>>> referenceLayer = lyrs[0]
>>> referenceLayer.name
u'cover'
```

The MoveLayer method requires a DataFrame object and the two Layer objects as arguments. The optional fourth parameter specifies the position relative to the reference layer. Enter the following code to move the 'centers' layer to the top of the table of contents:

```
>>> arcpy.mapping.MoveLayer(df, referenceLayer, layerToMove, 'BEFORE')
```

The code for removing and adding layers is similar. The difference comes in specifying the `Layer` objects (layers to be removed or added). Now that you have seen `MoveLayer` work in the ArcMap Python Window, you can become familiar with code for manipulating layers in stand-alone scripts, as shown in Examples 24.2 and 24.3.

Example 24.2 removes the first layer in the first data frame. To remove data from an open map by hand, you first need to right click on the layer in the table of contents and then you can select 'Remove'. Similarly, to remove a layer with a script, you first need to create a `Layer` object that is pointing to the layer you want to remove. Then you can call the `RemoveLayer` method on this object. The map in Example 24.2, 'layerManipExample2.mxd', has three layers, all in one data frame. The code gets a `MapDocument` object, lists the `DataFrame` objects, and lists the `Layer` objects from the first data frame. The script sets `layerToRemove` to the first `Layer` object in the list and then calls the `RemoveLayer` method. This method has two required arguments, a `DataFrame` object (for the data frame where the layer resides) and a `Layer` object (referencing the one to be removed). The script finally saves a copy of the map that now has two layers and releases the lock on the map by deleting the `MapDocument` object. The copy of the map only has the two remaining layers.

Example 24.2: Remove the first layer from a map.

```
# removeLayers.py
# Purpose: Remove the first layer in the table of contents.
# Input: No arguments required.
import arcpy

# Get a MapDocument object.
mxdName = 'layerManipExample2.mxd'
mapPath = 'C:/gispy/data/ch24/maps/'
mxd = arcpy.mapping.MapDocument(mapPath + mxdName)

# Get a list of the DataFrame objects.
dfs = arcpy.mapping.ListDataFrames(mxd)

# Get the first DataFrame object.
df = dfs[0]

# Get a list of Layer objects in this data frame.
lyrs = arcpy.mapping.ListLayers(mxd, '' , df)

# Get the first Layer object.
layerToRemove = lyrs[0]

# Remove the layer.
arcpy.mapping.RemoveLayer(df, layerToRemove)

# Save a copy of the map.
copyName = 'C:/gispy/scratch/' + mxdName[:-4] + '_V2.mxd'
mxd.saveACopy(copyName)
```

```
# Delete the MapDocument object to release the map.
del mxd
```

Example 24.3 adds data to a map. To add data to a map by hand, you click on the 'Add Data' button in the standard toolbar and browse to the data to add. To add data with a script, you need to specify the data path and create a `Layer` object pointing to the data. This layer is then passed into the AddLayer method. For adding layers, you again need to get a `MapDocument` and `DataFrame` objects to specify where the layer will be added. Since the layer doesn't exist yet, instead of pointing to an existing layer, you need to create a `Layer` object by calling the `Layer` class. The `Layer` class takes the name of a data file as an argument and returns a `Layer` object. Example 24.3 creates a `Layer` object named `layerObj` by pointing to the `fileName` variable. Next, it calls `AddLayer`, passing in two required arguments, `DataFrame` and `Layer` objects.

By default the `AddLayer` method adds a layer in the same way the 'Add Data' button works, using an automatic arrangement scheme. You can also specify 'TOP' or 'BOTTOM' of the data frame using the third optional argument. The `InsertLayer` method is similar to the `AddLayer` method, but it gives you tighter control on the placement by requiring a reference layer (like `MoveLayer`).

Example 24.2 and 24.3 manipulate vector data (a shapefile); The procedures for removing and adding vector and raster files are the same.

Example 24.3: Adding a shapefile layer to a map.

```
# addLayer.py
# Purpose: Add a data layer to a map.
# Input: No arguments required.
import arcpy

# Initialize data variables.
arcpy.env.workspace = 'C:/gispy/data/ch24/maps/'
fileName = '../USstates/MA.shp'
mapName = 'layerManipExample3.mxd'

# Instantiate MapDocument and DataFrame objects.
mxd = arcpy.mapping.MapDocument(arcpy.env.workspace + '/' + mapName)
dfs = arcpy.mapping.ListDataFrames(mxd)

# Get the first data frame.
df = dfs[0]

# Instantiate a Layer object.
layerObj = arcpy.mapping.Layer(fileName)

# Add the new layer to the map.
arcpy.mapping.AddLayer(df, layerObj)
# Save a copy of the map.
```

```
copyName = 'C:/gispy/scratch/' + mapName[:-4] + '_V2.mxd'
mxd.saveACopy(copyName)

# Delete the MapDocument object to release the map.
del mxd
```

24.3.2 *Working with Symbology*

The `mapping` module provides some capability for changing layer symbology using another layer management method, `UpdateLayer`. To manually change the symbology for a map layer, right click on a map layer name in the table of contents, select 'Properties' and click on the 'Symbology' table. Select a symbology type from the list on the left and then you can update the properties for your visualization preferences. For example, if you select 'Graduated colors' under 'Quantities', you can select a value field, a number of classes, and a color ramp. The available properties depend on the symbology type, e.g., the 'Graduated Symbols' type has a symbol size property but no color ramp, since it varies glyph size, instead of color, for each class. The `mapping` module allows you to modify some of these properties.

Python can only modify the properties associated with certain symbology types. A `Layer` object has a `symbologyType` property which has one of following values: `'GRADUATED_COLORS'`, `'GRADUATED_SYMBOLS'`, `'UNIQUE_VALUES'`, `'RASTER_CLASSIFIED'`, or `'OTHER'`. Each of the first four values has an `arcpy` class and properties that can be modified in Python. The `'OTHER'` symbology type is a catch-all for the remaining symbology types. Python can not modify the symbology of layers that have a symbology type of `'OTHER'`.

Modifying symbology in Python starts by getting the `Layer` object and checking the `symbologyType`, which tells you which symbology properties apply. To see how it works, open 'symbologyExample.mxd' in ArcMap and try the following code in the ArcMap Python window:

```
>>> mxd = arcpy.mapping.MapDocument('CURRENT')
>>> dfs = arcpy.mapping.ListDataFrames(mxd)
>>> df = dfs[0]
>>> lyrs = arcpy.mapping.ListLayers(mxd)
>>> layerToModify = lyrs[0]
>>> layerToModify.symbologyType
u'OTHER'
```

Data files that have a '.lyr' file extension are referred to as 'layer files'. These files store the symbology for a dataset. When you add a layer file to a map, it uses the stored `symbologyType`; Otherwise, when data is added to a map, the default `symbologyType` is `'OTHER'`. For example, when a shapefile is added to a map,

it is assigned the default symbology type. Depending on its `symbologyType`, a
`Layer` object can have a `symbology` class.

```
>>> layerToModify.symbology
NameError: The attribute 'symbology' is not supported on this
instance of Layer.
```

The `symbology` class is not supported for `layerToModify`. Map layers
with a `symbologyType` of `'OTHER'` do not support the `symbology` class.
Only map layers with one of the four other symbology types support the symbology
class. You can't change the `symbologyType` of a layer using an assignment statement, because this property is read-only:

```
>>> layerToModify.symbologyType = 'GRADUATED_COLORS'
RuntimeError: LayerObject: Set attribute symbologyType does not exist
```

Instead, modifying the `symbologyType` can be done using a previously created '.lyr' file. A '.lyr' file can act as training data that has the desired `symbolo-
gyType`. For our example, we'll use the sample dataset, 'gradColorNE.lyr', from
the 'C:\gispy\data\ch24\symbolTraining' directory. This dataset has a graduated
color symbology. To apply the symbology from this file, use the `UpdateLayer`
method, as in the following code:

```
>>> srcLay = 'C:/gispy/data/ch24/symbolTraining/gradColorsNE.lyr'
>>> srcLayObj = arcpy.mapping.Layer(srcLay)
>>> arcpy.mapping.UpdateLayer(df, layerToModify, srcLayObj)
>>> layerToModify.symbologyType
u'GRADUATED_COLORS'
```

Now we can modify graduated color symbology properties, such as the value
field for calculating the colors and the number of color classes:

```
>>> layerToModify.symbology.valueField
u'Category'
>>> layerToModify.symbology.valueField = 'AVE_FAM_SZ'
>>> arcpy.RefreshActiveView()
>>> layerToModify.symbology.numClasses
5
>>> layerToModify.symbology.numClasses = 3
>>> arcpy.RefreshActiveView()
```

Once the view is refreshed, you can see the resulting map that visualizes the
average family size in three classes (Figure 24.4). Though Python's access to symbology is somewhat limited, by developing a set of training layers, you can modify
a number of visual properties automatically.

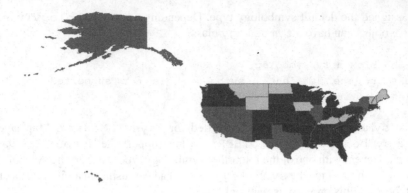

Figure 24.4   Graduate color symbology for average family size.

24.4   Managing Layout Elements

Map layout elements, such as legends, scale bars, and north arrows can also be repositioned and resized with the `mapping` module. The properties of a layout element can be changed manually in ArcMap by double-clicking on the element to launch the properties dialog. The element properties vary, depending on the type of element. Some (though not all) of these properties can be modified with a script.

The `mapping` module has a `ListLayoutElements` method to list the layout elements in the map. This returns a list of element objects, which have `name`, `type`, `position`, and `size` properties. Depending on the layout element, the object can have other properties or methods. For example, the title text can be modified using a `text` property, the style of a layer in the legend can be updated using the `updateItem` method, and a `sourceImage` property can be modified for picture elements. The 'layoutElemsExample.mxd' map document has layout elements visible to the left of the map in layout view. To show how this works, we'll modify the title element. First, close any ArcMap Python windows and then open 'layoutElemsExample.mxd' in ArcMap. Next, try the following code in the ArcMap Python window to get a list of layout element objects and print their names:

```
>>> mxd = arcpy.mapping.MapDocument('CURRENT')
>>> elems = arcpy.mapping.ListLayoutElements(mxd)
>>> for e in elems:
...     print e.name

Text Box
Scale Text
Alternating Scale Bar
North Arrow
Legend
Title
Layers
```

The loop prints the name of each element using the `name` property. Double-click on the title element in ArcMap to open its properties window and look at the 'Size and Positions' tab. The 'Element Name' is set to 'Title'. To get a specific element object, you need to know the element's name. For example, since the title layout element has the name `'Title'`, you can loop through the list to get this object as follows:

```
>>> for e in elems:
...       if 'Title' in e.name:
...             title = e
>>> title
<TextElement object at 0x2a1e8330[0x251d8a20]>
```

Alternatively, you can get a specific element by using the optional wild card parameter when you call the `ListLayoutElements` method, though this requires you to know both the name and the element type. The first parameter for this method, a `MapDocument` object, is required. The second parameter, the element type, is optional but will not accept a place holder, so you must specify one of the six predefined element types, `'DATAFRAME_ELEMENT'`, `'GRAPHIC_ELEMENT'`, `'LEGEND_ELEMENT'`, `'MAPSURROUND_ELEMENT'`, `'PICTURE_ELEMENT'`, or `'TEXT_ELEMENT'`. The third argument is a wild card for the element name. As an example, the north arrow is a `'MAPSURROUND_ELEMENT'` type element and it is named `'North Arrow'`. The following code uses these two pieces of information and indexes into the list to get the north arrow element:

```
>>> arrow = arcpy.mapping.ListLayoutElements(mxd,
        'MAPSURROUND_ELEMENT', '*Arrow*')[0]
>>> arrow.name
u'North Arrow'
```

Returning to our `title` variable, this is pointing to a text element object which has properties such as `elementWidth`, `elementHeight`, `elementPositionX`, `elementPositionY`, `fontSize`, `name`, `text`, and so forth. The following code uses the `text` and `fontSize` properties to change the map's title and refreshes the view to show the changes:

```
>>> title.text
u'Map_Title'
>>> title.text = 'USA'
>>> arcpy.RefreshActiveView()
>>> title.fontSize
42.0
>>> title.fontSize = 72
>>> arcpy.RefreshActiveView()
```

Currently, the map title is sitting in a staging area to the left of the main map view. We can move it to the center of the map using the position and dimensions of its data frame. `DataFrame` objects have `elementWidth`, `elementHeight`, `elementPositionX,` and `elementPositionY` properties which we can enlist for this task. Use the following code to get the `DataFrame` object for the map (which has only one data frame) and move the title to the center of the map:

```
>>> dfs = arcpy.mapping.ListDataFrames(mxd)
>>> df = dfs[0]
>>> title.elementPositionX = df.elementPositionX + \
    (df.elementWidth*0.5) - (title.elementWidth*0.5)
>>> arcpy.RefreshActiveView()
```

The title is now in the center (horizontally) of the frame. To move it to the top of the frame, modify the y-position:

```
>>> title.elementPositionY = df.elementPositionY + \
    df.elementHeight - title.elementHeight
>>> arcpy.RefreshActiveView()
```

You can not create new layout elements with the `mapping` module. However, you can use a template map document with map elements in a staging area, outside the data frame. Only the elements inside the data frame will be visible when the map is exported. For example, only the title of 'layoutElemsExample.mxd' has been moved inside the data frame. The rest of the elements are in a staging area, so they are not visible on export. Export 'layoutElemsExample.mxd' with the following code and confirm that the resulting document shows the title, but no other layer elements:

```
>>> myExport = 'C:/gispy/scratch/noSurrounds.pdf'
>>> arcpy.mapping.ExportToPDF(mxd, myExport)
```

For a template map document, you could start with something like the sample map, 'layoutElemsExample.mxd'. You may want to remove the data, move the title back to the staging area and add other elements (such as pictures or text boxes) in the staging area. Getting a layout element to modify its properties requires you to know the element name. When you add an element to a map document, it gets a name. However, for some elements, the default name is an empty string. For this reason, you need to make sure each element in your map template has a meaningful name. For example, add a text element to 'layoutElemsExample.mxd' (Insert > Text), double click on it and view the 'Size and Position' table The 'Element Name' box is empty; rename it as 'Text1', so that you can refer to this element by name. Once you have named all of the elements and placed them in a staging area, the template is ready. A script can save a copy of the template map, and then automatically add data

> **Note** New layout elements can not be created with the `mapping` module, but a pre-prepared map can be used as a template for automating layout element manipulation.

to the map and modify the symbology and map surrounds to streamline map production workflow.

24.5   Discussion

In this chapter, we used the `arcpy mapping` module to modify map, data frame, and layer properties, to add, move, and delete layers, and to modify symbology and layout elements. These operations involve working with map documents, data frames, layers, symbology, and layout element object properties and methods. Most mapping scripts begin by creating a `MapDocument` object, listing the `DataFrame` objects, and listing the `Layer` objects. To explore these objects further, reference the `'CURRENT'` map document in the ArcGIS Python window. When porting code to stand-alone applications, be sure to reference the full path file name of the map document and be aware of data locking conditions. This chapter demonstrated many of the `mapping` module capabilities, but not all. If you are working with temporal data attributes, you may want to investigate additional `mapping` module functionality, such as the `DataFrameTime` and `LayerTime` classes that enable dynamic map displays. One final tip, when you're using the `mapping` module extensively, it may be helpful to print the alphabetical lists of `mapping` classes, functions, and constants found in the ArcGIS Resources help pages.

24.6   Key Terms

`MapDocument` objects
`DataFrame` objects
`Layer` objects
`'CURRENT'` vs. map name
`RefreshActiveView` method

Layer file (.lyr extension)
`SymbologyType` property
`Layer` object symbology class
Layout element objects

24.7   Exercises

1. **exportToJPG.py** Write a script that exports a map to a JPEG image with the same base name as the map and places the image in 'C:/gispy/scratch'. The script should take one argument: the full path filename of a map document file.

   Example input: C:/gispy/data/ch24/maps/states.mxd

   Example output:
   ```
   >>>> C:/gispy/scratch/states.jpg created.
   ```

2. **listLayers.py** Write a script that takes the full path file name of a map document and prints the '.mxd' file path as well as its data frame and layer names of each layer. The script should take just one argument, the full path filename of the '.mxd' file. Use nested looping and format the output with tabs to achieve the indentation shown in the example.

   Example input: C:/gispy/data/ch24/maps/states.mxd

   Example output:
   ```
   >>> Map: C:/gispy/data/ch24/maps/states.mxd
   Frame 0: Northeast
        Layer 0: VT    C:\gispy\data\ch24\USstates\VT.shp
        Layer 1: RI    C:\gispy\data\ch24\USstates\RI.shp
        Layer 2: CT    C:\gispy\data\ch24\USstates\CT.shp
   Frame 1: Southeast
        Layer 0: VA    C:\gispy\data\ch24\USstates\VA.shp
        Layer 1: NC    C:\gispy\data\ch24\USstates\NC.shp
   ```

3. **addRaster.py** Write a script that adds a raster layer to an existing map and saves the results as a copy of the mxd in 'C:/gispy/scratch'. The script should take two arguments: the full path filename of the input '.mxd' file and the full path file name of the raster data to be added.

   Example input:
   C:/gispy/data/ch24/maps/testAdd.mxd C:/gispy/data/ch24/otherData/getty_rast

   Example output:
   ```
   >>> C:/gispy/scratch/testAdd.mxd created.
   ```

   The output map should have the same basename as the input map, but there should now be a 'getty_rast' layer in the table of contents.

4. **removeRastLayers.py** Write a script that removes any raster layers from a map and saves the results as a copy of the mxd in 'C:/gispy/scratch'. The script should take one argument, the full path filename of the '.mxd' file. The output map should have the same basename, but will have no rasters as layers in the table of contents.

Example input: C:/gispy/data/ch24/maps/testRemove.mxd

Example output:
```
>>> getty_rast_discrete layer removed.
getty_rast layer removed.
pic1.JPG layer removed.
C:/gispy/scratch/testRemove.mxd created.
```

5. **activeZoom.py** Practice using `mapping` properties and objects. Write a script that performs the following steps:

Step 1: Change the active view of a map to the data frame with the name specified by the user.

Step 2. Set only the layers in the active data frame to visible (This can by done by setting the visibility of all layers to `False` and then turning on only the needed layers.)

Step 3. Zoom to the extent of the first layer in the active data frame.

Step 4. Export a PDF of the map with the same base name as the data frame and in the same directory as the map document.

Do not save the changes to the map document within the script. The script should take three arguments, the full path filename of a map document, the name of a data frame in that document, and an output directory for the PDF.

Example input:
C:/gispy/data/ch24/maps/USregions.mxd Central C:/gispy/scratch

Example output:
```
>>> Map active view set to Central frame
southAtlantic layer hidden.
midAtlantic layer hidden.
newEngland layer hidden.
eastNorthCentral layer hidden.
eastSouthCentral layer hidden.
westSouthCentral layer hidden.
westNorthCentral layer hidden.
mountain layer hidden.
pacific layer hidden.
eastNorthCentral layer visible.
eastSouthCentral layer visible.
westSouthCentral layer visible.
westNorthCentral layer visible.
Data frame extent set to the extent of eastNorthCentral. C:/
   gispy/scratch/Central.pdf created.
```

The output '.pdf' document, C:/gispy/scratch/Central.pdf, looks something like Figure 24.5:

Figure 24.5   PDF exported by activeZoom.py for the sample input.

6. **modSymbol.py** Write a script that modifies the symbology of the first layer in the first data frame in a map and export the results using the following steps:

   Step 1. Use a training layer to modify the symbology to graduated symbols.
   Step 2. Change the field value of the graduated symbols to 'FEMALES', the number of classes to 3, and the normalization field to 'POP2012'.
   Step 3. Export a JPEG of the resulting map.

   Do not save the changes to the map document within the script. The script should take three arguments, a full path filename of the '.mxd' file, a full path training layer name, and a full path JPEG name.

   Example input: C:/gispy/data/ch24/maps/symbologyExample.mxd C:/gispy/data/ch24/symbolTraining/gradSymbolsNE.lyr C:/gispy/scratch/Females2010.jpg

   Example output:
   ```
   >>> C:/gispy/scratch/Females2010.jpg created.
   ```

The output image should look similar to this one:

7. **modLayout.py** Write a script that modifies the north arrow in a map. Specifically, it will move the north arrow to the bottom left corner of the data frame and triple the size of the arrow. Export a TIFF of the resulting map to the 'C:/gispy/scratch' directory. Do not save the changes to the map document within the script. The script should take two arguments, a full path map name and a TIFF base name, in that order. Assume the input map has an existing north arrow layout element named 'North Arrow'.

Example input:
C:/gispy/data/ch24 /maps/layoutElemsExample.mxd northArrow.tif

Example output:
```
>>> C:/gispy/scratch/northArrow.tif created.
```

The output image should look similar to this one:

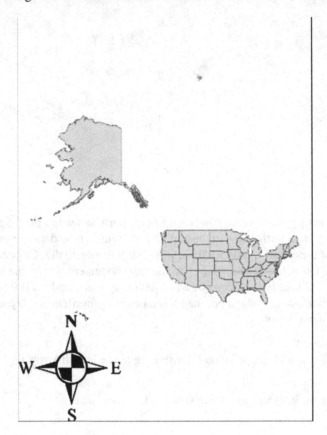

Index

\ (backslash) for line continuation, 40, 149, 154, 159, 219, 324, 465

\ (backslash) for \n, \t, and other escape sequences, 49–51, 57, 206, 358, 392, 396, 404, 421

+,-, *, /, **, and % (in math operations), 38, 100, 105, 211, 258, 276, 424, 426, 487

+ (plus sign) in concatenation, 43–45, 52–53, 60

* (asterisk) for import statements, 109

* (asterisk) forwild cards or placeholders, 191–193, 196, 219, 313–314, 516

/ and \\ (forward slash and double backslash) for file paths, 49–51

" or ' (double quotation mark or single quotation mark) for string literals, 39, 124, 151–152

" and ' (double quotation mark and single quotation mark) for sql queries, 150–154, 329

! (exclamation point) for ArcGIS Python expressions, 100, 101, 207, 316, 477, 497

= (equals sign) for assignment statements, 26–27, 148

== (double equal signs), checking for equality in Python, 55, 146–148

!= (exclamation point next to equals sign), checking for inequality in Python, 55, 146–148

(hash sign) for comments, 23, 173

(hash sign) as a default value, 465

'#' (hash sign string) for tool parameter placeholding, 102–103

<> (less than sign next to greater than sign) for checking inequality in SQL, 150–151, 329

""" or ''' (triple double quotation marks or triple single quotation marks) for multi-line string literals, 39, 54

""" or ''' (triple quotation marks) for docstrings, 265, 267, 418

" vs. ' (double quotation makes versus single quotation marks) for script arguments, 124, 130

: (colon) to start code blocks, 143, 174, 175, 177, 243, 267, 359, 418

: (colon) in slicing, 42–43, 60, 112, 127–128

: (colon) in dictionaries, 336–337

{} (curly braces) for dictionaries, 336, 337

() (parentheses) for tuples, 64

() (parentheses) for function calls, 24, 262, 267

[] (square braces) for dictionaries, 337–341, 350

[] (square braces) for indexing, 41, 60

[] (square braces) for lists, 59, 60, 145, 337, 340

[] (square braces) for slicing, 42, 60

. (period)
    for OOP dot notation, 45–46, 87, 97, 101, 155, 251, 296, 316, 394, 420, 428, 493
    for OOP double-dot notation, 107, 125, 311, 507

../ (dot dot slash) for relative path, 294

A

Absolute path. *See* Full path file name

abspath, os.path function.
    See os module functions

Access by key, 337–338

activeView MapDocument property, 512–513

Printed in the United States
By Bookmasters